MATHEMATICAL MODELING

No. 8

Edited by
William F. Lucas, Claremont Graduate School
Maynard Thompson, Indiana University

Birkhäuser
Boston • Basel • Berlin

Control and Chaos

Kevin Judd
Alistair Mees
Kok Lay Teo
Thomas L. Vincent

Editors

Birkhäuser
Boston • Basel • Berlin

Kevin Judd
Alistair Mees
Kok Lay Teo
Center for Applied Dynamics
and Optimization
Department of Mathematics
The University of Western Australia
Perth, 6907, Australia

Thomas L. Vincent
Aerospace and Mechanical Engineering
University of Arizona
Tucson, Arizona 85721
U.S.A.

Library of Congress Cataloging-in-Publication Data

Control and chaos / Kevin Judd ... [et al.] , editors.
    p.   cm. -- (Mathematical modeling ; no. 8)
  "Proceedings of the US-Australia Workshop on Control and Chaos
held in Honolulu, Hawaii from 29 June to 1 July 1995" --Pref.
  Includes bibliographical references.
  ISBN 0-8176-3867-9 (acid-free paper). -- ISBN 3-7643-3867-9 (Basel
: acid-free paper)
  1. Control theory--Congresses. 2. Chaotic behavior in systems-
-Congresses. I. Judd, Kevin, 1962- . II. US-Australia Workshop
on Control and Chaos (1995 : Honolulu, Hawaii) III. Series:
Mathematical modeling (Boston, Mass.) ; no. 8.
QA402.3.C618   1997   97-7728
003'.75--dc21   CIP

Printed on acid-free paper
© 1997 Birkhäuser Boston

Birkhäuser ®

Copyright is not claimed for works of U.S. Government employees.
All rights reserved. No part of this publication may be reproduced, stored in a retrieval system, or transmitted, in any form or by any means, electronic, mechanical, photocopying, recording, or otherwise, without prior permission of the copyright owner.

Permission to photocopy for internal or personal use of specific clients is granted by Birkhäuser Boston for libraries and other users registered with the Copyright Clearance Center (CCC), provided that the base fee of $6.00 per copy, plus $0.20 per page is paid directly to CCC, 222 Rosewood Drive, Danvers, MA 01923, U.S.A. Special requests should be addressed directly to Birkhäuser Boston, 675 Massachusetts Avenue, Cambridge, MA 02139, U.S.A.

ISBN 0-8176-3867-9
ISBN 3-7643-3867-9
Camera-ready copy provided by the editors.
Printed and bound by Quinn-Woodbine, Woodbine, NJ.
Printed in the U.S.A.

9 8 7 6 5 4 3 2 1

Contents

Understanding Complex Dynamics

Triangulating Noisy Dynamical Systems
    Stuart Allie, Alistair Mees, Kevin Judd and Dave Watson   1

Attractor Reconstruction and Control Using Interspike Intervals
    Tim Sauer   12

Modeling Chaos from Experimental Data
    Kevin Judd and Alistair Mees   25

Chaos in Symplectic Discretizations of the Pendulum
and Sine-Gordon Equations
    B.M. Herbst and C.M. Schober   39

Collapsing Effects in Computation of Dynamical Systems
    Phil Diamond, Peter Kloeden and Aleksej Pokrovskii   60

Bifurcations in the Falkner-Skan equation
    Colin Sparrow   91

Some Characterisations of Low-dimensional Dynamical Systems
with Time-reversal Symmetry
    John A. G. Roberts   106

Controlling Complex Systems

Control of Chaos by Means of Embedded Unstable Periodic Orbits
    Edward Ott and Brian R. Hunt   134

Notch Filter Feedback Control for k-Period Motion in a Chaotic System
    Walter J. Grantham and Amit M. Athalye   142

Targeting and Control of Chaos
    Eric J. Kostelich and Ernest Barreto   158

Adaptive Nonlinear Control: A Lyapunov Approach
    Petar V. Kokotović and Miroslav Krstić   170

Creating and Targeting Periodic Orbits
    Kathryn Glass, Michael Renton, Kevin Judd and Alistair Mees   183

Dynamical Systems, Optimization, and Chaos
    John B. Moore   197

Combined Controls for Noisy Chaotic Systems
    Mirko Paskota, Kok Lay Teo and Alistair Mees   207

Complex Dynamics in Adaptive Systems
  Iven M.Y. Mareels   226

Hitting Times to a Target for the Baker's Map
  Arthur Mazer   251

Applications

Controllable Targets Near a Chaotic Attractor
  Thomas L. Vincent   260

The Dynamics of Evolutionary Stable Strategies
  Yosef Cohen and Thomas L. Vincent   278

Nitrogen Cycling and the Control of Chaos in a Boreal Forest Model
  John Pastor and Yosef Cohen   304

Self-organization Dynamics in Chaotic Neural Networks
  Masataka Watanabe, Kazuyuki Aihara and Shunsuke Kondo   320

Preface

This volume contains the proceedings of the US-Australia workshop on Control and Chaos held in Honolulu, Hawaii from 29 June to 1 July, 1995. The workshop was jointly sponsored by the National Science Foundation (USA) and the Department of Industry, Science and Technology (Australia) under the US-Australia agreement.

Control and Chaos—it brings back memories of the endless reruns of "Get Smart" where the good guys worked for Control and the bad guys were associated with Chaos. In keeping with current events, Control and Chaos are no longer adversaries but are now working together. In fact, bringing together workers in the two areas was the focus of the workshop.

The objective of the workshop was to bring together experts in dynamical systems theory and control theory, and applications workers in both fields, to focus on the problem of controlling nonlinear and potentially chaotic systems using limited control effort. This involves finding and using orbits in nonlinear systems which can take a system from one region of state space to other regions where we wish to stabilize the system. Control is used to generate useful chaotic trajectories where they do not exist, and to identify and take advantage of useful ones where they do exist. A controller must be able to nudge a system into a proper chaotic orbit and know when to come off that orbit. Also, it must be able to identify regions of state space where feedback control will be effective. Several new methods were presented, including targeting algorithms and the use of Lyapunov functions for identifying controllable targets. There are many direct applications of the methods discussed in this workshop. These include mechanical and electrical systems which are to be controlled, or ecosystems and other biosystems that are to be managed.

The participants of the workshop were from the USA or Australia except for one from England and one from Japan. The participants were more or less balanced according to their interests in dynamics or control systems. Each participant gave a presentation in his or her field of expertise as related to the overall theme. The formal papers (contained in this volume) were written after the workshop so that the authors could take into account the workshop discussions, and relate their work to the other presentations. Each paper was reviewed by other participants who then wrote comments which are attached to the ends of the papers. We feel that the comments form a valuable part of this volume in that they give the reader a share of the workshop experience.

The papers are grouped, as they were presented, according to three classifications: Understanding Complex Systems, Controlling Complex Systems, and Applications. Part I, contains seven papers dealing with modeling, behavior, reconstruction, prediction, and numerics. Part II, contains nine papers on controlling complex systems by means of embedding unstable periodic orbits, targeting, filtering, optimization, and adaptive methods. Part III contains four applications papers including the control of a bouncing ball, evolutionary stability, chaos in ecosystems, and neural networks.

We would like to acknowledge the financial support of NSF and DIST which made this workshop possible. The cooperation and organizational skills of the East-West center where the workshop was held was also most appreciated. Jenny Harris gave invaluable assistance with the Australian end of the workshop organisation. AIM thanks the Institute of Nonlinear Science, U.C. San Diego, for hospitality.

Kevin Judd
Alistair I. Mees
Kok-Lay Teo
Thomas L. Vincent

Perth, Western Australia
Tucson, Arizona

August 1996.

Standing from left: John Pastor, Petar Kokotovic, John Moore, Katie Glass (*in front*), Masataka Watanabe (*behind*), Tim Sauer, Constance Schrober, Alistair Mees, Arthur Mazer, Ed Ott, Phil Diamond, Yosef Cohen, Tom Vincent, Eric Kostelich, Iven Mareels.

Seated from left: Kok-Lay Teo, Walt Grantham, Colin Sparrow, Kevin Judd, John Roberts.

Triangulating Noisy Dynamical Systems

Stuart Allie, Alistair Mees, Kevin Judd and Dave Watson
Centre for Applied Dynamics and Optimization
Department of Mathematics
The University of Western Australia
Perth, 6907, Australia

Abstract

Triangulation and tesselation methods have proven successful for reconstructing dynamical systems which have approximately the same dynamics as given trajectories or time series. Such reconstructions can produce other trajectories with similar dynamics, so giving a system on which one can conduct experiments, but can also be used to locate equilibria and determine their types, carry out bifurcation studies, estimate state manifolds and so on. In the past, reconstruction by triangulation and tesselation methods has been restricted to low-noise or no-noise cases. This paper shows how to construct triangulation models for noisy systems; the models can then be used in the same ways as models of noise-free systems.

1 Introduction

An important part of controlling systems is modeling them. The system identification and modeling problem is well-studied for linear systems, but there seem to be few results for strongly nonlinear systems. In this paper we present an approach that generalizes the triangulation method of modeling nonlinear dynamical systems, presented at the previous conference in this series [9].

The earlier work applied only to low-noise systems (roughly, where the noise level was small compared with the typical separation of the data points being used to build the model). This paper deals with the case where the noise is non-negligible. We still simplify matters somewhat, as will be explained below.

1.1 Embedding

It may be helpful to say a few words about embedding and reconstruction. Since this is a well-discussed area (see e.g. [1, 16]) we shall be brief. We start with a discrete-time dynamical system with state $x_t \in M$, where M is a d-dimensional manifold, with the dynamics described by

$$x_{t+1} = f(x_t) + \xi_t,$$

where f is unknown and ξ_t is the dynamical noise of the system. We assume that there is a measurement function $\phi : M \to \mathbb{R}$ such that

$$y_t = \phi(x_t) + \nu_t,$$

with ν_t being the measurement noise. The set $\{y_t\}$ is the scalar time series which is measured. This set is all that is available for constructing the model. From this, we make vectors in R^n of the form

$$z_t = (y_t, y_{t-\tau}, \ldots, y_{t-(n-1)\tau})$$

for an embedding dimension n and lag τ.

By a theorem due to Takens [17], we know that, in general, if n is sufficiently large, then such a mapping is an embedding $h : M \to \mathrm{R}^n$ with $z_t = h(x_t)$. Since h is an embedding, the following diagram commutes.

$$\begin{array}{ccc} x_t & \xrightarrow{f} & x_{t+1} \\ h \downarrow & & h \downarrow \\ z_t & \xrightarrow{F} & z_{t+1} \end{array}$$

It is the map F which we wish to model. However, because of the form of the embedded states z_t, it is sufficient to estimate the map $\rho : \mathrm{R}^n \to \mathrm{R}$ such that

$$y_{t+1} = \rho(z_t) + \epsilon_t$$

where ϵ_t represents the total noise present.

2 Triangulation approximations

2.1 Triangulating noise-free systems

Before looking at the proposed solution to the general modeling problem let us review the noise-free case. Notice that when $\epsilon = 0$ the problem is to find a map from R^n to R that fits the given data. One of us (AM) has discussed elsewhere [9, 11] how to optimally triangulate a set of points in R^n and then to produce a map $\hat{\rho}$ that is linear on each triangle (or simplex, if $n > 2$), is continuous across triangle boundaries, and is such that for each vertex v_i,

$$\hat{\rho}(v_i) = h_i$$

where h_i is the actual value (the "height") given by the data at v_i. We will see shortly what is meant by an optimal triangulation; probably the best method for constructing it is Watson's [18], and in this paper we will assume that construction of triangulations is a solved problem.

The definition of $\hat{\rho}$ is simple once the triangulation has been constructed: if $x \in T_j$ where triangle T_j has vertices $\{v_k\}$, where k is a set of $n+1$ labels to the set $\{v_i\}$, then there is a unique non-negative solution λ to the equations

$$\sum_{k \in N(x)} \lambda_k(x) v_k = x, \qquad (1)$$

$$\sum_{k \in N(x)} \lambda_k(x) = 1. \qquad (2)$$

Now define

$$\hat{\rho}(x) = \sum_{k \in N(x)} \lambda_k(x) h_k; \qquad (3)$$

that is, we express x as a convex combination of the vertices of the triangle it is in, then use the same convex combination of the vertex heights to approximate the height at x. It is easy to see that $\hat{\rho}$ is continuous and that if the heights did in fact satisfy
$$h = \rho(x)$$
for some smooth function ρ then
$$|\hat{\rho}(x) - \rho(x)| = O(\epsilon^2)$$
where ϵ is the diameter of triangle T_i. (This, incidentally, explains why we do not want to use just any triangulation: we should pick the one that minimizes diameters on average, which turns out to be the Delaunay triangulation [5, 19, 14, 11].)

We can now define an approximation $\hat{F}$ for the dynamical system from R^n to R^n; for example, in the two dimensional case we use
$$\hat{F}(y_t, y_{t-1}) = (\hat{\rho}(y_t, y_{t-1}), y_t) \tag{4}$$
where we are assuming a simple lag-1 embedding. This map takes triangles into triangles; for example, Fig 1 shows a triangulation of an embedding of the Henon

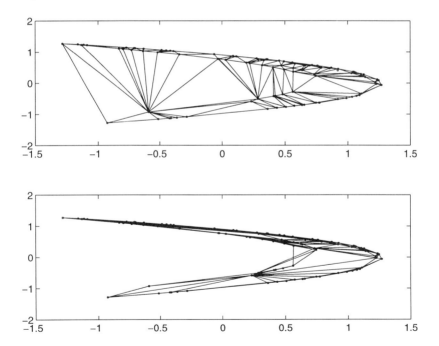

Figure 1: (a) The Delaunay triangulation of some points on an embedding of the x_1 coordinate of a trajectory from the the Henon map. (b) The image of the triangulation under the map $\hat{F}$ defined in (4).

map and the image of the triangulation. This picture alone gives us a good feel for the effect of the map, which is to take a region of the plane and bend and compress it onto a roughly horseshoe-shaped region. For more details of applications, see [9, 10, 11].

2.2 Triangulating noisy systems

The approximation defined by (3) assumes that we know the values of the h_i. In the noisy case, if our v_i are data points then we have noisy data values as an approximation to the h_i. If the v_i are not data points, then we have no a priori information about the $h_i = \rho(v_i)$. We *might* choose the set $\{v_i\}$ to be a subset of our (embedded) data points $\{z_t\}$, but there is no requirement to do so. Our modeling problem may now be expressed as the problem of selecting the v_i and the h_i.

Assuming we have chosen the v_i in some way, we wish to choose the h_i so as to give the best approximation to f given $\{z_t\}$ and $\{y_t\}$. To do this, we note that we can define a matrix Λ with elements $\Lambda_{ij} = \lambda_j(x_i)$ and we define the λ by (1) for $j \in N(x_i)$ and $\lambda_j(x_i) = 0$ for $j \notin N(x_i)$. We can then write the approximation as a matrix equation thus:

$$\hat{\rho}(x_i) = \sum_{j \in N(x_i)} \lambda_j(x_i) h_j \quad (5)$$

$$= \sum_j \lambda_j(x_i) h_j$$

$$= (\Lambda h)_i \quad (6)$$

where $h = (h_1, \ldots, h_{n_v})^T$ for n_v vertices. We solve for the h_i by solving the least squares problem defined by

$$\text{minimise}: \quad (y - \Lambda h)^T (y - \Lambda h) \quad (7)$$
$$\text{over}: \quad h \quad (8)$$

where $y = (y_1, \ldots, y_n)^T$.

All that remains then is to describe a method for choosing the $\{v_i\}$. To do this, we utilize a heuristic based on expansion and contraction of the model. We allow the model to grow by, say, k vertices, by selecting the k data points with the worst fitted data values, that is, the largest values of $|y_t - \hat{f}(x_t)|$. We then remove the k vertices which are (or are closest to) the data points with the *best* fitted data values. This process of growing and shrinking the model continues until either their is no change in the set of vertices before and after the expansion and contraction or the MSSE fails to improve after expansion and contraction (in which case, the previous set of vertices with the better MSSE is kept.)

In brief then, the final algorithm is as follows:

1. Construct an initial affine model using $(n+1)$ vertices which form a single simplex containing the entire data set.

2. Select the m data points with the worst fitted data values. Bring these points into the triangulation.

3. Select the m data points with the best fitted data values. Remove these points from the triangulation.

4. If the triangulation has changed and the MSSE has improved, go back to 2.

5. Calculate the minimum description length for the current model size.

6. Have we found a local minimum of description length (as a function of number of vertices)? If so, stop now.

7. Increase the model size by one by bringing the worst fitted point into the triangulation and goto 2.

We check for the existence of a (local) minimum of description length as a function of model size (number of vertices) simply by storing the model with the smallest description length to date and declaring it to be a local minimum if we have not found a smaller value after a certain number of increases in model size.

3 Which model is best?

3.1 Model Selection Criteria

Once we are able to construct a model of a given size (that is to say, of a given number of parameters), we must decide how to choose between different sizes. There are various criteria in general use, such as the Akaike Information Criterion (AIC) and the Bayesian Information Criterion (BIC). In this work however, we utilize the criterion of Minimum Description Length (MDL) as described by Rissanen [15]. The description length of a data string is essentially the length of code required, in some optimal coding system, to completely describe the data in terms of a given model. The application of MDL to model selection is discussed in some detail in [7].

3.2 Description Length of Triangulation Models

In order to encode our model, we must transmit sufficient information for the receiver to be able to reconstruct the data completely. One way of doing this is by using what Rissanen calls a "two-part code". Firstly, we encode the information about the residuals of the model as defined by $\epsilon_t = y_t - \hat{\rho}(x_t)$. It is a well-known result of coding theory that one can encode this information in a code of length bounded below by $-\log_2 P(\epsilon)$ bits, where $P(\epsilon)$ is the probability distribution of the residuals, and this length is approachable to better than 1 bit [4]. Secondly, we must encode information about our model. In our case, this means that we must transmit the values of the vertex positions $v_i \in R^n$ and their heights $h_i \in R$. The values of these parameters are, in principle, known to arbitrary precision and we must truncate them to transmit them in a code of finite length. The important feature of the application of the MDL criterion is that we can optimize over the truncation. That is, we can choose the parameter precisions which give the smallest description length of the data.

In the present case, we *assume* that the residuals have a Gaussian distribution, in order to simplify the calculation of description length. This is a reasonable assumption in light of the fact that we choose the heights h_i by solving a least squares problem as stated above. However, in practice, we find that the residuals are often not sufficiently close to Gaussian and we could improve our models by using a more sophisticated model of the noise.

It is worth noting how the MDL criterion selects the "best" model in some well-defined sense. The term in the description length which describes the residuals will, in general, decrease as the model size increases, due to more accurate fitting of the data. However, as the model size increases, the code needed to describe the model parameters will tend to increase. It is in this way that the MDL criterion selects the model which best achieves a balance between the competing aims of fitting the data as accurately as possible and the desire for parsimony in the model.

4 Results

In this section, we present the results of some preliminary applications of the modeling techniques described here.

4.1 Example of triangulated surface: the Ikeda map

The Ikeda map [3] is a map $\theta : R^2 \to R^2$ defined by

$$\begin{aligned} z &= 0.4 - 6/\left(1 + x^2 + y^2\right) \\ \theta(x,y) &= \left(a + b \times (x\cos(z) - y\sin(z)), b \times (x\sin(z) + y\cos(z))\right), \end{aligned} \quad (9)$$

where a and b are parameters. For the data used, we have $a = 1$, $b = 0.7$. For this example, we are modeling the x-component of the map. This then gives us a map $\rho : R^2 \to R$ which we model by triangulation of the data. The data used to generate this figure was a trajectory of some 500 points with Gaussian noise of standard deviation 0.05 added. Figures 2 and 3 show the surface defined by ρ and an example of the piecewise linear surface which is defined by the approximation $\hat{\rho}$. The model constructed used 25 vertices selected from the data points. Note that although the approximated surface appears to differ considerably from the correct surface, the approximation produces dynamics which are very similar to the original system.

4.2 The Rössler System

The data set chosen was generated by the numerical integration of the equations of motion which produce the Rössler attractor. These equations are

$$\begin{aligned} \dot{x} &= -z - y, \\ \dot{y} &= x + ay, \\ \dot{z} &= b + z(x - c), \end{aligned} \quad (10)$$

where a, b and c are parameters. The original time series consisted of 1000 points of such data. A small amount of normally distributed noise (with a standard deviation of 0.04) was added to this. The data was embedded in 3 dimensions, with a lag of 3, and the prediction "lag" was for one time step. Hence, we are modeling the map f defined by.

$$y_t = f(x_t),$$

where

$$x_t = (y_{t-1}, y_{t-4}, y_{t-7}).$$

The model selected by the MDL principle contained only 7 vertices chosen from the data points, plus 4 vertices placed well outside the convex hull of the data to give

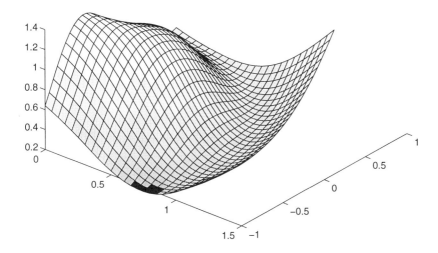

Figure 2: The surface defined by the x-coordinate of the Ikeda map.

an enclosing simplex. Despite this small model (in terms of number of parameters), the model seems to encompass much of the important dynamics of the system. Sample trajectories of the model are shown in figure 4 alongside the trajectory from which the model was constructed. There are two types of trajectories shown, those run without any "dynamical" noise, and those with. The noise added was chosen by random selection from the residuals of the model, hence this type of orbit represents a form of "dynamical bootstrapping." Note that the model reproduces the important folding of the attractor which is the major nonlinearity of this system and the source of the interesting dynamics. In figure 5, we see the accuracy of the model in making one-step predictions.

5 Conclusions

The present paper can best be seen as a research announcement. It is the beginnings of applying to triangulation modeling the methods developed by two of the authors (AM and KJ) for radial basis modeling, as described elsewhere in this volume [6]. A fuller description of this work and, in particular, more detailed applications, will be found in [2].

There are two parts to the modeling process: an easy part, involving linear least-squares fitting (at least when the noise is Gaussian) and a hard part, involving

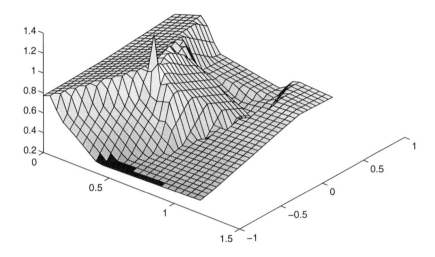

Figure 3: The surface defined by the approximation $\hat{p}$ to the x-coordinate of the Ikeda map.

selection of vertex locations. The problem fits into the format we have described elsewhere as *pseudo-linear* modeling [7] and our solution is of the same kind as we used for radial-basis models: we reduce the hard nonlinear part of the problem to a linear (but still NP-hard) subset selection problem which we then solve by a heuristic method which has a history of performing well [13, 12, 7]. In the present case the nonlinear part is the selection of vertices in a triangulation of given size, and we choose some set of candidate vertices such as the data points themselves, then select the subset as described in Section 2.2. The result is a function from R^n to R which models the dynamics of one or more embedded time series.

Selecting the best model from among candidates of different sizes is done using minimum description length, which is described in detail elsewhere. This approach has worked well with other pseudo-linear models of a number of physical systems [8] and we expect it to work well in the present case, though further research is necessary.

The main limitations of the present approach are the inaccuracy of the approximation and the limitations of the noise model used. The former problem is understood, at least in principle, and the main gains are likely to arise from a more sophisticated noise model.

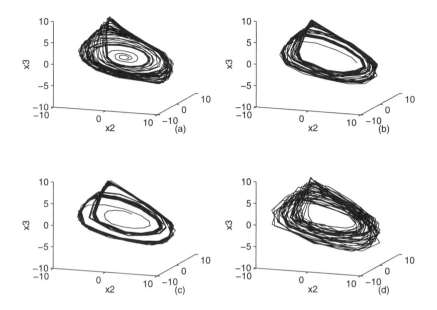

Figure 4: (a) The original time series (embedded). (b) and (c) Free run trajectories of the model. (d) A "bootstrapped" trajectory of the model.

References

[1] A. M. Albano, A. I. Mees, G. C. deGuzman, and P. E. Rapp. Data requirements for reliable estimation of correlation dimensions. In A. V. Holden H. Degn and L. F. Olsen, editors, *Chaos in biological systems*, pages 207–220. Plenum, New York, 1987.

[2] S. P. Allie, A. I. Mees, K. Judd, and D. F. Watson. Triangulations, noise and minimum description length. *In preparation*, 1995.

[3] M. Casdagli. Nonlinear prediction of chaotic time series. *Physica D*, 35:335–356, 1989.

[4] C. M. Goldie and R. G. E. Pinch. *Communication Theory*, volume 20. Cambridge University Press, Cambridge, 1991.

[5] P. J. Green and R. Sibson. Computing Dirichlet tessellations in the plane. *The Computer Journal*, 21:168–173, 1978.

[6] K. Judd and A. I. Mees. Pseudo-linear models and subset selection.

[7] K. Judd and A. I. Mees. On selecting models for nonlinear time series. *Physica D*, 82:426–444, 1995.

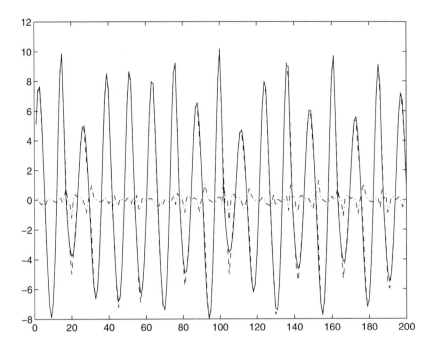

Figure 5: One step predictions of the model for the Rössler system. The solid line is the test data series, the dashed line is the one-step prediction of the model and the dash-dot line is the difference between the test set and the model. Note that even with a very small model (only 11 vertices altogether), the one-step predictions are very good.

[8] K. Judd, A. I. Mees, T. Molteno, and N. Tufillaro. Modeling chaotic motions of a string from experiment data. *in preparation*, 1995.

[9] A. I. Mees. Modelling complex systems. In A. I. Mees T. Vincent and L. S. Jennings, editors, *Dynamics of Complex Interconnected Biological Systems*, volume 6, pages 104–124. Birkhauser, Boston, 1990.

[10] A. I. Mees. Modelling dynamical systems from real-world data. In A. Passamente N. B. Abraham, A. M. Albano and P. E. Rapp, editors, *Measures of Complexity and Chaos*, volume 208, pages 345–349. Plenum, New York, 1990.

[11] A. I. Mees. Dynamical systems and tesselations: Detecting determinism in data. *International Journal of Bifurcation and Chaos*, 1:777–794, 1991.

[12] A. I. Mees. Nonlinear dynamical systems from data. In F. P. Kelly, editor, *Probability, Statistics and Optimisation*, pages 225–237. Wiley, Chichester, England, 1994.

[13] A. I. Mees. Reconstructing chaotic systems in the presence of noise. In M. Yamaguti, editor, *Towards the Harnessing of Chaos*, pages 305–321. Elsevier, Tokyo, 1994.

[14] A. Okabe, B. Boots, and K. Sugihara. *Spatial Tesseslations: Concepts and Applications of Voronoi Diagrams*. Wiley, Chichester, England, 1992.

[15] J. Rissanen. *Stochastic Complexity in Statistical Inquiry*, volume 15. World Scientific, Singapore, 1989.

[16] T. Sauer, J. A. Yorke, and M. Casdagli. Embedology. *J. Stat. Phys.*, 65:579–616, 1992.

[17] F. Takens. Detecting strange attractors in turbulence. In D. A. Rand and L. S. Young, editors, *Dynamical Systems and Turbulence*, volume 898, pages 365–381. Springer, Berlin, 1981.

[18] D. F. Watson. Computing the n-dimensional Delaunay tesselation with application to Voronoi polytopes. *The Computer Journal*, 24:167–172, 1981.

[19] D. F. Watson. *Contouring: A guide to the analysis and display of spatial data*. Pergamon Press, 1992.

Commentary by I. Mareels

The paper discusses a universal modelling technique based on the idea of triangulating noisy data obtained from the output of a dynamical system, whose state may also be perturbed by noise.

The paper really proposes a method on how to go about constructing a model in a very general framework. It leaves the reader to speculate on how the analysis may be performed and possible success may be quantified.

It is nevertheless a very thought provoking paper and its emphasis on modelling is most appreciated.

Commentary by C. Sparrow

I find the triangulation method of making models from data, as described in this paper, attractive. In contrast to some other methods they are really trying to model the underlying geometry, rather than just attempting to model the data. In this paper the data is assumed to be noisy, but to come from a system where the behaviour is to a great extent determined by a low-dimensional deterministic system. In this case there is serious question about how to pick a best model; if the model is too crude it will be simple to state, but will not reproduce the data very well, whilst if it is too fine it will produce better results but may be almost as complicated to describe as the original data set. The minimal description length criterion for choosing a model is interesting, even though in practical applications it is very unlikely that your motivation is to provide an optimal coding of the data set. I would like to know how this criterion relates to the geometric virtue of triangulation methods – i.e. if an MDL model is chosen, does this model also come close to giving you the best geometric insight into the underlying dynamics?

Attractor Reconstruction and Control Using Interspike Intervals

Tim Sauer
Department of Mathematics
George Mason University
Fairfax, VA 22030, USA

Abstract

Takens showed in 1981 that dynamical state information can be reconstructed from an experimental time series. Recently it has been shown that the same is true from a series of interspike interval (ISI) measurements. This method of system analysis allows prediction of the spike train from its history. The underlying assumption is that the spike train is generated by an integrate-and-fire model. We show that it is possible to use system reconstruction based on spiking behavior alone to control unstable periodic trajectories of the system in two separate ways: by making small (subthreshold) changes in a system parameter, and by using an on-demand pacing protocol with superthreshold pulses.

1 Introduction

Prediction of the future behavior of a system is possible to the extent that the system is deterministic. When it is possible to predict the future with a deterministic model, predictability can be exploited to characterize and control the system.

The discovery of chaos in simple nonlinear deterministic systems has led to a change in modes of system analysis, in particular for the description of the time evolution of systems. The new ingredient due to chaos is the possibility that nonlinear systems with few degrees of freedom, while deterministic in principle, can create output signals that appear complex, and mimic stochastic signals from the point of view of conventional time series analysis. The reason for this is that trajectories that have nearly identical initial conditions will separate from one another at an exponentially fast rate. This exponential separation causes chaotic systems in the laboratory to exhibit much of the same medium to long-term behavior as stochastic systems. However, the key fact is that short-term prediction is not ruled out for chaotic systems, if there are a reasonably low number of active degrees of freedom.

State-space reconstruction methods for nonlinear systems began with the pioneering work of Packard et al. [7] and Takens [12]. This work showed that vectors representing a number of delayed versions of some observable measurement can be put in a one-to-one correspondence with the states of the system being observed. We summarize these ideas, and their later extensions [10], in Section 2 of this article.

The fact that the current system state can be identified using a vector of time series measurements leads to a number of applications, since analysis of the geometry of the chaotic attractor underlying the time series can be performed in the state-space reproduced from the delay-coordinate vectors. Applications include noise-filtering and prediction of chaotic time series, and control of unstable periodic orbits solely from a time series record [6].

At present, increased attention is being paid to the dynamical properties of physiological systems, and particularly to the nonlinear aspects of the systems. Given the nonlinearity of the subsystems involved, it is often argued that the existence of deterministic chaotic behavior in physiological systems is likely. Although irregular signals are being produced, useful information is being processed by the system. However, little direct evidence for chaos exists within the research literature at present, possibly due to the difficulty of acquiring precise data and the potential high-dimensionality of the dynamics.

A characteristic feature of dynamical processing in physiological systems is the capacity for information to change in form while being passed between individual units. Perhaps some information-passing in nervous systems depends primarily on spike trains, or temporal patterns of firing of action potentials, and is largely independent of the amplitudes of the potentials. In these systems, one expects the information content to be contained in the temporal pattern of firings, rather than a time series.

Recent work [9] has shown that state-space reconstruction is possible using vectors of successive spike train measurements, under fairly general assumptions on the dynamical source of the spikes. Specifically, it is shown that if spikes are produced by an integrate-and-fire model from a generic signal from some dynamical system, then there is a one-to-one correspondence between the system states and interspike interval vectors. We will review these ideas in Section 3. See [11] for an application of interspike interval analysis to experimental neurological data.

Section 4 discusses a practical consequence of the correspondence: that spike trains of moderate length are predictable. Section 5 shows that the correspondence allows us to control chaos, in the sense that an unstable orbit of an attractor can be stabilized either with small or large perturbations, using knowledge of the system dynamics learned through spike trains. Section 6 discusses the limitations of this study and directions for further work.

2 State representation from time series

When studying phenomena through laboratory experiment, it is seldom possible to measure all relevant dynamical variables. In many cases the dimension of the state space relevant to the system may be infinite. How can the dynamical behavior be studied in such a situation?

A key element in resolving this general class of problems is provided by the mathematics of embedding theory. In typical situations, points on the dynamical attractor in the full system phase space have a one-to-one correspondence with measurements of a limited number of variables. This is a powerful fact. By definition, a point in the phase space carries complete information about the current system state. If the equations defining the system dynamics are not explicitly known, this phase space is not directly accessible to the observer. A one-to-one correspondence means that the phase space state can be identified by measurements.

Assume we can simultaneously measure m variables $(y_1(t), \ldots, y_m(t))$, which we denote by the vector $\mathbf{y}(t)$. The m-dimensional vector $\mathbf{y}$ can be viewed as a function of the system state $\mathbf{x}(t)$ in the full system phase space:

$$\mathbf{y} = \mathbf{F}(\mathbf{x}) = (f_1(\mathbf{x}), \ldots, f_m(\mathbf{x})). \tag{1}$$

We call the function $\mathbf{F}$ the *measurement function*, and we call the m-dimensional vector space in which the vectors $\mathbf{y}$ lie the *reconstruction space*.

We have grouped the measurements as a vector-valued function $\mathbf{F}$ of $\mathbf{x}$. The fact that $\mathbf{F}$ is a function is a consequence of the definition of state – information about the system is determined uniquely by the state, so each measurement is a well-defined function of $\mathbf{x}$.

As we discuss in detail below, as long as m is taken sufficiently large, the measurement function $\mathbf{F}$ generically defines a one-to-one correspondence between the attractor states in the full phase space and m-vectors $\mathbf{y}$. By one-to-one, we mean that for a given $\mathbf{y}$ there is a *unique* $\mathbf{x}$ on the attractor such that $\mathbf{y} = \mathbf{F}(\mathbf{x})$. When there is a one-to-one correspondence, each vector formed of m measurements is a proxy for a single system state, and the fact that the entire system information is determined by a state $\mathbf{x}$ is transferred to be true as well for the measurement vector $\mathbf{F}(\mathbf{x})$. In order for this to be so, it turns out that it is enough to take m larger than twice the box-counting dimension of the phase space attractor.

The one-to-one property is useful because the state of a deterministic dynamical system, and thus its future evolution, is completely specified by a point in the full phase space. Suppose that when the system is in a given state $\mathbf{x}$ one observes the vector $\mathbf{F}(\mathbf{x})$ in the reconstruction space, and that this is followed one second later by a particular event. If $\mathbf{F}$ is one-to-one, each appearance of the measurements represented by $\mathbf{F}(\mathbf{x})$ will be followed one second later by the same event. This is because there is a one-to-one correspondence between the attractor states in phase space and their image vectors in reconstruction space. Thus there is predictive power in measurements $\mathbf{y}$ that are matched to the system state $\mathbf{x}$ in a one-to-one manner.

A key element in such an analysis is provided by embedding theory. In typical situations, points on the dynamical attractor in the full system phase space have a one-to-one correspondence with vectors of time-delayed amplitude measurements of a generic system variable. System analysis of this type is suggested in [7], and an embedding theorem due to Takens [12] (later extended in [10]) established its validity.

3 State representation from spike trains

A statistical process in which the dynamical information is carried by a series of event timings is called a *point process*. See [4] for an introduction to the statistical literature on this subject. The main question of this section is the following: If a point process is the manifestation of an underlying deterministic system, can the states of the deterministic system be uniquely identified from the information provided by the point process?

In particular, neurobiological systems are often marked by measurable pulses corresponding to a cell reaching a threshold potential, which triggers rapid depolarization, followed by repolarization, which restarts the cycle. The times of these discrete pulses can be recorded. This type of data differs from a time series of an observable measured at regular time intervals. Many hypotheses and models for the description of the time variability of these pulses have been proposed [14].

Our approach is meant to be a simple paradigm that shares qualitative characteristics with real systems, and that demonstrates feasible ways in which complicated information about system states can be communicated and transformed in type. In particular, we will focus on the linkage between continuous dynamics and the interspike intervals (ISI's) produced by them, rather than modeling the detailed mechanism of any single system.

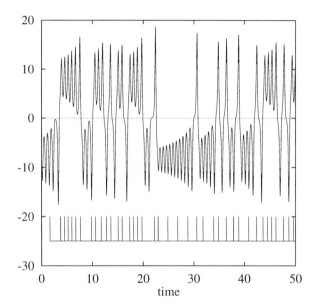

Figure 1: The upper trace is a solution of the Lorenz equations graphed as a function of time. The lower trace shows the times at which spikes are generated according to equation 2, with $S(t) = (x(t) + 2)^2$ and $\Theta = 60$.

For concreteness, we make a simple hypothesis connecting an underlying continuous dynamical system to the point process. The time series from the dynamical system is integrated with respect to time; when it reaches a preset threshold, a spike is generated, after which the integration is restarted. This integrate-and-fire model is chosen for its simplicity and potential wide applicability. In our simulations we hypothesize that the input to be integrated is a low-dimensional chaotic attractor. We use the Mackey-Glass equation and the Lorenz equations as simple representative examples.

Let $S(t)$ denote the signal produced by a time-varying observable of a finite-dimensional dynamical system. Assume that the trajectories of the dynamical system are asymptotic to a compact attractor X. Let Θ be a positive number which represents the firing threshold. After fixing a starting time T_0, a series of "firing times" $T_1 < T_2 < T_3 < \ldots$ can be recursively defined by the equation

$$\int_{T_i}^{T_{i+1}} S(t) \, dt = \Theta \tag{2}$$

From the firing times T_i, the interspike intervals can be defined as $t_i = T_i - T_{i-1}$.
Figure 1 shows a time trace of the x-coordinate of the Lorenz [2] equations

$$\begin{aligned}
\dot{x} &= \sigma(y-x) \\
\dot{y} &= \rho x - y - xz \\
\dot{z} &= -\beta z + xy
\end{aligned} \qquad (3)$$

where the parameters are set at the standard values $\sigma = 10, \rho = 28, \beta = 8/3$. In the lower part of the figure, the spiking times generated by (2) with threshold $\Theta = 60$ are shown as vertical line segments.

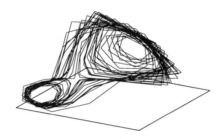

Figure 2: Reconstruction of Lorenz dynamics from a spike series. Vectors of three successive interspike time intervals are plotted, connected by line segments. The series of time intervals used to make the recontruction was created using integrate-and-fire model (2), with an input signal from the Lorenz attractor.

It turns out that if the interspike intervals are finite, and under certain genericity conditions on the underlying dynamics, signal and threshold, the series $\{t_i\}$ of ISI's can be used to reconstruct the attractor X. In other words, there is a one-to-one correspondence between m-tuples of ISI's and attractor states, which associates each vector $(t_i, t_{i-1}, \ldots, t_{i-m+1})$ of ISI's with the corresponding point $x(T_i)$ on the attractor. In analogy with the original Takens' theorem [12] and its generalization [10], the condition $m > 2D_0$ is sufficient, where D_0 is the box-counting dimension of the attractor X. As with Takens' theorem, smaller m may be sufficient in particular cases.

Figure 2 shows a delay plot of interspike intervals produced by the x-coordinate of the Lorenz attractor [2]. Since the box-counting dimension of the Lorenz attractor is about 2.05, the theorem stated above says that generic reconstructions in $m = 5$ dimensions are topologically equivalent to the original Lorenz attractor, although its orientation in five dimensional Euclidean space could be rather twisted. Figure 2 can be viewed as a projection from five dimensional space to three dimensional space of this topologically equivalent attractor.

4 Nonlinear prediction of spike timings

One practical consequence of such a theorem is that the ISI vectors $(t_i, \ldots, t_{i-m+1})$ can be used to reconstruct the attractor X sufficiently to make measurements of dynamical invariants of X possible, and to do short-term prediction on the series $\{t_i\}$. Thus the possibility exists of predicting future interspike intervals from past history, which has practical applications. To explain the meaning of short-term, it is instructive to recall the time series case. For chaotic time series, the length of time horizon for which prediction is effective depends on the separation time of nearby trajectories, and so inversely on the largest Lyapunov exponent. For ISI series, the horizon depends both on the Lyapunov exponents of the underlying process and on the threshold Θ. As Θ increases, the predictability decreases, as seen below.

We will apply a simple version of a nearest-neighbor prediction algorithm. The version used here is sufficient to measure the level of determinism in the series. In order to quantify the predictability of the series of ISI's we will use the concept of surrogate data [13] to produce statistical controls.

The prediction algorithm works as follows. Given an ISI vector $V_0 = (t_{i_0}, \ldots, t_{i_0-m+1})$, the 1% of other reconstructed vectors V_k that are nearest to V_0 are collected. The values of the ISI for some number h of steps into the future are averaged for all k to make a prediction. That is, the average $p_{i_0} = \langle t_{i_k+h} \rangle_k$ is used to approximate the future interval t_{i_0+h}. The difference $p - t_{i_0+h}$ is the h-step prediction error at step i_0. We could instead use the series mean m to predict at each step; this h-step prediction error is $m - t_{i_0+h}$. The ratio of the root mean square errors of the two possibilities (the nonlinear prediction algorithm and the constant prediction of the mean) gives the normalized prediction error

$$\text{NPE} = \frac{\langle (p_{i_0} - t_{i_0+h})^2 \rangle^{1/2}}{\langle (m - t_{i_0+h})^2 \rangle^{1/2}} \qquad (4)$$

where the averages are taken over the entire series. The normalized prediction error is a measure of the (out-of-sample) predictibility of the ISI series. A value of NPE less than 1 means that there is linear or nonlinear predictability in the series beyond the baseline prediction of the series mean.

Our goal in predicting ISI's is to verify that the nonlinear deterministic structure of the dynamics that produced the intervals is preserved in the ISI's. Linear autocorrelation in the time series, present for example in correlated noise, can cause NPE to be less than 1. In order to control for this effect, we calculate the NPE for the original series of ISI's as well as for stochastic series of the same length, called surrogate data, which share the same linear statistical properties with the original series.

We use two types of surrogate data in our analysis. The first, called a random phase (RP) surrogate, is a series with the same power spectrum as the original series, but is the realization of a stochastic process. The autocorrelation of the original series is preserved in the surrogate series, while the nonlinear deterministic structure is eliminated. If the predictability of the original can be shown to be statistically different from the predictability of such surrogates, the null hypothesis that the original series was produced by a stochastic process can be rejected, which is evidence that the nonlinear structure of the underlying dynamical system is present in the interspike intervals. The second type of surrogate, called a Gaussian-scaled shuffle (GS) surrogate [13], is a shuffle (rearrangement) of the original series. The GS surrogate (called an amplitude-adjusted surrogate in [13]) corresponds to the null hypothesis that the series is a monotonically-scaled version of amplitudes produced by a Gaussian random process with similar autocorrelation.

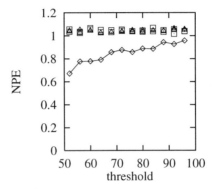

Figure 3: The graph shows normalized one-step prediction error for 12 ISI series created using the Lorenz attractor input signal $S(t) = (x+2)^2$, and varying thresholds Θ along the horizontal axis. Diamonds denote the normalized prediction error of the ISI series; squares (RP) and triangles (GS) denote the NPE of the surrogate series.

The result of applying the prediction algorithm to interspike intervals from the Lorenz equations is shown in Figure 3. For these calculations we fixed the embedding dimension $m = 3$, and predicted one step ahead ($h = 1$). Several surrogate series of each of the two types were generated. Prediction results for the original ISI series, the random-phase and Gaussian-scaled surrogates are shown in the Figure. For low threshold there is a statistically significant difference between the original series and its surrogates. The conclusion is that there is predictability in the ISI series caused by the underlying deterministic dynamics. (More precisely, there is predictability not explained by any of the null hypotheses controlled for by the surrogate data.)

The theory implies predictability, but gives no estimate for the length of spike train necessary to detect it. Our findings indicate that data sets of lengths accessible in physiological settings may be sufficient in principle.

A second important point shown by Figure 3 regards the reconstruction quality as a function of threshold. As the threshold θ in (2) is increased, the effect on the interspike interval reconstruction is analogous to increasing the sampling interval in the case of time series reconstruction. We expect the quality of the reconstruction to degrade when the threshold is high enough that the intervals become comparable to the reciprocal of the largest Lyapunov exponent, the decorrelation time of the dynamics. Whether predictability can be found in a particular set of experimental data may depend on this issue.

5 Control from interspike intervals

Stabilization of unstable periodic orbits in chaotic systems using small controls was described by Ott, Grebogi, and Yorke [5], and is commonly referred to as OGY control. Several experimental examples of the success of this type of stabilization are documented in [6]. The fundamental idea is to determine some of the low period unstable orbits embedded in the attractor, and then use small feedback to control the system. The OGY method is model-free, in that it relies on time series information to reconstruct the linear dynamics in the vicinity of the unstable orbit.

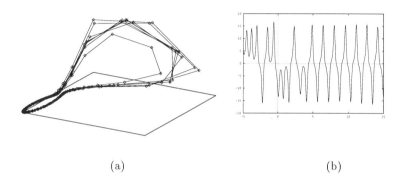

(a) (b)

Figure 4: Subthreshold control of the unstable "figure-eight" orbit of the Lorenz attractor, using the interspike interval history only. (a) A plot in reconstructed state space. The three dimensions of the space are three consecutive time intervals. After an initial transient, the systems becomes trapped in the figure-eight orbit. (b) A plot of the time series from the Lorenz attractor, which is being integrated to form the spikes. The vertical line shows the time at which the control protocol is turned on.

In this section we would like to prove the viability of this concept when the dynamical information is carried by interspike intervals instead of a time series. We will demonstrate two different types of control protocols. The first will be the analogue of OGY control for spike trains. Using small changes in a system parameter, an unstable orbit will be stabilized. This type of control is called subthreshold control, since the control signal is not large enough on their own to make the system fire. For this demonstration we will use spike intervals produced by an observable from the Lorenz system, and stabilize the figure-8 trajectory.

The second protocol will use superthreshold pulses to stabilize an unstable equilibrium. This technique, often called on-demand pacing, will be shown to work for deterministic as well as random integrate-and-fire systems. Therefore, although the success of subthreshold control can be viewed as evidence of deterministic dynamical behavior in an unknown system, the same cannot be said for superthreshold control.

We begin with subthreshold control. The standard version of the Lorenz attractor exhibits trajectories cycling erratically around two unstable equilibrium points.

The number of times a typical trajectory cycles around one point before moving to the other is known to be unpredictable on moderate to long time scales. We will choose to stabilize the unstable periodic trajectory which cycles exactly once around one equilbrium, then exactly once around the other, before returning to the first and repeating the process. This "figure-eight" trajectory of the Lorenz attractor is unstable, in the sense that a trajectory started near this behavior will tend to move away from it.

In this feasibility study we used the Lorenz equations (3) to generate an input signal $S(t) = x(t) + 25$. The parameters $\sigma = 28$ and $\rho = 10$ are set at the standard values, and the parameter β is nominally set at the standard value 8/3, but will be used as control parameter.

Using the reconstruction of Lorenz dynamics solely from the interspike intervals produced by (2), local linear models for a pair of points along the figure-eight orbit were constructed numerically. This step involves careful use of interpolation, since as Figure 4(a) shows, any given point on the reconstructed orbit may be reached only rarely by an interspike interval vector.

Figure 4(a) shows the result of the control procedure. Using small perturbations to the nominal value of β, after an initial transient (the single inside loop), the stabilization is achieved, and the system will continue the figure-eight behavior indefinitely. A time trace of the x-coordinate of the underlying Lorenz attractor being used to drive the integrate-and-fire model is shown in Figure 4(b). It was not used in the controlling process, and is shown here simply to verify success of the control procedure.

Superthreshold control of spike train dynamics is conceptually simpler. In this scenario, the goal is to change an erratic firing pattern to a regular pattern. The controller has the capability of causing the system to fire at any time, say through a large pulse applied to the system. We choose a desired fixed interspike interval and then ask the contoller to apply superthreshold spikes to regulate the system to spiking rhythmically at that interval. In the on-demand pacing protocol, the controller applies the external spike whenever the interspike interval has exceeded the desired value. Obviously, this will cause the spike intervals to be capped at the desired level. What we will show is that in addition, under realistic assumptions on the dynamics, the spike times will be bounded from below as well, resulting in an evenly-timed spike sequence.

To make a model which is perhaps closer to a neurophysiological process, we add to the integrate-and-fire model (2) some relative refractoriness. For this experiment we use a time series from the Mackey-Glass equation as the signal to be integrated. The Mackey-Glass [3] delay differential equation is

$$\dot{x}(t) = -0.1x(t) + \frac{0.2x(t-\Delta)}{1+x(t-\Delta)^{10}}. \tag{5}$$

When $\Delta = 30$, the attractor for this system has dimension around 3.5, according to numerical estimation. For $\Delta = 100$, the correlation dimension is approximately 7.1. (See [1].)

The relative refractoriness is included by changing the constant threshold Θ in the integrate-and-fire hypothesis (2) to a simple "leaky potential" function. The threshold Θ will decrease at a constant rate ($\dot{\Theta} = -0.5$ in our example) until firing, at which time Θ is replaced by $\Theta + 50$, representing the repolarization stage.

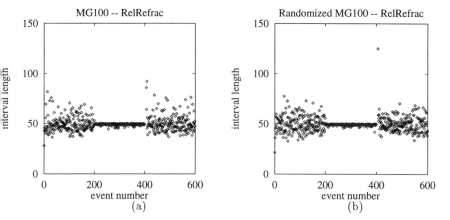

Figure 5: Application of superthreshold pulses to control an integrate-and-fire model with relative refractoriness. (a) The input signal to the integrator is a Mackey-Glass time series. Intervals are determined according to (2), where Θ is represented by a leaky potential. On-demand pacing control is applied between event numbers 200 and 400 only. (b) Input signal to the integrator is random noise with the same power spectrum as in (a). Similar control can be achieved for this random spike train.

The results of control are shown in Figure 5(a). A signal from equation (5) with $\Delta = 100$ is integrated until the threshold is met as in (2), with the assumption that the threshold Θ evolves as discussed above. For spike (event) numbers 1 to 200, the interspike interval lengths of the free-running system is graphed on the vertical axis. Since the integrated signal is 7-dimensional, no pattern can be ascertained by visual inspection. After event 200, the on-demand control protocol is turned on. The system immediately relaxes into a rhythmical behavior, with interspike intervals held near a fixed value. Note that not only are there no intervals above the nominal value (which is trivially enforced by on-demand pacing), but intervals below the nominal value have been largely eliminated as well. After spike 400, when the on-demand pacing is suspended, the system re-establishes its prior erratic behavior.

The same effect can be seen when the deterministic Mackey-Glass input is replaced by a random noise input. We manufacture the noise by creating a random-phase surrogate of the Mackey-Glass time series. That is, the phases of a Fourier-transformed version of the time series are randomized, and then the inverse Fourier transform is applied, resulting in a random signal with the same spectral characteristics as the deterministic Mackey-Glass signal used above. Figure 5(b) shows the same on-demand control technique as applied above. The control has an effect similar to that in the deterministic case.

6 Conclusions

We have suggested methods of studying and exploiting interspike interval series recorded from experiment, in the absence of known model equations. If they are generated according to hypothesis (2), the ISI vectors can be used to represent states of the underlying attractor in a similar manner as delay coordinate vectors from amplitude time series. No attempt has been made thus far to minimize description length for an ISI reconstruction (see Judd and Mees, this volume), but that would presumably lead to significantly increased efficiency in applications.

Although equation (2) was chosen to be as free of assumptions as possible, it is a hypothesis, and the primary features of any particular physiological system may not be well represented by this hypothesis. This study is therefore a kind of feasibility study. Further work with other integrate-and-fire models for specific contexts, such as the leaky integrator of section 5, is needed.

One strong motivation for this work is to advance understanding of information processing in neural systems. The possibility exists that much of the communication and processing that transpires in the central nervous system, for example, is coded in the temporal firing patterns of neural circuits. Our approach is designed to provide fundamental knowledge toward solving the so-called temporal coding problem. The article by Watanabe et. al (this volume) studies the problem of learning in neural systems; see that article and the references therein for current work on this very important problem.

This study shows the theoretical possibility of using the OGY control procedure, presented by Ott in this volume, with obvious modifications for systems measured only by spike trains. In systems where deterministic but aperiodic spike trains are measured, and where a global parameter is accessible for small time-dependent perturbations, this technique may be used to entrain the system into periodic spiking behavior.

Moreover, it was shown that the on-demand control procedure can be used to enforce equilibrium spiking behavior in a neurophysiologically motivated spiking model system. This capability was shown to exist for a deterministic system as well as a similar system that was principally random. We conclude that ability to control is not necessarily a reliable diagnostic for deterministic dynamical mechanisms.

Acknowledgments

This research was supported in part by the U.S. National Science Foundation (Computational Mathematics).

References

[1] M. Ding, C. Grebogi, E. Ott, T. Sauer, J.A. Yorke, Estimating correlation dimension from a chaotic time series: when does plateau onset occur? Physica D **69**, 404 (1993).

[2] E. Lorenz, Deterministic nonperiodic flow, J. Atmos. Sci. **20**, 130 - 141 (1963).

[3] M.C. Mackey, L. Glass, Oscillations and chaos in physiological control systems, Science **197**, 287 (1977).

[4] M. Miller, D. Snyder, *Random Point Processes in Time and Space*. Springer-Verlag, New York (1991). P.A.W. Lewis, ed. *Stochastic Point Processes: Statistical Analysis, Theory, and Applications*. Wiley-Interscience, New York (1972).

[5] E. Ott, C. Grebogi, E. Ott, Controlling chaos. Phys. Rev. Lett. **64**, 1196 (1990)

[6] E. Ott, T. Sauer, J.A. Yorke, *Coping with Chaos: Analysis of Chaotic Data and the Exploitation of Chaotic Systems*. Wiley Interscience, New York, 418 pp. (1994).

[7] N. Packard, J. Crutchfield, J.D. Farmer, R. Shaw, Geometry from a time series, Phys. Rev. Lett. **45**, 712 (1980).

[8] D. Prichard, J. Theiler, Generating surrogate data for time series with several simultaneously measured variables, Phys. Rev. Lett. **73**, 951 (1994).

[9] T. Sauer, Reconstruction of dynamical systems from interspike intervals, Phys. Rev. Lett. **72**, 3811 (1994).

[10] T. Sauer, J.A. Yorke, and M. Casdagli, Embedology, J. Stat. Phys. **65**, 579 (1991).

[11] S.J. Schiff, K. Jerger, T. Chang, T. Sauer, P.G. Aitken, Stochastic versus deterministic variability in simple neuronal circuits. II. Hippocampal slice, Biophys. J. **67**, 684 (1994).

[12] F. Takens, Detecting strange attractors in turbulence, Lecture Notes in Math. **898**, Springer-Verlag (1981).

[13] J. Theiler, S. Eubank, A. Longtin, B. Galdrakian, J.D. Farmer, Testing for nonlinearity in time series: the method of surrogate data, Physica D **58**, 77 (1992).

[14] H. Tuckwell, *Introduction to Theoretical Neurobiology*, vols 1,2. Cambridge University Press (1988).

Commentary by T. Pastor

This paper explores the dynamics of a system by examining the vector of interspike intervals derived from a time series of system output. The interspike interval is similar to the differential time interval suggested by Babloyantz (1989) and used by us in investigating the attractor of a forest ecoystem model (Pastor and Cohen, this volume). However, this paper by Sauer shows the power of using the vector of interspike intervals to reconstruct the time dynamics of a system and thereby gain some understanding of system structure when only the signal is known. As an example, the Lorenz attractor is reconstructed from its vector of interspike intervals. Sauer then shows how OGY control of this reconstructed system can be achieved by small perturbations to β in the Lorenz

$$\dot{z} = -\beta z + xy$$

This technique has potential value for the understanding and management of ecological systems, provided that there is a long-enough time series to properly reconstruct the attractor from the vector of interspike intervals. Some candidate data sets for which this might be possible include pollen records sampled at extremely fine intervals (MacDonald et al. 1993) and trapping records for hare and lynx of the Hudson's Bay Company. The dynamics of the latter might be chaotic (Schaffer

1984). The application of this technique to the lynx-hare system opens some intriguing possibilities. β in the above equation is analogous to the mortality terms in Lotka-Volterra equations for the lynx-hare system; some of this mortality can be achieved by hunting or trapping. The control of the Lorenz attractor by small perturbations to β in the example by Sauer suggests that similar control might be achieved for a chaotic predator-prey system by regulated hunting or trapping. The possibility that such a control mechanism for an ecological system can be designed only from data on the interspike intervals without knowing the underlying model offers some practical tools for wildlife managers.

References

[1] Babloyantz, A. 1989. Some remarks on nonlinear data analysis. Pages 51-62 in N.B. Abraham, A.M. Albano, A. Passamante and P.E. Rapp (editors). Measures of Complexity and Chaos. New York, Plenum Press.

[2] MacDonald, G.M., T.W.D. Edwards, K.A. Moser, R. Pienitz, and J.P. Smol. 1993. Rapid response of treeline vegetation and lakes to past climate warming. Nature 361: 243-245.

[3] Schaffer, W. 1984. Stetching and folding in lynx fur returrns: evidence for a strange attractor in nature? The American Naturalist 124: 798-820.

Commentary by M. Watanabe and K. Aihara

The Tim Sauer's article forwards some intriguing possibilities and promise. It deals with spike trains generated by an integrate-and-fire model and shows that information is well contained in these spike trains and can be used for system control. This approach is quite interesting because in the field of neuroscience, whether the brain uses coding by spatio-temporal spike patterns or firing-rates has been a subject of controversy. If we assume the former, there may be a place in the brain where coding is changed from firing-rate based coding to spatio-temporal spike train based coding and vice versa, since some experimental data suggest that information is firing rate coded in sensory and motor neurons. The Tim's results give an important insight to this question of coding translation in the brain.

Modeling Chaos from Experimental Data

Kevin Judd and Alistair Mees
Centre for Applied Dynamics and Optimization
Department of Mathematics
The University of Western Australia
Perth, 6907, Australia

Abstract

There are several robust methods for controlling nonlinear, chaotic systems when the dynamics are known. When the dynamics are not known a model could be constructed from time-series data and a controller attached to the model. Recent research has shown that the minimum description length principle provides a good model without over-fitting. An algorithm for constructing minimum description length radial basis models and an example illustrating the modeling a complex system will be described. Specifically, we describe the successful modeling of a mechanical system with time varying parameters. The variation of the parameters is exploited to construct a parametric model that we can consider using to control a system by adjusting its parameters.

1 Introduction

One method of controlling a chaotic system is to first build a model of the dynamics, which includes the dependence upon parameters, then use this model to predict and control the system's behavior. This paper concentrates on the modeling of nonlinear systems from time-series data and describes some new methods for constructing such models. These new methods employ the minimum description length principle to identify good radial basis models that avoid over-fitting data.

This paper begins with a brief description of a modeling procedure, which uses the minimum description length principle. The details of the modeling procedure and the theory behind it appear elsewhere [7, 4, 5, 6]. We then describe an experimental system with time varying parameters and the techniques used to construct a parametric model the dynamics from a time series of measurements taken from the system. Finally, we comment on the relevance of these modeling methods to control of chaotic systems.

2 Modeling time series

The modeling procedures we employ assumes that the system under analysis, with the view to controlling, has dynamics described by

$$x_t = f(x_{t-1}, \beta_t) + \nu_t,$$
$$y_t = c(x_t) + \xi_t,$$

where $x_t \in \mathrm{R}^d$ is the state of the system at time t, $\beta_t \in \mathrm{R}^k$ are time varying parameters of the system, $y_t \in \mathrm{R}$ is our measurement of the system via a read-out function c, and ν_t and ξ_t are i.i.d. random variates representing the dynamical noise and measurement noise respectively. We assume there is no knowledge of the system other than the time series of measurements y_t and perhaps limited knowledge of β_t, for example, that β is monotonically increasing over the period the measurements were made. Our aim is to build a model of the dynamics even when the dynamics are governed by a nonlinear function $f(x, \beta)$.

If the dynamics were linear, then a traditional method of modeling the dynamics constructs an autoregressive model of the form

$$Y_{t+1} = a_0 + b_0 Y_t + b_1 Y_{t-1} + \cdots + b_k Y_{t-k} + \epsilon_t.$$

Sophisticated techniques exist to estimate the parameters of this model for fixed order k. The order of the model cannot usually be determined in advance and must instead be estimated from the time-series. The optimal order of the model for a time series is of length n is usually taken to be where one attains the minimum of the Akaike Information Criteria (AIC) [2]

$$AIC(k) = n \ln \left(\sum_{i=1}^n e_t^2 / n \right) + 2k,$$

or the minimum of the Schwarz Information Criterion (SIC) [11]

$$SIC(k) = n \ln \left(\sum_{i=1}^n e_t^2 / n \right) + k \ln(n),$$

where e_t is the residual error of the model fit, that is, the difference between the fitted value of y_t and the actual value.

When the system is nonlinear there are no longer well established techniques for modeling the system; although there is the beginnings of a theoretical basis which we will now describe. A celebrated theorem of Takens states that, generically, one can reconstruct the state space and dynamics of a nonlinear system by the method of time delay embedding. This entails constructing a vector time series $z_t = (y_{t-T}, y_{t-2T}, \ldots, y_{t-dT}) \in \mathrm{R}^d$ from the scalar time-series y_t using suitably chosen d and T. There is a growing body of theory and practical knowledge on how toselect d and T to guarantee an adequate reconstruction [1]. Having obtained an embedding, the dynamics of the system can be modeled as

$$y_t = F(z_t, \beta_t) + \epsilon_t,$$

for an some unknown nonlinear function F and unknown i.i.d. random variates ϵ_t. The aim of the modeler is to estimate F. Unfortunately, when F in nonlinear, there is no unique way to estimate this function, which is unlike the case with linear autoregressive models, described above, where there is an ordered hierarchy of models from which one can select the first model that describes the system with sufficient accuracy as determined by an information criterion. With nonlinear models there is an unordered, infinitely branching hierarchy of models, so that the question of selecting an optimal model is not clear cut or even answerable. The best one can do at present is to restrict attention to a, more or less, ordered family of models and select the optimal model in ways analogous to that used for linear models employing an analogous information criterion.

Our modeling procedure has three components: a suitable hierarchal class of nonlinear models; methods to select optimal models of a prescribed order from the model class; and criteria that determines when a model is sufficiently accurate and not over-fitting the time series.

3 The model class

The parameterized family of nonlinear models $G(z, \beta, \lambda)$ we have chosen to use are *pseudo-linear* models, which are linear combinations of some arbitrary set of *basis functions* $f_i(x)$, that is,

$$G(z, \beta, \lambda) = \sum_{i=1}^{m} \lambda_i \ f_i(z, \beta).$$

The set of basis functions should include a large variety of nonlinear functions including linear functions. The major advantage of pseudo-linear models is the parameters λ are easily estimated once the basis functions are chosen. The aim is to approximate $F(z, \beta)$ with $G(z, \beta, \lambda)$, that is, we are assuming

$$\begin{aligned} y_t &= G(z_t, \beta_t, \lambda) + \epsilon_t \\ &= \sum_{i=1}^{m} \lambda_i \ f_i(z_t, \beta_t) + \epsilon_t \end{aligned}$$

To estimate the parameters λ define

$$\begin{aligned} V_i &= (f_i(z_1, \beta_1), \ldots, f_i(z_n, \beta_n))^T, i = 1, \ldots, m \\ y &= (y_1, \ldots, y_n)^T, \\ \lambda &= (\lambda_1, \ldots, \lambda_m)^T, \end{aligned}$$

and so if we were to assume that modeling errors have a Gaussian distribution, then minimizing the sum of squares of errors with respect to λ,

$$\min_{\lambda} \|y - V\lambda\|, \tag{1}$$

will obtain the optimal (maximum likelihood) model for the chosen basis functions. The minimization is easily accomplished using the singular value decomposition of V. Generally pseudo-linear models with many basis functions—indeed any nonlinear model having many parameters—over-fit time series resulting in models with unstable dynamics, because they incorporate properties of the original data that should be attributed to noise. One must use an information criterion, analogous to AIC and SIC, that will prevent over-fitting, by limiting the size of the model. The appropriate criterion for nonlinear models is *description length*.

4 Description length and model selection

Our modeling procedures rest upon the minimum description length principle, which is Occam's Razor applied in a modeling context, that is, the best model is the smallest model. Suppose one has collected, from an experiment, a time series y_t, $t = 1, \ldots, n$ measured to 12 bits, and wishes to communicate this data to a colleague. One could send the raw data. Alternatively, one could construct a dynamical model from the data, which enables one to predict a value of y_t from earlier values. If you and your colleague have previously agreed on a class of models, then you could communicate the data by sending the parameters of a model, enough initial data to start predicting future values of the time series and the errors between the true time series and the values predicted by the model. Given this information your colleague can reconstruct exactly the experimental data to its measured accuracy. An important point to appreciate is the parameters and errors need only be specified to finite accuracy. Furthermore, if the model is any good, then the total number of bits required to transmit parameters, initial values and errors will be less than the number of bits of raw data. The minimum description length principle of modeling states, in effect, that the best model of a time series is the model that allows the greatest compression of the data. The concept of description length was introduced by Rissanen [9]. Description length is a generalization of AIC and SIC to nonlinear models, in fact, SIC is an asymptotic approximation to the description length of a linear model.

Usually one does not calculate the description length of a model exactly, but rather calculates a convenient approximation of it. For pseudo-linear models with k basis functions, it can be shown [4] that the description length is bounded by

$$S_k = (\frac{n}{2} - 1)\ln\frac{e^\mathsf{T} e}{n} + (k+1)(\frac{1}{2} + \ln\gamma) - \sum_{j=1}^{k} \ln\delta_j,$$

where $e = y - V\lambda$ are the fitting errors, γ is related to the scale of the data, and δ solves $Q\delta = 1/\delta$ where

$$Q = nV^\mathsf{T} V/e^\mathsf{T} e.$$

The variables δ can be interpreted as the precision to which the parameters λ are specified.

Our procedure for constructing pseudo-linear models, which incorporates the minimum description length principle, is:

- Select a large population of basis functions $f_i(z, \beta)$.
- For $k = 0, 1, 2, \ldots$
    - Find the best model using only k of the basis functions (i.e., the maximum likelihood model.)
    - Calculate the description length S_k of the model.
- Stop when S_k stops decreasing.

A typical variation of the description length with model size k is shown in figure 1.

5 Radial basis models

There is a great variety of potential basis functions that could be used in building nonlinear models. We have concentrated on radial basis functions:

$$f_i(x) = \phi(|x - c_i|/r_i)$$

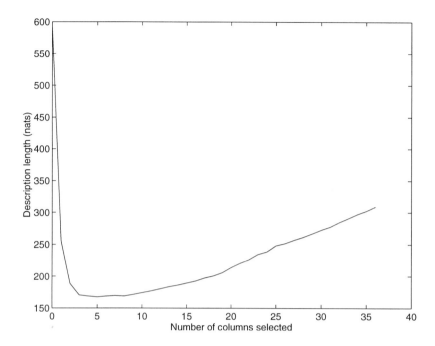

Figure 1: This figure shows a typical variation of description length with model size, that is, number of basis functions used. There is an initial rapid decrease in description length as model size increases, indicating that there is dynamical structure in the data and a model will effect a compression of the data. However, as the model size increases the gain from an additional basis function is marginal and the cost of transmitting the extra coefficient for the basis function exceeds the overall contribution to a decrease in modeling error.

for arbitrary *centre* c_i, *radii* r_i and function ϕ. It is almost always advantageous to have available constant and linear functions too. There have been many methods suggested for the selection of centres. We have introduced the ideas of "chaperons" as centres and find them the most flexible and effective; chaperons are centres generated by adding a random variate to the embedded data points [4].

For very complex problems one may need to search over a greater population of potential basis functions than can be realistically handled; recall that the main computation is the solution of equation 1. Under these circumstances an iterative method (akin to simulated annealing) of selecting an optimal model from a population of basis functions by the above algorithm, then generating a new population that includes the previous best selections and takes into account where the model has the greatest errors. It appears that such an algorithm converges to an optimal model. Figure 2 show an example of the fitting of a complex nonlinear function by the above

method when
$$f_i(x) = \exp\left(\frac{1-\nu_i}{\nu_i}\left|\frac{x-c_i}{r_i}\right|^{\nu_i}\right),$$
where r_i and ν_i are also adjustable parameters—the new parameter ν_i transforms a sombrero into a top-hat as ν_i increases.

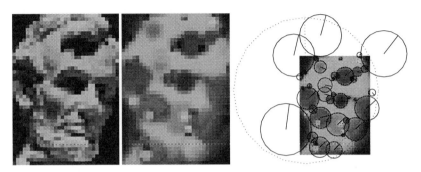

Figure 2: A well-known American. (a) The original data. (b) The image fitted by a radial basis model. Unfocusing one's eyes and standing back the likeness is there. The fit does not look good to our eyes because the model minimizes the sum of squares error, but our eye uses a different measure of closeness that puts more emphasis on edges. (c) The positions and radii of the centers used. Solid circles correspond to positive λ_s and dotted circles to negative λ_s. The radius marked within circles indicates the value of ν_i as on a linear dial increasing anti-clockwise with down the minimum value of 1 and up equaling 10.

6 Experimental data: nonlinear oscillations

The remaining part of this paper describes an application to a nonlinear mechanical system, the planar-forced vibrations of a stretched, uniform string. Bajaj and Johnson [3] have analyzed the weakly nonlinear partial differential equations describing the forced vibrations of stretched uniform strings. The equations take into account motions transverse to the plane of the forcing, which are induced by a coupling with longitudinal displacements, and changes in tension that occur for large amplitude motions. The averaged equations of a resonant system can be reduced by successive approximation to a four-dimensional system of ordinary differential equations. These equations have a complex bifurcation structure exhibiting Hopf, saddle-node, period-doubling and homoclinic (Shil'nikov) bifurcations, although it is the Shil'nikov mechanism that underlies most of this complexity.

Molteno and Tufillaro have performed experiments with guitar strings and have recorded periodic and apparently chaotic motions [8]. They generously provided us with some their data. We constructed a model of the dynamics from a single time-series by the above methods, and from an analysis of the model were able to infer a Shil'nikov mechanism as being the source of chaotic motions in the experimental system. This is an involved and technical argument that we will not enter into here; instead we concentrate on a simpler problem of identifying details of bifurcations, not immediately apparent in the experimental data, which is of more obvious importance to the control of chaotic systems.

7 Modeling experimental data

Molteno and Tufillaro conducted a forced string experiment [8], collecting copious data of which we have taken a single experiment's recording. During this experiment there was a monotone drift in the parameters, due to heating of the apparatus. Figure 3 shows Takens embedding plots of the beginning and end of a long time series that exhibits a transition from apparently chaotic to periodic motion. The mechanism of this transition was shown by us to be generated by a Shil'nikov mechanism.

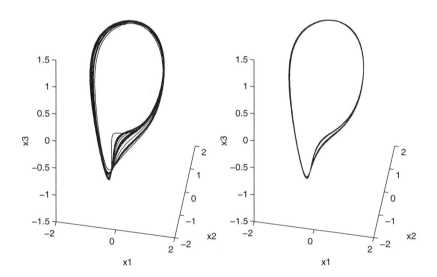

Figure 3: The complete time series consisted of an approximately 2 minute run sampling at 1.34016 kHz giving over 128 thousand samples. Shown here are segments from the beginning and end of the times series each of 5000 samples embedded in 3 dimensions with a lag of 28. The data shows a transition from apparently chaotic motion to periodic motion.

Here we analyze the local bifurcations in greater detail and establish that there is a period-doubling cascade, which includes a period three window, although none of this is immediately observable in the data. Using an interactive 3-D viewer[1] it

---
[1] The investigators used SceneViewer, an Open Inventor application available on Silicon Graphics

was found that the embedded time series seems to lie on a two-dimensional ribbon with a half twist and the suggestion of a fold. The system is rather like the Rossler system [10] but with a half twist. The half twist occurs at the bottom of figure 3 and the fold at the top of the attractor, within an arc of about one fifth of the attractor length. The two-dimensional ribbon structure suggests modeling the system with a one-dimensional Poincare return map. If this were to fail, then we would try a two-dimensional return map. As will be seen, a one-dimensional model is very successful.

A Poincare return map can be obtained directly from a time series of a flow, in some instances, by simply constructing a time series of suitable local minima or local maxima. If $\dot{x} = f(x)$ and $y = g(x)$ is a scalar measurement, then $\dot{y} = Dg(x)\dot{x} = Dg(x)f(x)$ and local extrema of y occur where $Dg(x)f(x) = 0$. If the readout function $g(x)$ is non-degenerate, then $Dg(x)$ is everywhere non-vanishing. Consequently, $Dg(x)f(x) = 0$ defines codimension 1 submanifolds on which the vector field $f(x)$ is transverse, in fact normal, except where $f(x) = 0$.

In practice, one needs a time series sampled frequently enough to accurately determine local extrema, and one must take care to ensure that local extrema all live on the same piece of submanifold and that the vector field is crossing in the same direction in all cases. For the system considered we chose local minima y_t such that $y_s > y_t$ for all $|s - t| < T/2$ where T is the approximate period of the time series (e.g., $T/2$ is the first minimum of the autocorrelation function.)

By the above methods we obtained two scalar time series, the time of the nth local minimum t_n and the system state at the minimum $z_n = y(t_n) \in \mathrm{R}$. We then assumed that the z_n are realizations of a stochastic process Z_n where

$$Z_{n+1} = F(Z_n, \beta_n) + \epsilon_n, \qquad (2)$$

β is a bifurcation parameter and ϵ_n are i.i.d. Gaussian random variates. Our aim is to model $F(z, \beta)$. We assume that the bifurcation parameter β is a monotonic function of time and hence define[2], $\beta_n = \alpha t_n$, and build a model of $z_{n+1} = \hat{F}(z_n, \beta_n) + \epsilon_n$, $(z_n, \beta_n) \in \mathrm{R}^2$. This is a somewhat unusual practice because we are building a model of a "non-stationary" system, that is, it has time varying parameters. However, this can be used to our advantage because knowing that the parameters have a monotone variation enables us to build a model of both the dynamics and the parameterization of the dynamics from a single time series.

The function $F(z, \beta)$ was modeled with a radial basis model using cubics with chaperons as centres and linear functions. Interactive viewing of this surface $\hat{F}(z, \beta)$ showed it to be unimodal for fixed β, very nearly quadratic in z and very nearly linear in β; figure 4 is a two dimensional projection of the surface. Figure 5 shows a numerically computed bifurcation diagram of $\hat{F}$, which is seen to have a period doubling bifurcation that only vaguely corresponds to an equivalent plot of the experimental data shown in figure 6. Notable is the lack of a period doubling cascade and period three window in the experimental data; we will now search for reasons why these are not seen.

---

and other platforms.

[2]The scale parameter α is chosen so that the range of β_n is the same as that of z_n, because this scaling is advantageous to the radial basis model.

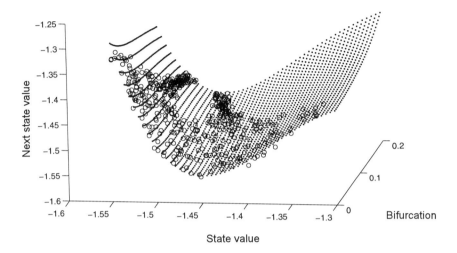

Figure 4: A projection from three dimensions of the embedded return map data with bifurcation parameter dependence (circles) and the surface fit of the return map model (dots).

Analysis of model

Figure 7 shows the empirically derived[3] expectation of z_n for the system (2) when there is a small dynamical noise component and no simulation of measurement noise. The dynamical noise chosen had a standard deviation $\sigma = 0.003$, which is half of the standard deviation of the residual errors of the fitted model. The choice of σ is somewhat arbitrary. Using the full standard deviation of the residuals gave too great a dispersion, which indicates some measurement error is present, for example, from the method of calculating the z_n. On the other hand, we expect some significant level of dynamical noise because additive measurement error would give a uniform blurring of the bifurcation diagram, but this is not consistent with the dispersion of the experimental data. A maximum likelihood estimate of the relative contributions of measurement and dynamical noise could be obtained through boot-strapping, but we have not done this.

As can be seen, the period three orbits of the model system are sufficiently sensitive to perturbation that this small amount of dynamic noise is enough to almost completely mask their existence. The expectation of the state variable z_n shown in figure 7 agrees well with the experimental sample shown in figure 6, although the

---
[3] That is a Monte Carlo simulation.

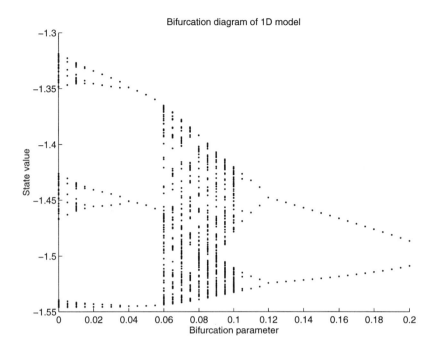

Figure 5: A numerically determined bifurcation diagram of the model (2). It is seen to be a classic period doubling bifurcation complete with period three window.

model predicts a more prominent central ridge in the period three window and a hollow above it. There are four possible sources of discrepancy between predicted expectation of the state variable and apparent experimental distribution: small errors in the model to which the location of the middle point of the period three orbit is sensitive; the assumption that the stochastic terms in (2) are normally distributed; the assumption that the experimental system attains its asymptotic state; and additional blurring due to measurement error that accounts for the remaining residual component. With regard to the first source of discrepancy, when our algorithms to generate models were run repeatedly to generate many models and their bifurcation diagrams compared it was found that they differed only slightly, but most notably in the onset of the period three window and the location of the middle point of the period three orbit. This could partly account for the noted discrepancies. With regard to the second source of discrepancy, the assumption of normality of errors is most likely false. There are several sources of error in the experimental time series z_n, notably errors in determining local minima, which are likely to be nearly normally distributed, and the intrinsic dynamical noise of the system. The latter is really an accumulation of small perturbations through a complete cycle of the flow. The system certainly flattens these errors onto the two-dimensional sheet of its attractor and this implies that on return to the Poincare section there is a larger perturbation across the sheet

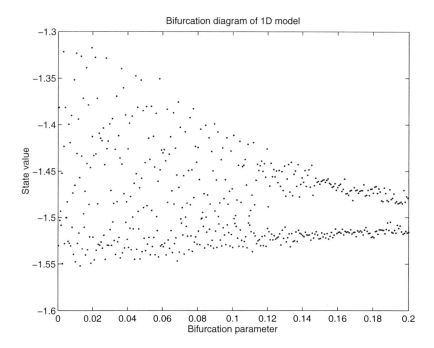

Figure 6: A plot a the time series z_n where time axis has been rescaled to be bifurcation parameter β. If the system were noise free and the model accurate, then this plot should resemble figure 5. It does so only vaguely; there is no clear period doubling or period three window.

than perpendicular to the sheet. This would imply greater levels of noise than the model might predict and furthermore this blurring would be more noticeable of an sensitive period three orbit than it would be of a more stable period two orbit. The third source of discrepancy arises because figure 7 shows the asymptotic distribution of states given noise, whereas in figure 6 the bifurcation parameter is constantly increasing and the system does not necessary attain its asymptotic distribution for any parameter value, in particular it is unlikely to do so in sensitive regions such as near bifurcation points and in the period-three window. The final source of discrepancy might account for some of the small discrepancies, but does not account for the main discrepancies because it is a uniform blurring.

8 Control and chaos

In this presentation we have briefly overviewed a nonlinear modeling procedure that uses the minimum description length principle to avoid over-fitting of data. We have also shown how one can, under some circumstances, build parametric models of sys-

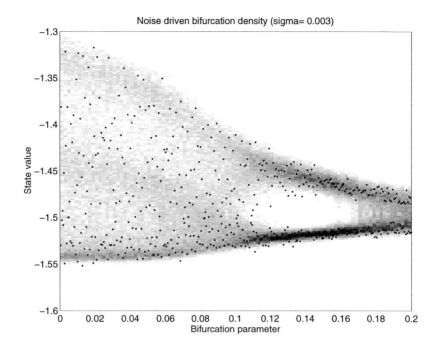

Figure 7: The empirically calculated expectation of the state variable z_n for the model system when the noise has a standard deviation of 0.003. The density plot shows the asymptotic probability of Z_n falling in a narrow range when β is held fixed.

tems from a single time series of a non-stationary system. One aspect of modeling that we have passed over here the need for good reconstruction methods such described by Sauer in this publication. Embedding is a preliminary step to modeling and our methods assume that it has be successfully accomplished.

In this book Glass, Kostelich, Ott and Teo describe new methods that control of chaotic systems with small perturbations of control parameters, although Vincent uses more traditional methods. Some of methods require knowledge of the nonlinear maps and flows that govern the dynamics of a system, but in many applications such detailed knowledge is not available. The modeling methods we have described provide useful parameterized models of system dynamics from time series measurements. This implies that the control methods can be applied in a much broader range of application. It should be emphasized that the modeling methods assume very little about the system under study and in fact can exploit chaotic dynamics and nonstationarity to explore the phase and parameter space of the dynamics.

References

[1] H. D. I. Abarbanel, R. Brown, J. J. Sidorowich, and L. S. Tsimring. *Analysis of Observed Chaotic Data*, volume 65(4). 1993.

[2] H. Akaike. A new look at the statistical identification model. *IEEE Trans. Ann. Control*, 19:716–723, 1974.

[3] A. K. Bajaj and J. M. Johnson. On the amplitude dynamics and crisis in resonant motion of stretched strings. *Phil. Trans. R. Soc. Lond.*, 338A:1–41, 1992.

[4] K. Judd and A. I. Mees. A model selection algorithm for nonlinear time series. *Physica D*, 82:426–444, 1995.

[5] K. Judd and A. I. Mees. Modeling chaotic motions of a string from experimental data. *Submitted for publication*, 1995.

[6] K. Judd and A. I. Mees. A simulated annealling approach model selection for highly nonlinear systems. *in preparation*, 1995.

[7] A. I. Mees. Parsimonious dynamical reconstruction. *International Journal of Bifurcation and Chaos*, 3:669–675, 1993.

[8] T. Molteno and N. B. Tufillaro. *J. Sound Vib.*, 137:327, 1990.

[9] J. Rissanen. *Stochastic Complexity in Statistical Inquiry*, volume 15. World Scientific, Singapore, 1989.

[10] O. E. Rossler. An equation for continuous chaos. *Physics Letters*, 57A:397–398, 1976.

[11] G. Schwarz. Estimating the dimension of a model. *The Annals of Statistics*, 6(2):461–464, 1977.

Commentary by E. Ott

The theoretical basis for the procedure in this paper seems to be the hypothesis that the delay coordinate system can be modeled by

$$y_t = F(z_t, \beta_t) + \epsilon_t, \tag{3}$$

where ϵ_t is an unknown i.i.d. random variable representing the noise. For a nonlinear system, it would seem that a more reasonable model is

$$y_t = F(z_t, \beta_t) + \epsilon_t f(z_t, \beta_t). \tag{4}$$

Would this effect the best way to choose a description length criterion?

In many cases the measurement may be a vector rather than a scalar. In such cases use of the full vector information would seem to be much preferable (as compared to reduction of the measurement to a scalar). This would presumably allow fewer delays to be used and hence less time for noise errors to build up in the constructed delay vector. Can you comment on whether the methods of this paper have a natural generalization to this case?

Commentary by E. Kostelich

Radial basis functions are one popular class of functions for modeling the dynamics of time series data. In contrast to methods based on the computations of Jacobian matrices (see for instance J.-P. Eckmann and D. Ruelle, Rev. Mod. Phys. 57 (1985), 617), radial basis functions provide a straightforward way to interpolate data points that may not lie exactly on the observed attractor.

I think that the most interesting aspect of this paper is the attempt to model nonstationary data, as described in Sections 7 and 8. The underlying process exhibits low dimensional chaos, and a bifurcation parameter varies monotonically with time. This kind of problem is often faced in laboratory experiments, but most existing analysis methods for chaotic data assume an underlying dynamical process that is stationary. Despite the rather poor agreement with the observed bifurcation diagram in Fig. 6 with the predicted one in Fig. 5, the differences can be attributed to measurement errors.

However, attempts to model the dynamics of chaotic but nonstationary data may be very difficult if the system exhibits sufficiently long transients. In such a case, the estimation of the model and the parameters is problematic, because one never sees a "steady state" attractor for any parameter value, and it is not clear that the notion of "dynamical noise" is sensible in the context of chaotic transients. Chaotic transients can persist for a very long time after a parameter is varied; however, the details of the behavior depend heavily on the system under investigation. In the data examined here, the transients probably were sufficiently short that the dynamics could be estimated with good accuracy.

Reply to commentaries from Judd and Mees

The reviewers raise interesting issues and ask questions that are phrased in the expectation of a reply.

Ott asks whether equation (4) would not be a better model, that is, one where the variance of the noise ϵ_t is state dependent, and whether the description length criterion can be applied in this case. This is sometimes advantageous and furthermore description length tells you whether it is. The authors have modelled systems as Ott suggests, using radial basis model for both F and f. Description length criteria must now take account of the parameters used in both F and f; note that the addition complexity of f is only worthwhile if it significantly reduces the number of bits required to code the errors ϵ_t, otherwise one is trading bits from F, which is not likely to be helpful. There is a procedural complexity with the scheme, since the sum of squares of errors that one wishes to minimize is now

$$\sum_{t=1}^{n} \left(\frac{y_t - F(x_t, \beta_t)}{f(x_t, \beta_t)} \right)^2. \tag{5}$$

It would be prohibitively complex to solve this nonlinear problem. Instead, we used an iterative procedure: fit F with f as fixed weights (initially constant), then estimate f on the residuals. For some data we found that a nontrivial f was selected with marginal improvement of the model, but for others description length found there was no advantage over a constant variance.

Chaos in Symplectic Discretizations of the Pendulum and Sine-Gordon Equations

B.M. Herbst and C.M. Schober
Program in Applied Mathematics
University of Colorado
Boulder, Colorado, 80309, USA

Abstract

Homoclinic orbits play a crucial role in the dynamics of perturbations of the pendulum and sine-Gordon equations. In this paper we examine how well the homoclinic structures are preserved by symplectic discretizations. We discuss the property of exponentially small splitting distances between the stable and unstable manifolds for symplectic discretizations of the pendulum equation. A description of the sine-Gordon phase space in terms of the associated nonlinear spectral theory is provided. We examine how preservation of the homoclinic structures (i.e. the nonlinear spectrum) depends on the order of the accuracy and the symplectic property of the numerical scheme.

1 Introduction

In this paper we investigate the effectiveness of symplectic integrators in preserving the phase space structure of the sine-Gordon equation,

$$u_{tt} - u_{xx} + \sin u = 0 \qquad (1)$$

when periodic boundary conditions, $u(x,t) = u(x+L,t)$, are considered. For these boundary conditions, the sine-Gordon phase space contains tori and homoclinic orbits. A fundamental question is what happens to these structures under small perturbations. The answer depends on a variety of factors; e.g. the type of perturbation and the geometry and dimension of the phase space.

The homoclinic orbits are very sensitive to the perturbations caused by numerical discretizations and, from a numerical point of view, this sensitivity is troublesome. When initial data is chosen close to a homoclinic orbit, truncation errors can excite instabilities not present initially and quasi-periodic solutions can rapidly degenerate into solutions exhibiting temporal chaotic behavior [3]. Depending on the form of the discretization, other numerical artifacts include the development of high frequency oscillations or the collapse onto nearby periodic orbits. For example, the results of using a doubly discrete discretization of sine-Gordon due to Hirota are shown in Figure 1. One might expect this scheme, which is in fact completely integrable, to be very efficient – it is local and explicit and has exact quasi-periodic N-phase and homoclinic solutions. Even so, homoclinic crossings as well as grid scale oscillations occur (see [3] for a complete description).

The sine-Gordon equation is a completely integrable Hamiltonian system. Symplecticness is a characteristic property possessed only by solutions of Hamiltonian systems. The question then is whether the use of symplectic integrators can ameliorate some of the difficulties encountered when numerically solving the sine-Gordon equation.

Symplectic integrators have proven to be very successful in handling planar Hamiltonian systems; i.e. they preserve the integrability of the continuous system with exponential accuracy in the discretization parameter. In section 2 a brief overview of symplectic discretizations of Hamiltonian systems is provided. Although symplectic discretizations of planar Hamiltonian systems generically produce chaotic solutions in the vicinity of homoclinic orbits, in section 3, using Melnikov's function, we show that the width of the chaotic layer about the homoclinic orbit is exponentially small in the discretization parameter. These ideas are illustrated in section 4 using the pendulum equation, which is relevant to our study since it arises as a spatially uniform reduction of the sine-Gordon equation.

Determining the importance of the preservation of symplecticness in the integration of high dimensional Hamiltonian systems which result from discretizing the PDE's of soliton theory is an open problem. In particular, in section 5 we address the following questions: **i)** Do symplectic integrators preserve the structure of the sine-Gordon phase space appreciably better than nonsymplectic methods? More specifically, do trajectories explore a smaller region of phase space when symplectic methods are used? **ii)** Does the relative performance of symplectic integrators improve if initial values are chosen away from the the unstable homoclinic manifolds? If very long time integrations are considered? **iii)** How important is the order of accuracy of the numerical scheme in the preservation of the qualitative properties of the phase space?

To answer these questions we need a simple description of the geometry of sine-Gordon phase space; this is provided in terms of the main spectrum of an associated operator (see section 5.2). The sine-Gordon spectrum is invariant; however, since the integrable system undergoes perturbations induced by numerical truncation errors or even roundoff, the spectrum is no longer invariant. By monitoring the evolution of the spectrum under the numerical flow we correlate irregularities in the numerical solution to changes in the geometry of the phase space. Thus, the schemes are evaluated in terms of the accuracy with which the nonlinear spectrum of sine-Gordon is preserved.

2 Hamiltonian Systems.

Let $\mathbb{R}^{2n}$ be an *even* dimensional oriented Euclidean space with coordinates $(p_1, \ldots, p_n, q_1, \ldots, q_n)$ and let $H(\boldsymbol{p}, \boldsymbol{q})$ be a function defined on a (open, connected) domain, $\Omega \subset \mathbb{R}^{2n}$. Then a Hamiltonian system with n degrees of freedom can be defined by,

$$\frac{dp_i}{dt} = -\frac{\partial H}{\partial q_i}, \quad \frac{dq_i}{dt} = \frac{\partial H}{\partial p_i} \quad i = 1, \ldots, n, \qquad (2)$$

or more conveniently as,

$$\frac{d\boldsymbol{z}}{dt} = J^{-1}\mathbf{grad}H(\boldsymbol{z}), \qquad (3)$$

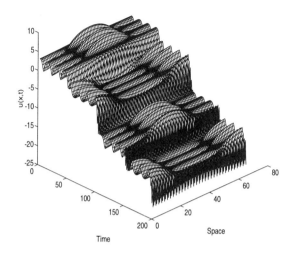

Figure 1: The doubly discrete integrable discretization.

where $z := (p, q)^T$ and J is the *symplectic matrix*,

$$J := \begin{pmatrix} 0_n & I_n \\ -I_n & 0_n \end{pmatrix} \quad (4)$$

with 0_n and I_n denoting the zero and unit matrices of dimension n respectively. System (2) has two important properties: **(i)** The Hamiltonian is a constant of motion (in fact, for $n = 1$ the system is integrable) **(ii)** The phase flow preserves the sum of the oriented areas obtained by projecting $\Sigma \subset \Omega$ onto the n 2-dimensional planes (p_i, q_i), $i = 1, ..., n$. This is simply a generalization of the area preservation property which planar Hamiltonian systems possess. The second property is a fundamental geometric property of Hamiltonian systems—it holds *only* for Hamiltonian systems, see e.g. [6]. It is our purpose to discretize (2), retaining as much of these two properties as possible.

However, consistent discretizations in general are not able to preserve both the area and the Hamiltonian. Since area preservation (in particular its higher dimensional analogue) is the more fundamental of the two properties, much effort has gone onto the construction of *symplectic* discretizations.

2.1 Symplectic Discretizations

We begin by considering planar Hamiltonian systems. In this case, it is straightforward to verify that a map is area preserving. Let

$$(P, Q) = \phi(p, q) \quad (5)$$

define a smooth transformation in some domain Ω. This transformation preserves area if and only if its Jacobian is identically one, i.e.

$$\frac{\partial P}{\partial p}\frac{\partial Q}{\partial q} - \frac{\partial P}{\partial q}\frac{\partial Q}{\partial p} = 1, \quad \forall (p,q) \in \Omega. \tag{6}$$

We simplify by assuming that the Hamiltonian is separable, i.e. that $H(p,q) = T(p) + V(q)$. The following is a first order discretization of (2),

$$\begin{aligned} p_{n+1} &= p_n - kV'(q_n) \\ q_{n+1} &= q_n + kT'(p_{n+1}) \end{aligned} \tag{7}$$

where a prime denotes a derivative with respect to the argument, and k the time step. To check that the transformation $(p_n, q_n) \to (p_{n+1}, q_{n+1})$ is symplectic, it is easy to check that its Jacobian is indeed one. As a consequence of the separability of the Hamiltonian, the scheme is explicit. Although (7) is only first order, there are several procedures to construct higher order symplectic schemes. When the Hamiltonian is separable, a general form of higher order schemes may be given by (see e.g. [11]),

$$\begin{aligned} p_{i+1} &= p_i - C_i k V'(q_i) \\ q_{i+1} &= q_i + D_i k T'(p_{i+1}), \end{aligned} \tag{8}$$

where $i = 1, \ldots, m$ and the coefficients C_i and D_i are determined in order for the scheme to be symplectic and of $O(k^m)$. Here p_1 and q_1 are the numerical approximation at the time t and p_{m+1} and q_{m+1} are the approximations at the next time level, $t + k$. For example, the first order scheme (7) is given by $m = 1$ and

$$C_1 = 1, \quad D_1 = 1.$$

A second order scheme is given by $m = 2$ and,

$$C_1 = 0, \quad C_2 = 1, \quad D_1 = \tfrac{1}{2} = D_2$$

Note that the schemes defined by (8) are explicit since the Hamiltonian is separable. This is only one way of constructing higher order symplectic discretizations and the one used in this study; for other constructions, see [10].

In the planar case, symplectic simply means a consistent, area preserving discretizations of (2). The higher dimensional analogue is a little more complicated. If we denote the transformation from one time step to the next by ϕ (cf (5)), the appropriate quantity to preserve for higher dimensional systems is given by,

$$\phi'^T J \phi' = J, \quad \forall (p,q) \in \Omega \tag{9}$$

where ϕ' is the Jacobian matrix of the transformation and J is the 2×2 dimensional symplectic matrix defined by (4). Again, several procedures for constructing discretizations with exactly these properties have been developed. As in the planar case, one can ensure that the scheme is explicit if the Hamiltonian is separable. The Hamiltonian of the sine-Gordon equation and the discretization of it which we use are separable (c.f. eqn(49)). This allows for very efficient explicit implementations of the symplectic schemes.

3 Measuring the splitting distance

In general, symplectic discretizations loose the integrability of the continuous planar Hamiltonian systems; i.e they do not preserve the Hamiltonian exactly. A simple geometric argument shows that symplectic schemes generically produce chaotic solutions in the vicinity of homoclinic orbits, regardless of the size of the time step k or the order of the discretization. In this section, we argue that symplectic discretizations remain very close to being integrable. We take an heuristic approach using Mel'nikov's function, to estimate the splitting distance between the stable and unstable manifolds. In this way we obtain an estimate of the width of the chaotic layer about the homoclinic orbit and a measure of the deviation from integrability.

3.1 The Mel'nikov function

Consider a planar symplectic mapping given by

$$Q_\epsilon : \boldsymbol{x} \to \boldsymbol{F}(\boldsymbol{x}) + \epsilon \boldsymbol{G}(\boldsymbol{x}, \epsilon), \ \boldsymbol{x} \in \mathbb{R}^2, \tag{10}$$

and $\boldsymbol{F}, \boldsymbol{G} : \mathbb{R}^2 \to \mathbb{R}^2$. Assume that the unperturbed map ($\epsilon = 0$) admits a constant of motion, $H(\boldsymbol{x}_n) = \text{const}$, for all n where $\boldsymbol{x}_{n+1} = \boldsymbol{F}(\boldsymbol{x}_n)$, i.e.

$$H(\boldsymbol{F}(\boldsymbol{x}_n)) = H(\boldsymbol{x}_n). \tag{11}$$

Also assume that the homoclinic orbit of the unperturbed map is obtained from $H(\hat{\boldsymbol{x}}_n(\xi)) = 0$, where ξ is a phase factor indicating the position of the initial condition on the homoclinic orbit. For this map the splitting distance between the stable and unstable manifolds at the phase point, ξ, is given to first order in ϵ by the Mel'nikov function (see for example [13]),

$$M(\xi; \epsilon) = \sum_{n=-\infty}^{\infty} \boldsymbol{G}(\hat{\boldsymbol{x}}_{n-1}(\xi), \epsilon) \wedge \hat{\boldsymbol{v}}_n(\xi) \tag{12}$$

where $\hat{\boldsymbol{v}}_n(\xi)$ denotes the n-th iterate of the tangent vector to the unperturbed homoclinic orbit from the phase point ξ. Writing the vectors in component form as $\boldsymbol{u} := (u^{(1)}, u^{(2)})^T$, the wedge product is defined by $\boldsymbol{u} \wedge \boldsymbol{v} = u^{(1)}v^{(2)} - u^{(2)}v^{(1)}$. Note that $\hat{\boldsymbol{x}}_n(\xi)$ is evaluated on the homoclinic orbit of the unperturbed problem.

It is useful to give another interpretation of the Mel'nikov function. The change in the constant of motion $H(\boldsymbol{x}_n)$ under the perturbed system (10) is calculated from,

$$\begin{aligned} H(\boldsymbol{x}_{n+1}) &= H(\boldsymbol{F}(\boldsymbol{x}_n) + \epsilon \boldsymbol{G}(\boldsymbol{x}_n)) \\ &= H(\boldsymbol{F}(\boldsymbol{x}_n)) + \epsilon DH(\boldsymbol{F}(\boldsymbol{x}_n)) \cdot \boldsymbol{G}(\boldsymbol{x}_n) + O(\epsilon^2) \\ &= H(\boldsymbol{x}_n) + \epsilon DH(\boldsymbol{F}(\boldsymbol{x}_n)) \cdot \boldsymbol{G}(\boldsymbol{x}_n) + O(\epsilon^2), \end{aligned} \tag{13}$$

where we have made use of (11) and DH denotes the gradient of H. Rewriting (13) as

$$\begin{aligned} \Delta H_n &= \epsilon DH(\boldsymbol{F}(\boldsymbol{x}_n)) \cdot \boldsymbol{G}(\boldsymbol{x}_n) + O(\epsilon^2) \\ &= \epsilon \boldsymbol{G}(\hat{\boldsymbol{x}}_n(\xi), \epsilon) \wedge \hat{\boldsymbol{v}}_{n+1}(\xi) + O(\epsilon^2) \end{aligned} \tag{14}$$

(recall that $\boldsymbol{x}_{n+1} = \boldsymbol{F}(\boldsymbol{x}_n)$ on the homoclinic orbit), it follows that the Mel'nikov function is a first order estimate of total change in the constant of motion, $H(\hat{\boldsymbol{x}}_n)$, over the homoclinic orbit.

Assume that the Mel'nikov function given by (12) can be written as

$$M(\xi, k) = \sum_{n=-\infty}^{\infty} m(n\tau - \xi), \qquad (15)$$

where $m(s)$ is determined by the homoclinic orbit and the perturbation and τ is somehow related to the discretization parameter. Since we sum over all values of n, it follows that M is periodic of period τ, $M(\xi + \tau, \tau) = M(\xi, \tau)$. Since $m(s)$ is evaluated on the unperturbed homoclinic orbit, it decays exponentially fast as $s \to \pm\infty$. The Mel'nikov sum therefore converges uniformly for small but finite k.

The Fourier coefficients of M are given by

$$\hat{M}_n(\tau) = \frac{1}{\tau} \int_0^\tau M(\xi, \tau) \exp(-i\mu_n \xi) d\xi, \qquad (16)$$

where $\mu_n = 2\pi n/\tau$. Inserting expression (15) for $M(\xi, \tau)$ into (16), and interchanging the summation and integration, we find

$$\hat{M}_n(\tau) = \int_{-\infty}^{\infty} m(s) \exp(-i\mu_n s) ds. \qquad (17)$$

If the perturbation and homoclinic orbit are analytic in a strip around the real axis, $m(s)$ is analytic in a strip surrounding the real axis. This allows one to conclude (see for instance [12]) that the leading order behavior of the integral is determined by the distance of the nearest singularities of $m(s)$ to the real axis. More specifically, if the nearest singularity is located at a distance $i\rho$ from the real axis, then

$$\hat{M}_n(\tau) \propto \exp(-2\pi\rho n/\tau). \qquad (18)$$

Thus we have established that all modes with $n \neq 0$ decay exponentially fast in τ. In fact, we know that the stable and unstable manifolds either intersect or are identical in the case of planar Hamiltonian systems. Therefore, the average splitting distance, $\hat{M}_0(\tau)$, decays at least as fast as the higher order coefficients, proving the exponential decay for the $n = 0$ mode. More precisely, a simple geometric argument shows that $\hat{M}_0$ is in fact, zero.

We illustrate these ideas with the following prototypical example.

4 The pendulum equation.

We consider the pendulum equation in the form

$$q'' + \sin q = 0. \qquad (19)$$

Note that (19) can also be written as the Hamiltonian system,

$$\begin{aligned} q' &= p \\ p' &= -\sin q, \end{aligned} \qquad (20)$$

with the Hamiltonian function given by

$$H(q, p) = \tfrac{1}{2} p^2 - \cos q. \qquad (21)$$

The homoclinic orbit to the origin is given by

$$p^2 - \cos q = 1,$$

or,

$$q(t) = \pi + 4\arctan[\exp(t+\gamma)]. \quad (22)$$

The nearest singularities in the complex plane are located at

$$t_0 = \pm \tfrac{1}{2} i\pi. \quad (23)$$

4.1 Discretizations

The first discretization to be considered is the second order scheme given by,

$$Q_{n+1} - 2Q_n + Q_{n-1} + k^2 \sin Q_n = 0, \quad (24)$$

which can be rewritten as a first order system of the form

$$\begin{aligned} Q_{n+1} &= Q_n + kP_n \\ P_{n+1} &= P_n - k\sin Q_{n+1}. \end{aligned} \quad (25)$$

Note that this scheme is indeed area preserving, i.e. it is symplectic. Thus we expect to see some unusual, nonintegrable behavior, in the numerical solution. The solution obtained from this discretization for $k = 0.2$ is shown in Figure 2(a). Apparently all is well; no abnormal behavior is observed. However, by changing scale and taking a close up look as in Figure 2(b), the situation changes dramatically. Now the familiar KAM features—resonant islands, chaos and invariant curves—become visible.

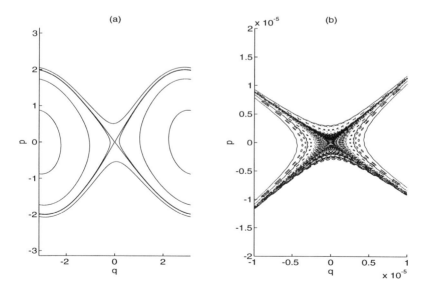

Figure 2: The solution of the pendulum equation. (b) is a close up view of (a).

To understand the reasons for this, we turn to a different discretization,

$$\begin{aligned} Q_{n+1} &= Q_n + kP_n \\ P_{n+1} &= P_n + \frac{2}{k} i \ln\left[\frac{1 + \frac{1}{3}k^2 \exp(iQ_{n+1})}{1 + \frac{1}{3}k^2 \exp(-iQ_{n+1})}\right]. \end{aligned} \quad (26)$$

This scheme is also area preserving. In fact, it is *integrable* with a constant of motion given by (see [7]),

$$H := \tfrac{1}{2}(\cos Q_n + \cos Q_{n-1}) + (k^2/16)\cos(Q_n + Q_{n-1}) + \frac{1}{k^2}[\cos(Q_n - Q_{n-1}) - 1]. \quad (27)$$

The expression for the homoclinic orbit associated with (26) is given by (see [3]),

$$Q_n = \pi + 4\arctan(\exp(pnk + \gamma), \quad (28)$$

where

$$\cosh pk = \frac{1 + \frac{1}{3}k^2}{1 - \frac{1}{3}k^2}. \quad (29)$$

Comparison with the analytical homoclinic orbit, (22) shows no difference, except for a phase shift, determined by (29).

The symplectic discretization (25) can be written as a perturbation of the integrable discretization (26). A standard Taylor expansion shows that the perturbation is of the form,

$$\frac{1}{8}k^3 \sin(2Q_{n+1}) + O(k^5)$$

This enables us to estimate the splitting distance as a function of k, as explained in the previous section. According to (28) the nearest singularities are located at $\pm i\pi/2$ and it follows from Mel'nikov's method that the splitting distance to first order is given by,

$$D \propto \exp(-\pi^2/k), \text{ as } k \to 0. \quad (30)$$

In [14], a more rigorous analysis is carried out and yields the same leading order estimate. Note: The estimate (30) depends on the area preservation property and not on the order of the discretization. This suggests that the qualitative behavior, determined by the magnitude of the splitting distance, is independent of the order of accuracy of the discretization. We do not expect to observe significant qualitative differences between symplectic schemes of different orders. This has been verified numerically; see [5].

For higher dimensional Hamiltonian systems, it is not known if similar estimates can be obtained. It is of interest then to observe how symplectic discretizations of the sine-Gordon equation, a PDE analogue of the pendulum equation, perform.

5 The sine-Gordon equation.

The sine-Gordon equation can be viewed as an infinite dimensional Hamiltonian system,

$$q_t = \frac{\delta H}{\delta p}, \quad p_t = -\frac{\delta H}{\delta q} \quad (31)$$

with
$$H(p,q) = \int_0^L \left[\tfrac{1}{2}p^2 + \tfrac{1}{2}(q_x)^2 + 1 - \cos q\right] dx \tag{32}$$
and where $q := u$ and $p := u_t$ are the conjugate variables and δ denotes the variational derivative.

The Poisson bracket of any two functionals F and G is defined to be
$$\{F, G\} = \int_0^L \left[\frac{\delta F}{\delta q}\frac{\delta G}{\delta p} - \frac{\delta F}{\delta p}\frac{\delta G}{\delta q}\right] dx, \tag{33}$$
and the evolution of any functional F under the sine-Gordon flow is governed by
$$\frac{dF}{dt} = \{F, H\}. \tag{34}$$
Obviously, the Hamiltonian H is conserved by the sine-Gordon flow. Moreover, the sine-Gordon equation is a completely integrable system as there exists an infinite family of conserved functionals in involution with respect to the Poisson bracket (33). This allows the sine-Gordon equation to be solved with the inverse scattering transform.

An important distinction between the pendulum and sine-Gordon equation is that the latter possesses much more complicated homoclinic structures [1]. Here we simply provide numerical evidence of the presence of homoclinic orbits in the sine-Gordon equation and their analytical description without going into a derivation of the formula.

The simplest solutions of the sine-Gordon equation are the fixed points, $u(x,t) = (n\pi, 0)$ for any integer n. Through a linearized analysis, it can be shown that the fixed points for even n are stable (centers) and that the fixed points for odd n are unstable. The seperatrices originate from the the unstable saddle points. Because of the symmetry, $u \to u \pm 2\pi$ is also a solution, we only consider the fixed point $u = (\pi, 0)$. For $u(x,t) = \pi + \epsilon(x,t)$, $|\epsilon(x,t)| << 1$, with $\epsilon(x,t) = \hat{\epsilon}_n(t)\exp(i\mu_n x) + \hat{\epsilon}_n^*(t)\exp(-i\mu_n x)$, $\mu_n = 2\pi n/L$, n an arbitrary integer, it follows that
$$\frac{d^2}{dt^2}\hat{\epsilon}_n + \omega_n^2 \hat{\epsilon}_n = 0, \tag{35}$$
(and similarly for $\hat{\epsilon}_n^*(t)$) where $\omega_n^2 = \mu_n^2 - 1$. Hence, the n-th mode grows exponentially, if $0 \le \mu_n^2 < 1$. It is worth noting that the zeroth mode, $n = 0$, is the most unstable in the sense that it grows at the fastest rate (it is therefore the dominant mode).

It is convenient to rewrite (35) as a system
$$\frac{d}{dt}\begin{pmatrix}\hat{\epsilon}_n \\ \hat{\eta}_n\end{pmatrix} = \begin{pmatrix} 0 & 1 \\ 1-\mu_n^2 & 0 \end{pmatrix}\begin{pmatrix}\hat{\epsilon}_n \\ \hat{\eta}_n\end{pmatrix}.$$
The eigenvalues are given by $\lambda = \pm\sqrt{1-\mu_n^2}$ and the eigenvectors, in the case of an unstable mode, by $(1, \lambda)^T$.

The eigenvectors translate into initial conditions as,
$$\begin{aligned} u(x,0) &= \pi + \epsilon_0 \cos(\mu_n x) \\ u_t(x,0) &= \epsilon_0\sqrt{1-\mu_n^2}\cos(\mu_n x). \end{aligned} \tag{36}$$

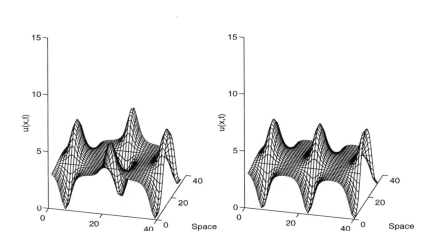

Figure 3: (a) Outside and (b) Inside the homoclinic orbit.

To demonstrate the significance of the unstable "direction", we choose an initial condition on either "side" of the unstable direction. The simplest case of interest (i.e. the solution has spatial structure) is obtained by letting $L = 2\sqrt{2}\pi$, $\mu = 2\pi/L$, which allows two unstable modes. The initial conditions are

$$\begin{aligned} u(x,0) &= \pi + 0.1\cos(\mu x) \\ u_t(x,0) &= (0.1+p)\sqrt{1-\mu^2}\cos(\mu x), \end{aligned} \qquad (37)$$

where p is a small real parameter which moves the inital condition across the homoclinic orbit. Figures 3(a), (b) show the solutions with (a) $p = +0.01$ and (b) $p = -0.01$. Different qualitative behavior is readily observed—the temporal period shown in Figure 3(a) is about twice that of Figure 3(b)—this suggests, which we confirm below, that the orbit initiated by (36) (with $p = 0$) forms a separatrix, (cf. Figure 4). The solution in Figure 3(a) can be thought of as 'outside' and the solution in Figure 3(b) as 'inside' the homoclinic orbit.

The homoclinic orbit shown in Figure 4 is a member of a family of homoclinic orbits which has the following explicit representation [3]:

$$u(x,t) = \pi + 4\tan^{-1}\left[\frac{b_1}{b_2}\cos(p_1 x + \gamma_1)\operatorname{sech}(p_2 t + \gamma_2)\right], \qquad (38)$$

where

$$p_1^2 + p_2^2 = 1, \quad b_1^2 p_1^2 = b_2^2 p_2^2. \qquad (39)$$

One can think of each $p_1 = 2\pi n/L$, n an integer such that $p_1^2 < 1$ (in order to satisfy (39)), as a separate homoclinic orbit. Figure 2 shows the situation for $n = 1$. Note that the case $p_1 = 0$, reduces to the homoclinic orbit of the pendulum equation (28).

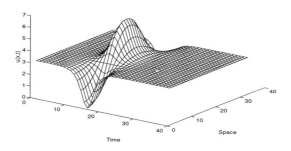

Figure 4: The surface $u(x,t)$ of a homoclinic solution.

5.1 A Motivational Experiment

Before we systematically examine the numerical schemes, we briefly present some numerical results obtained using a Fourier pseudospectral spatial discretization (50) and a semi-discrete integrable discretization described by Faddeev and Takhtajan [15] of the sine-Gordon equation. In both cases N = 32 elements are used and the time integration is performed using the 4th-order Runge-Kutta integrator D06BBf in the NAG library (specifying a relative error of 10^{-10}). We consider the initial data

$$u(x,0) = \pi + 0.1\cos(\mu x), u_t(x,0) = 0, \qquad (40)$$

with parameters $\mu = 2\pi/L$ and $L = 2\sqrt{2}\pi$.

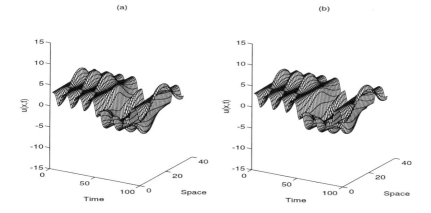

Figure 5: (a) Integrable semi-discretization, (b) Pseudospectral discretization

Figures 5(a) and (b) show the solutions obtained using the integrable and Fourier pseudospectral spatial discretization respectively. The results are an improvement over the completely integrable scheme shown in Figure 1; at least no high frequency oscillations are observed. In addition both schemes give very similar results, normally an indication that the numerical solution has converged to the true solution (we note that the situation does not change in a qualitative manner if the grid is refined). The solution remains quasi-periodic until a homoclinic crossing is observed and a rolling motion starts. Homoclinic crossings are not usually associated with integrable problems. Since we are using a spatial discretization that adequately resolves the spatial structure, one might infer that the homoclinic crossings are due to the nonsymplectic time integrator. Since the Fourier pseudospectral method can be written as a finite dimensional Hamiltonian system, see eqn (49), it is natural to use a symplectic method to discretize the time variable.

Figure 6(a) shows the result if the Fourier pseudospectral method is combined with a first order symplectic method using a time step $\Delta t = 0.01$ (see section 2). The regularity of the solution is exactly what one might expect from an integrable problem such as the sine-Gordon equation. Given the efficiency of symplectic integrators for planar problems (c.f. 30) one might conclude that, (i) Figure 6(a) is the true qualitative behavior of the sine-Gordon equation and, (ii) symplectic methods are amazingly efficient; a first order routine with a large time step has outperformed the very sophisticated NAG routine. However, the results of a second order symplectic integrator (see Figure 6b) complicates the issue. Although it takes a little longer before a homoclinic crossing is observed than in Figures 5(a) and (b), homoclinic crossings are not avoided as with the first order scheme, Figure 6(a).

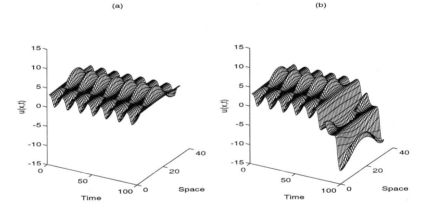

Figure 6: (a) First order symplectic, (b) Second order symplectic.

A correct interpretation of these results can be obtained using the associated spectral theory. We now show how the geometric structure of the infinite dimensional phase space may be described in terms of the Floquet discriminant. More specifically, it implicitly defines the homoclinic orbits and allows one to measure the width of the chaotic layer which appears about the homoclinic orbits when the system is perturbed.

5.2 Spectral Theory of the sine-Gordon equation

The phase space of the sine-Gordon equation with periodic boundary conditions can be described in terms of the spectrum of the following linear operator (the spatial part of the associated Lax pair; for a detailed description see [1]):

$$\mathcal{L}(u, \lambda) = \left[A\frac{d}{dx} + \frac{i}{4}B(u_x + u_t) + \frac{1}{16\lambda}C - \lambda I \right], \tag{41}$$

where

$$A = \begin{pmatrix} 0 & -1 \\ 1 & 0 \end{pmatrix}, \quad B = \begin{pmatrix} 0 & 1 \\ 1 & 0 \end{pmatrix} \tag{42}$$

$$C = \begin{pmatrix} \exp(iu) & 0 \\ 0 & \exp(-iu) \end{pmatrix}, \quad I = \begin{pmatrix} 1 & 0 \\ 0 & 1 \end{pmatrix}, \tag{43}$$

$u := (u(x,t), u_t(x,t))$ is the potential and $\lambda \in \mathbb{C}$ denotes the spectral parameter. The spectrum of $\mathcal{L}^{(x)}$ is defined as

$$\sigma(\mathcal{L}) := \left\{ \lambda \in \mathbb{C} | \mathcal{L}^{(x)}v = 0, |v| \text{ bounded } \forall x \right\}. \tag{44}$$

Since the potential u solves the sine-Gordon equation, and is of spatial period L, the spectrum is obtained using Floquet theory. The fundamental matrix, $M(x, x_0; u, \lambda)$, of the spectral operator (41) is defined by

$$\mathcal{L}(u, \lambda)M = 0, \quad M(x_0, x_0; u, \lambda) = \begin{pmatrix} 1 & 0 \\ 0 & 1 \end{pmatrix} \tag{45}$$

and the Floquet discriminant $\Delta(u, \lambda) := \mathrm{tr} M(x_0 + L, x_0; u, \lambda)$. The spectrum of $\mathcal{L}(u, \lambda)$ is given by the following condition on Δ:

$$\sigma(\mathcal{L}^{(x)}) := \{ \lambda \in \mathbb{C} | \Delta(u, \lambda) \text{ is real and } -2 \leq \Delta(u, \lambda) \leq 2 \}. \tag{46}$$

The discriminant is analytic in both its arguments. Moreover, for a fixed λ, Δ is invariant along solutions of the sine-Gordon equation:

$$\frac{d}{dt}\Delta(u(t), \lambda) = 0. \tag{47}$$

Since Δ is invariant and the functionals $\Delta(u, \lambda)$, $\Delta(u, \lambda')$ are pairwise in involution, Δ provides an infinite number of commuting invariants for the sine-Gordon equation.

When discussing the numerical experiments, we monitor the following elements of the spectrum which determine the nonlinear mode content of solutions of sine-Gordon equation and the dynamical stability of these modes:

(i) Simple periodic/antiperiodic spectrum

$$\sigma^s = \{\lambda_j^s | \Delta(\lambda, u) = \pm 2, \, d\Delta/d\lambda \neq 0\}.$$

(ii) Double points of the periodic/antiperiodic spectrum

$$\sigma^d = \{\lambda_j^d | \Delta(\lambda, u) = \pm 2, \, d\Delta/d\lambda = 0, \, d^2\Delta/d\lambda^2 \neq 0\}.$$

The periodic/antiperiodic spectrum provides the actions in an action-angle description of the system. The values of these actions fix a particular level set. Let λ denote the spectrum associated with the potential u. The level set defined by u is then given by,

$$\mathcal{M}_u \equiv \{v \in \mathcal{F} | \Delta(v, \lambda) = \Delta(u, \lambda), \lambda \in \mathbb{C}\}. \tag{48}$$

Typically, $\mathcal{M}_u$ an infinite dimensional stable torus. However, the sine-Gordon phase space also contains degenerate tori which may be unstable. If a torus is unstable, its invariant level set consists of the torus and an orbit homoclinic to the torus. These invariant level sets, consisting of an unstable component, are represented in general by complex double points in the spectrum. A complete and detailed description of the sine-Gordon phase space structure is provided in [1]; we illustrate the main ideas by means of a simple example.

Again, consider the spatially and temporally uniform solution, $u(x,t) = (\pi, 0)$. For this solution, the Floquet discriminant is given by $\Delta(u, \lambda) = 2\cos\left(\lambda + \frac{1}{16\lambda}\right)L$ and the spectrum by $\sigma(\mathcal{L}) = \mathbb{R} \bigcup (|\lambda|^2 = 1/16)$. The periodic spectrum is located at $\lambda_j = \frac{1}{2}\left[\frac{j\pi}{L} \pm \sqrt{\frac{j^2\pi^2}{L^2} - \frac{1}{4}}\right]$, j integer. Each of these points is a double point embedded in the continuous spectrum and becomes complex if $0 \leq \left(\frac{2\pi j}{L}\right)^2 < 1$. Note that the condition for complex double points is exactly the same as the condition for unstable modes. This is a special case of the general result mentioned above which relates complex double points of the linear spectral problem to homoclinic solutions of the sine-Gordon equation.

What is the spectrum of solutions that are small perturbations of $u(x,t) = (\pi, 0)$? Figures 3(a),(b) and 4 are three possible states nearby $(\pi, 0)$ in the "two unstable mode regime". These different states are obtained simply by varying the parameter p a little in initial data (37). As we saw, the waveforms are quite distinctive. With a perturbation analysis [3], it is found that these different states also have distinctive spectral configurations (see Figure 7). Figure 7(a) corresponds to the homoclinic orbit of Figure 4 and shows that both eigenvalues are double. Figures 7(b) and (c), which are the spectral representations of the waveforms shown in Figures 3(b) and (a) respectively, show how the complex double point at 45° has split into two simple points; it has opened either into a "gap" in the spectrum (Figure 7(b)) or has formed a "cross state" in the spectrum (Figure 7(c)).

Under the sine-Gordon flow, the spectral configurations remain invariant. However, due to perturbations induced by numerical discretization, the spectrum evolves in time. We now monitor the evolution of the spectrum under the different numerical flows. This will provide a quantitative measure of the qualitative properties of symplectic and nonsymplectic discretizations of sine-Gordon.

5.3 The sine-Gordon Numerical Experiments.

We discretize the Hamiltonian (32) as follows:

$$H = \frac{1}{2} \sum_{n=-\frac{1}{2}N}^{\frac{1}{2}N-1} \left[|\dot{A}_n|^2 + \mu_n^2 |A_n|^2\right] - \frac{1}{N} \sum_{j=-\frac{1}{2}N}^{\frac{1}{2}N-1} \cos\phi_j. \tag{49}$$

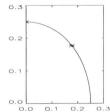 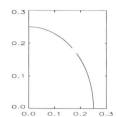 

Figure 7: The nonlinear spectrum. (a) Homoclinic orbit, (b) Inside the homoclinic orbit ('gap state'), (c) Outside the homoclinic orbit ('cross state')

This yields the exponentially accurate (for analytic solutions) Fourier pseudospectral scheme,
$$\ddot{A}_n + \mu_n^2 A_n + F_n\{\sin\phi\} = 0, \tag{50}$$
where
$$A_n = F_n\{\phi\} := \frac{1}{N}\sum_{j=-\frac{1}{2}N}^{\frac{1}{2}N-1} \phi_j \exp(-2\pi i n j/N)$$
and
$$\phi_j = F_n^{-1}\{A\} := \sum_{n=-\frac{1}{2}N}^{\frac{1}{2}N-1} A_n \exp(2\pi i n j/N).$$

We compare the higher dimensional counterparts of the first-, second- and fourth order symplectic discretizations defined by (8) with the Runge-Kutta methods of the same orders of accuracy.

5.3.1 Numerical Results

For the numerical experiments in the unstable regime, we use initial data (40). This initial data is in the "effectively" chaotic regime as the zeroth double point remains closed, i.e. the initial data is on the level set containing the homoclinic manifold [3]. Closed double points cannot be preserved by the numerical schemes and in the following experiments one observes that the zeroth mode is immediately split into a gap state by the numerical scheme. This splitting distance tells us how well the homoclinic structure is preserved by the numerical schemes.

To interpret the evolution of spectrum plots, recall that under perturbations the complex double points can split in two ways—either into a gap along an arc of the circle, or into a cross along the radius (cf. Figure 7). For each set of experiments, we show a signed measure of the splitting distance for each complex double point as a function of time. Positive and negative values represent gap and cross states respectively. Homoclinic crossings occur when the splitting distance passes through zero. Note that we measure the nonlinear spectrum throughout with an accuracy of about 10^{-6}.

It is worth mentioning the performance of a finite difference scheme implemented with Runge-Kutta (2nd and 4th order) and symplectic (1st, 2nd and 4th order) integrators. Interesting lattice dynamics occur such as collapse onto nearby (stable) periodic orbits, but as a numerical scheme it performs quite poorly. It does not preserve either component of the spectral configuration. More importantly, there is *no* substantial difference in the performance the Runge-Kutta and symplectic integrators [4].

One might argue that the similarities between the the symplectic and nonsymplectic schemes are due to an inadequate spatial resolution provided by the finite difference spatial discretization. Accordingly we oncentrate on the exponentially accurate Fourier pseudospectral method (50), implemented with Runge-Kutta (2nd and 4th order) and symplectic (1st, 2nd and 4th order) integrators. These schemes are denoted PS2RK, PS4RK, PS1SY, PS2SY and PS4SY respectively.

We use N=32 Fourier modes and a fixed time step $\Delta t = L/512$. For initial data (40), the deviations in the spectrum corresponding to the first two nonlinear modes under the PS1SY, PS2RK, PS2SY PS4RK, PS4SY flows are given in Figures 8-12 for $0 \leq t \leq 500$. This method is exponentially accurate in space which allows for a very accurate initial approximation of the spectral configuration. The evolution of the spectrum under the numerical flow is primarily due to the time integrators.

The splitting distance for both modes obtained with PS1SY (Figure 8) is $O(10^{-2})$ and is larger than that obtained with the Runge-Kutta schemes. This result highlights the fact that symplecticness is *not* enough to preserve the phase space geometry. The instability (rolling motion) observed in the wave form coincides with the first homoclinic crossing, i.e. when the splitting distance changes sign in Figure 6. This takes place later for PSY1SY than for the other schemes. However, once the crossings start they are larger in amplitude for PSY1SY, indicating that the trajectory explores a larger region of the sine-Gordon phase space.

Using PS2RK (Figure 9) and PS2SY (Figure 10), the spectrum for the first mode does not execute any homoclinic crossings and so the torus component is much more accurately preserved than with the previous spatial discretizations. The zeroth mode still displays homoclinic crossings which occur earlier than with the lower order PS1SY. Since the initial data is chosen on the homoclinic manifold, it is to be expected that there will be an earlier onset and higher density of homoclinic crossings when a more accurate scheme is used. Refinement can accentuate the frequency of homoclinic crossings as the numerical trajectory is trapped in a narrower band about the homoclinic manifold. With PS2RK there is a $O(10^{-3})$ drift in the zeroth mode, but no strong growth. The drift is eliminated when using PS4RK (Figure 11). The spectral deviations are $O(10^{-4})$ for PS4RK and PS4SY (Figure 12). Again, there is not an appreciable difference between the Runge-Kutta and symplectic integrators in their ability to preserve the integrable structure. An accurate representation of the global structures appears to be more a function of accuracy than symplecticness. In this effectively chaotic region of phase space, the chaotic width about the unstable torus is only slightly more sharply defined with the symplectic integrator. Note that the regular behavior of the first order symplectic scheme is due to larger error in the actions and not superior performance.

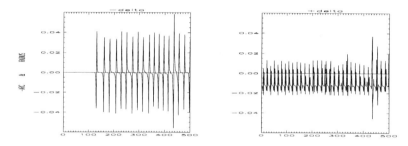

Figure 8: ps1sy: $u(x, 0) = \pi + 0.1 \cos \mu x$, $u_t(x, 0) = 0$, $N = 32$, $t = 0 - 500$

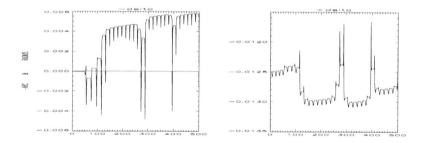

Figure 9: ps2rk: $u(x, 0) = \pi + 0.1 \cos \mu x$, $u_t(x, 0) = 0$, $N = 32$, $t = 0 - 500$

In long time studies of low dimensional Hamiltonian systems, symplectic integrators have been reported as superior in capturing global phase space structures since standard integrators may allow the actions to drift[10]. To investigate this issue for sine-Gordon, we examine a time slice $10,000 \leq t \leq 10,500$. For PS4RK the deviations in the actions associated with the zeroth mode oscillates about 1.2×10^{-4} whereas for PS4SY it oscillates about 5×10^{-5} [4]. However, the Runge-Kutta integrator can be made more efficient using variable time steps. We apply the Runge-Kutta code, D02DDf, of the NAG library which is a fully adaptive time stepping method, to the pseudospectral method (PSNAG). PSNAG does provide an improvement as it oscillates about 5×10^{-5}. For a complete description of these experiments, the reader is referred to [4]. The chaotic width (amplitude of the splitting distance) obtained with the fixed time step has been diminished with this adaptive method. Consequently, for the long timescale regime, the slight advantage obtained with PS4SY has been eliminated using variable time steps.

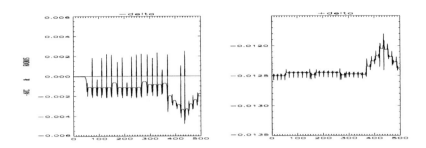

Figure 10: ps2sy: $u(x,0) = \pi + 0.1\cos\mu x$, $u_t(x,0) = 0$, $N = 32$, $t = 0 - 500$

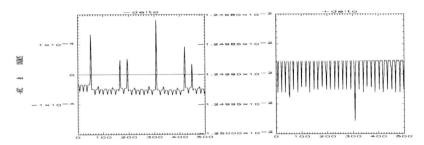

Figure 11: ps4rk: $u(x,0) = \pi + 0.1\cos\mu x$, $u_t(x,0) = 0$, $N = 32$, $t = 0 - 500$

To determine if an improvement in the performance of symplectic integrators occurs when dealing with stable structures, we have used the following initial data :

$$u_0 = 3.1 + 0.1(\cos(\mu x) + \cos(2\mu x)), \quad L = 4\sqrt{2}\pi. \tag{51}$$

which is a finite distance away from the homoclinic manifold. Initially we are nearby a solution ($u = \pi$) with 3 unstable modes— however all the double points have been split into gaps of magnitude 10^{-2} by the *initial* values and we refer to this as the stable case. In order to obtain homoclinic crossings, the perturbations in the λ need to be more than 10^{-2}. Therefore, provided the numerical schemes are sufficiently accurate,

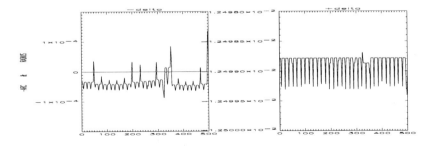

Figure 12: ps4sy: $u(x,0) = \pi + 0.1\cos\mu x$, $u_t(x,0) = 0$, $N = 32$, $t = 0 - 500$

no homoclinic crossings are expected. Although the results are not presented here, we note that our studies in this regime confirm our previous observations: the first order symplectic scheme executes deviations large enough for homoclinic crossings to occur. A comparison of the second and fourth order symplectic and Runge-Kutta schemes shows that the symplectic schemes capture most of the phase space structures of interest, although not substantially more accurately than the standard schemes which do not take into account the Hamiltonian nature. In fact, a drift in the actions occurs even when using the symplectic integrators and is eliminated using a smaller time step.

6 Conclusions

For planar Hamiltonian systems, there is no doubt that symplectic integrators lead to remarkable improvements; they are able to preserve the homoclinic orbits with exponential accuracy. However, no theoretical results are available on the deviation from integrability of symplectic discretizations of the Hamiltonian PDE's of soliton theory. Using the same fixed time step in the symplectic and standard integrators, we find very little difference between the qualitative properties (precisely defined) of symplectic and nonsymplectic schemes as applied to spatial discretizations of the sine-Gordon equation. The ability to preseve the phase space structure appears to depend more upon the accuracy of the numerical scheme rather than the property of symplecticness. *It is not only the symplectic property that is important.*

Smaller step sizes were not necessary for the standard integrators to capture the same features as those obtained with the symplectic integrators. In addition, the Runge-Kutta integrators we employ are not state of the art. If a sophisticated variable time step method is used, the very small improvement seen using the symplectic method may be lost. Even for long time integrations, where symplectic integrators perform better in low dimensional problems because the standard schemes sometimes permit a drift in the actions, no significant difference was detected, either in the stable or unstable regimes. Consequently, there does not appear to be a clear advantage in using symplectic integrators for numerical implementation of the sine-Gordon equation.

Throughout this volume the sensitive dependence on initial data which a chaotic process displays is exploited to control the chaos and maintain the system about a final desired state. It would be interesting to see if any of the ideas on targeting etc. could be adapted and incorporated in numerical studies of nonlinear PDE's.

Acknowledgements: This work was partially supported by AFOSR grant F49620-94-0120, ONR grant N00014-94-0194 and NSF grants DMS-9024528, DMS-9404265.

References

[1] N. Ercolani, M.G. Forest and D.W. McLaughlin. Geometry of the Modulational Instability Part III: Homoclinic Orbits for the Periodic Sine-Gordon Equation. Physica D, **43**, pp349-384 (1990).

[2] M.J. Ablowitz, B.M. Herbst and Constance M. Schober. Numerical Simulation of quasi-periodic solutions of the sine-Gordon equation. Physica D, **87**, pp37-47 (1995).

[3] M.J. Ablowitz, B.M. Herbst and C.M. Schober. On the Numerical Solution of the sine-Gordon equation. I. Integrable Discretizations and Homoclinic Manifolds. Accepted J. Comput. Phys. (1995).

[4] M.J. Ablowitz, B.M. Herbst and C.M. Schober. On the Numerical Solution of the sine-Gordon equation. II. Numerical Schemes. Submitted J. Comput. Phys. (1995).

[5] B.M. Herbst, Numerical evidence of exponentially small splitting distances in symplectic discretizations of planar Hamiltonian systems, Preprint. (1991)

[6] V.I. Arnold. Mathematical methods of classical mechanics. Springer-Verlag, New York (1978).

[7] A. Bobenko, N. Kutz and U. Pinkall. The discrete quantum pendulum. SFB 288 Preprint 42, Berlin (1992).

[8] P.J. Channel and C. Scovel. Symplectic integration of Hamiltonian systems. Nonlinearity, 3, pp231- 259 (1990).

[9] J.M. Sanz-Serna. Runge-Kutta schemes for Hamiltonian systems. BIT, 28, pp877-883 (1988).

[10] J.M. Sanz-Serna and M.P. Calvo. Numerical Hamiltonian problems. Chapman and Hall, London (1994).

[11] H. Yoshida. Construction of higher order symplectic integrators. Phys. A., 150, pp262-268 (1990).

[12] J.D. Murray. Asymptotic analysis. Springer-Verlag, New York(1984).

[13] R.W. Easton. Computing the Dependence on a Parameter of a Family of Unstable manifolds: Generalized Mel'nikov Formulas. Nonlinear Analysis, Theory, Methods and Appl., 8, pp1-4 (1984).

[14] Gelfreich, V.G., Lazutkin, V.F. and Tabanov, M.B., Exponentially small splitting distances in Hamiltonian systems, Chaos 1, pp5-9 (1991).

[15] L.D. Faddeev and L.A. Takhtajan. Hamiltonian methods in the theory of solitons. Springer-Verlag, Berlin (1987).

Commentary by A. Mazer

This paper performs numerical experiments on the Sine Gordon equation to evaluate the performance of different numerical techniques. There is a particular interest in comparing the performance of symplectic integrators with nonsymplectic integrators. The Sine Gordon equations are chosen because the their dynamic structure (a complete set of first integrals) provides a way for meaningful performance evaluation. The experiments have implications for any Hamiltonian system.

This paper shows that discretization of the Sine Gordon Equations causes a perturbation resulting in the splitting of homoclinic orbits. The splittings then intersect resulting in chaos. The paper's objective is to evaluate the performance of different numerical schemes in evolving along the original homoclinic orbit before chaotic dynamics take the trajectory noticeably away.

The criteria used in performance evaluation is maintanence of the constants of motion of the system. The paper demonstrates that symplectic mappings perform no better than other numerical schemes such as Runge Kutta.

The paper gives a nice introduction to the onset of chaos first described by Melnikov. Additionally, it nicely sets the Melnikov integral in the Hamiltonian framework. The bibliography includes references for those interested in more details.

Commentary by K. L. Teo

The symplectic discretization is studied for dynamic systems. It is examined how well the homoclinic structures are preserved after symplectic discretizations. This is done by applying the symplectic discretization technique of the Hamiltonian system. In particular, two important equations: pendulum equation and sino-Gordon equation, are studied.

The topic of the paper looks interesting. The paper is fairly well written.

Collapsing Effects in Computation of Dynamical Systems

Phil Diamond
Department of Mathematics
University of Queensland,
Brisbane, Qld 4072, Australia

Peter Kloeden and Aleksej Pokrovskii
School of Comp & Maths
Deakin University,
Geelong, Vic 3217, Australia

Abstract

Computer simulations of dynamical systems contain discretizations, where finite machine arithmetic replaces continuum state spaces. In some circumstances, complicated theoretical behaviour has a tendency to collapse to either trivial and degenerate behaviour or low order cycles as a result of discretizations. Characteristics of such collapsing effects often seem to depend on the corresponding discretization only in a random way. We describe a procedure to construct a stochastic process with similar statistical characteristics. Results of computer modelling with one and two dimensional systems are discussed.

1 Introduction

Let f be a smooth unimodal function of $[0, 1]$ onto $[0, 1]$, such that $f(0) = f(1) = 0$. A well known example is the logistic mapping with Malthusian parameter $r = 4$,

$$F(x) = 4x(1 - x) = 1 - (2x - 1)^2, \quad x \in [0, 1].$$

Dynamical systems generated by many such mappings are classical chaotic one-dimensional systems.

Denote by $\mathbf{L}_\nu$ the uniform $1/\nu$ lattice on $[0, 1]$:

$$\mathbf{L}_\nu = \{0, 1/\nu, 2/\nu, \ldots, 1\}, \quad \nu = 1, 2, \ldots .$$

For $x \in [0, 1]$ and $k/\nu \leq x < (k + 1)/\nu$, for some $0 \leq k \leq \nu - 1$, denote the roundoff operator $[x]_\nu$ by

$$[x]_\nu = \begin{cases} k/\nu & \text{if } k/\nu \leq x \leq (k + 1/2)/\nu, \\ (k+1)/\nu & \text{if } (k + 1/2)/\nu < x < (k + 1)/\nu. \end{cases}$$

By f_ν denote the mapping $\mathbf{L}_\nu \mapsto \mathbf{L}_\nu$ defined by

$$f_\nu(\xi) = [f(\xi)]_\nu, \quad \xi \in \mathbf{L}_\nu.$$

The mapping f_ν is a $\mathbf{L}_\nu$-*discretization* [23] of the mapping f.

The behaviour of a discretized system can differ sharply from that of the original system. For a start, chaotic trajectories cannot arise because $\mathbf{L}_\nu$ is a finite set and so every trajectory of f_ν is periodic. Thus, the properties of a simulation are precisely those of the cycles of f_ν. Consequently, if computed orbits are to simulate a chaotic system closely, these cycles must be very long and their statistical character should be close to the invariant measure of the underlying theoretical system.

This paper studies the asymptotic behaviour of some characteristics of systems f_ν as $\nu \to \infty$. Some properties of long cycles of mappings of such type were investigated in some detail in [1]. It appears that these properties are different from theoretical estimates obtained by standard methods suggested, for instance, in [2], [13], [18]. Below we concentrate on analysis of some specific characteristics. One is the proportion of collapsing points, that is, the proportion of those points $x \in \mathbf{L}_\nu$ with the property that $f_\nu^m(x) = 0$ for some positive integer m. Alternatively, trajectories may collapse, not to the origin, but by entering a periodic orbit of reasonably low frequency. In this case, the average length of the transient part of the trajectory and of the length of the cyclic part, are important. It will be shown that the statistical properties of these features can be analyzed. Moreover, methods which arise naturally in this analysis should be useful in broader areas.

A point $\xi \in \mathbf{L}_\nu$ will be called ν-*collapsing* if $f_\nu^n(\xi) = 0$ for some finite n. Note, that $f_\nu^n = 0$ implies $f_\nu^m = 0$ for all $m > n$, because $f(0) = 0$. That is, 0 is an *absorbing state* for each discretization f_ν. Denote the set of ν-collapsing points by $Y(\nu; f)$. Let $p_\nu(f)$ denote the proportion of collapsing points from the lattice $\mathbf{L}_\nu$,

$$p_\nu(f) = \frac{\#(Y(\nu; f))}{\nu}.$$

This paper will study statistical properties of the sequence

$$\mathbf{P}(f) = p_1(f), p_2(f), \ldots, p_\nu(f), \ldots .$$

The starting point of this investigation was a simple numerical experiment with the logistic mapping F. Let $\tilde{p}_\nu(F)$ be the proportion of collapsing elements from a random sample of 100 initial values $\xi \in \mathbf{L}_\nu$. The value of $\tilde{p}_\nu(F)$ is a reasonable statistical approximation to $p_\nu(F)$. Figure 1 graphs $\tilde{p}_{N+n}(F)$ for the interval $N = 2^{27}$, $1 \leq n \leq 500$, a different random sample of initial values being taken for each n. Interesting features to note are that

- dependence on ν is highly irregular and the sequence does not appear to be correlated;
- quite high proportions of collapsing points occur at frequent ν;
- the graph is quite typical and does not depend on the actual interval chosen.

These features are an artifact of the interaction between the dynamical system induced by F and the discretization $[\ \cdot\ ]_\nu : \mathbb{R} \mapsto \mathbf{L}_\nu$. This has been discussed in further detail elsewhere [10].

Instead of collapse to the trivial cycle $\{0\}$, computed mappings may collapse to cycles of low order. The characteristics of interest here are the length of the cycle, the *basin of attraction* of the cycle, that is the proportion of lattice points attracted to

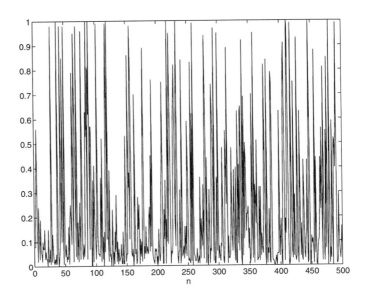

Figure 1: The proportion of collapsing elements $\tilde{p}_{N+n}$ for $N = 2^{27}$, $1 \leq n \leq 500$, for the logistic mapping $F(x) = 4x(1-x)$.

the cycle, and the length of the transient part of computed trajectories. Again, there is a marked lack of correlation of the sequence with ν of any of these characteristics [9, 10]. For example, Figure 2 graphs the size of a basin for $10^6 \leq \nu \leq 10^6 + 500$

The lack of correlation has also been noticed elsewhere when studying periodic cycles as a function of ν [13]. It is desirable to find a simple model to explain this interaction both qualitatively and quantitatively. Any such model should allow straightforward theoretical analysis and so provide a means of predicting the severity of such collapsing effects in numerical simulation of chaotic dynamical systems.

Some authors have used sequences of completely random mappings ([3] and references therein) as an heuristic model to describe sequences not unrelated to $\mathbf{P} = \{p_\nu(f)\}$. These models have not been completely successful for smooth one-dimensional unimodal mappings (see [1] and references therein). Instead of uniformly random mappings, we suggest *random mappings with a single attracting centre* [22].

These mappings can be described informally as follows: suppose that we have distinct point masses $x_0, x_1, \ldots, x_K$ forming an ensemble X. Let all but the first have unit mass, but suppose that x_0 has much higher mass $\Delta \gg 1$. Now imagine that it is possible to connect each point of the ensemble with every other point of X, and itself, by an arrow. Suppose that the probability that an arrow goes from x_i to x_j is *proportional to the mass of the end point of an arrow*, that is the mass of x_j. A *realization* of the random function is a random selection of such arrows, with respect to this probability distribution, with the constraint that only one arrow leaves each point of X. Any realization is thus a function on the finite set X. This is the principal property which distinguishes random mappings on X from a Markov chain on X.

A more formal description of random mappings with a single attracting centre

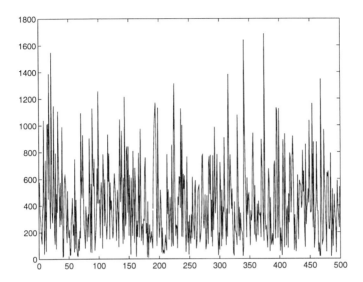

Figure 2: The length of attractive cycles of ν–discretizations of $3.96x(1-x)$ with $\xi_* = [0.5]_\nu$ in the basin of attraction, graphed against ν, $10^6 \leq \nu \leq 10^6 + 500$.

is given in the next section, along with our main hypothesis that they provide good models of collapsing effects. Section 3 discusses theoretical estimates of statistical characteristics of the model. Section 4 shows that these estimates are in close agreement with computer experiments for the logistic mapping F. Sections 5, 6 briefly discuss applications of the model to some other mappings.

2 Random mappings with a single attracting centre

Let $X = x_0, x_1, \ldots, x_K$ be a finite set and $\Gamma = \lambda_0, \lambda_1, \ldots, \lambda_K$ be a set of positive real numbers with
$$\sum_{i=0}^{K} \lambda_i = 1.$$
Introduce a random mapping $T(\,\cdot\,) = T(\,\cdot\,|\,\Gamma\,)$ of the set X into itself as follows:
$$P(T(x_i|\,\Gamma\,) = x_j) = \lambda_j, \qquad 0 \leq i,j \leq K$$
and the image of an element x_i is chosen independently of those of other elements of X. It is convenient to treat the λ_i, $i = 0, \ldots K$ as *weights* of corresponding elements x_i. An important parameter of the random mapping T is
$$H(T) = \sum_{i=0}^{K} \lambda_i^2.$$
This is the mathematical expectation of the weight of the random image of any element $x_i \in X$. The mapping $T(\,\cdot\,|\,\Gamma\,)$ is called *completely random* if it has the uniform distribution $\lambda_j = 1/(K+1)$, $j = 0, 1, 2, \ldots, K$.

Distinguish another particular type of random mapping T. Let $X = \{0, 1, \ldots, K\}$ and let Δ be a positive number with $1 < \Delta < K$. Consider a set of weights Λ where

$$\lambda_0 = \frac{\Delta}{K + \Delta} \quad \text{and} \quad \lambda_i = \frac{1}{K + \Delta}, \quad i = 1, \ldots, K.$$

The corresponding mapping $T(\,\cdot\,|\,\Lambda\,)$ is called a *random mapping on the set* $\{0, 1, \ldots, K\}$ *with a single attracting centre* 0 [22] and is usually denoted by $T_{\Delta,K}$. In other words, the random mapping $T_{\Delta,K}$ is defined by

$$P(T_{\Delta,K}(i) = j) = \begin{cases} \dfrac{\Delta}{K + \Delta} & \text{if} \quad j = 0, \\[6pt] \dfrac{1}{K + \Delta} & \text{otherwise}. \end{cases} \tag{1}$$

It should be noted that, in general, neither $T(\,\cdot\,|\,\Lambda\,)$ nor $T_{\Delta,K}$ have 0 as an absorbing state. Clearly, for $1 \ll \Delta \ll K$ there is an approximate representation

$$H(T_{\Delta,K}) \approx K^{-1}, \quad \Delta \approx \lambda_0 K. \tag{2}$$

Define the *collapsing component* Z of $T_{\Delta,K}$ to be the random subset of $\{0, 1, \ldots, K\}$,

$$Z_{\Delta,K} = \{i \in E(K) : T_{\Delta,K}^n i = 0 \text{ for some } n\}.$$

The collapsing component is essentially the set of eventually infected elements in some mathematical epidemiology models [12].

Define the random variable

$$Q_{\Delta,K} = \frac{\#(Z_{\Delta,K})}{K + 1}.$$

That is, $Q_{K,\Delta}$ is the proportion of elements of $\{0, 1, \ldots, K\}$ belonging to the collapsing component of the mapping $T_{\Delta,K}$.

Now we come to the main claim of this paper. Let f be a unimodal C^3 function of $[0,1]$ onto $[0,1]$ and $f(0) = f(1) = 0$. Suppose also that f has a negative Schwartzian, that is,

$$\frac{f'''}{f'} - \frac{3}{2}\left(\frac{f''}{f'}\right)^2 < 0$$

and that f has a non–degenerate critical point c, that is $f''(c) \neq 0$. Further assume that $|f'(0)|, |f'(1)| > 1$. Then $c_1 = f(c) = 1$, so $Df^n(c_1) = Df^{n-1}(0)$ is exponential with multiplier > 1. Such conditions are commonly used in ergodic theory of one–dimensional mappings [21] and are sufficient for the existence of a unique absolutely continuous invariant measure of f such that there exists $K > 0$ so that

$$\mu(A) \leq K|A|^{1/2}$$

for any measurable set $A \subseteq [0,1]$.

The fundamental hypothesis linking the proportion of collapsing points and the collapsing component of the random mapping (1) is as follows:

There exist positive constants $a = a(f)$, $b = b(f)$ with the following property. For large N and $1 \ll n \ll N$, the statistical properties of the sequence

$$\mathbf{P}(N, n; f) = p_N(f), p_{N+1}(f), \ldots, p_{N+n}(f) \tag{3}$$

are similar to those of the random variable

$$Q_{\Delta(\nu;f),K(\nu;f)}, \qquad \nu = N, N+1, \ldots, N+n,$$

where

$$\Delta(\nu;f) = a(f)\frac{\sqrt{\nu}}{\ln \nu}, \qquad K(\nu;f) = \left[b(f)\frac{\nu}{\ln \nu}\right]. \tag{4}$$

We have found no rigorous justification for this hypothesis. However the close agreement between its theoretical consequences and computational experiments strongly suggests that it does hold or, at least, some close modification is valid.

Our path to a link between statistical properties of collapsing effects and this particular class of random mappings was not completely straightforward. On the one hand, computational experiments for many functions showed clearly that some probabilistic description would have to be found (see Figures 1, 4). However, given a starting point, the computer orbit is completely determined and will always repeat in subsequent experiments. Consequently, Markov chain models did not match the circumstances in theory, nor did they give the statistical characteristics that were observed in practice. Another possible model was that of a random graph, where once a choice of starting point is made, the realization of the graph is then determined for all subsequent iterations on the computer lattice. As a model, uniform random graphs did not really explain the experimental results, but the class of random graphs which are those with single attracting centres gave excellent correspondence with observations. Moreover, under the assumption that the deterministic system had an absolutely continuous invariant measure, a strong motivational argument could be made that these random models were indeed what should be studied. However, an axiomatic description of finite discretization processes that would give rise to such random functions is not immediately obvious.

This hypothesis can be used in a number of different forms. On the one hand, these parameters can be estimated by some reasonable heuristics — see steps 1 and 2 below. On the other hand, standard procedures can be used to identify the parameters $a(f)$ and $b(f)$, or both these approaches might be combined.

Now let us introduce some heuristic reasoning which gives some informed guesses about $a(f)$ and $b(f)$ for any given unimodal mapping f satisfying the assumptions. This discussion will also provide the motivation for the hypothesis.

Let f be a symmetric unimodal mapping of $[0,1]$ onto $[0,1]$, with negative Schwartzian and satisfying the assumptions. Then f has a unique absolutely continuous invariant measure $\mu = \mu_f$ ([21], p. 376). The density of the measure μ_f is positive and has no singularities except at 0 and 1. Moreover, ([16], p. 211), there exist limits

$$\alpha_0 = \lim_{\gamma \to 0} \frac{\mu([0,\gamma])}{\sqrt{\gamma}}, \qquad \alpha_1 = \lim_{\gamma \to 1} \frac{\mu([1-\gamma,1])}{\sqrt{\gamma}}. \tag{5}$$

These limits are discussed in the appendix.

Step 1. First introduce an intermediate construction. Let $\mathbf{V}_{\nu,f} = f_\nu(\mathbf{L}_\nu)$. Note that the cardinality of $\mathbf{V}_{\nu,f}$ is less than ν for large ν. Denote $W_{\nu,f}(\eta) = \{x \in [0,1] : f_\nu([x]_\nu) = \eta\}$, $\eta \in \mathbf{V}_{\nu,f}$. By ergodic theorems, the proportion of elements from a typical trajectory

$$\mathbf{x} = [x_0]_\nu, f_\nu([x_0]_\nu), \ldots, f_\nu^n([x_0]_\nu), \ldots$$

which belong to $W_{\nu,f}(\eta)$ is equal to $\mu_f(W_{\nu,f}(\eta))$. In other words:

Lemma 1. *The proportion of elements from a typical trajectory* **x** *which satisfy* $f_\nu([x]_\nu) = \eta$, $\eta \in \mathbf{V}_{\nu,f}$, *is* $\mu_f(W_{\nu,f}(\eta))$.

From Lemma 1, it is natural to consider as a model of f_ν the random mapping $T_\nu = T(\,\cdot\,|\,\Lambda_{\nu,f})$ of the set $\mathbf{V}_{\nu,f}$ to itself, where the corresponding set of weights $\Lambda_{\nu,f} = \{\lambda_\zeta : \zeta \in \mathbf{V}_{\nu,f}\}$ are defined by

$$\lambda_\zeta = \mu_f(W_{\nu,f}(\zeta)).$$

In particular, it is possible to compute the weight λ_0 of the element 0,

$$\begin{aligned}\lambda_0 = \mu_f(W_{\nu,f}(\zeta)) &= \mu_f(\{x \in [0,1] : f_\nu([x]_\nu) = 0\}) \\ &= \mu_f(\{x \in [0,1] : 0 \leq f([x]_\nu) \leq 1/(2\nu)\}) \\ &= \mu_f(\{x \in [0,1] : [x]_\nu \in f^{-1}([0, 1/(2\nu)])\}).\end{aligned}$$

But $f^{-1}([0, 1/(2\nu)]) \subseteq [0, 1/(2\nu)] \bigcup [1 - 1/(2\nu), 1]$. So because of roundoff,

$$\lambda_0 = \mu_f([0, 1/(2\nu)] \bigcup [1 - 1/(2\nu), 1]) \approx \frac{\alpha_0 + \alpha_1}{\sqrt{2}} \frac{1}{\nu^{1/2}}. \tag{6}$$

Now consider $H(T(\,\cdot\,|\,\Lambda_{\nu,f}))$. Since the Schwartzian derivative is negative, by the Minimum Principle ([21], Section II.6), [0,1] can be partioned into three disjoint intervals, $[0,1] = [0, d_0] \bigcup (d_0, d_1) \bigcup [d_1, 1]$ such that $|Df(x)| \geq 1$ on $[0, d_0] \bigcup [d_1, 1]$ and $|Df(x)| < 1$ on (d_0, d_1). By symmetry, $d_1 = 1 - d_0$ and there exists $z \in [0,1]$ such that $f([0, d_0]) = f([d_1, 1]) = [0, z]$ and $f((d_0, d_1)) = (z, 1)$. Since f is expanding and symmetric on $[0, d_0] \bigcup [d_1, 1]$,

$$\begin{aligned}W_{\nu,f}(\zeta) &= \{x \in [0,1] : f([x]_\nu) \in (\zeta - 1/(2\nu), \zeta + 1/(2\nu)]\} \\ &= \{x \in [0,1] : [x]_\nu \in f^{-1}(\zeta - 1/(2\nu), \zeta + 1/(2\nu)]\}\end{aligned}$$

for all $\zeta \in [0, z] \bigcap \mathbf{L}_\nu$ such that $f^{-1}(\zeta) \subset [0, d_0] \bigcup [d_1, 1]$. For preimages $x \in [0, d_0]$, there exists an integer $k = k(\zeta)$, $0 \leq k \leq [\nu/2]$ such that $W_{\nu,f}(\zeta) = [k/\nu, (k+1)/\nu]$. Similarly, if the preimage $x \in [d_1, 1]$, $W_{\nu,f}(\zeta) = [(\nu - k)/\nu, (\nu - k + 1)/\nu]$. For $x \in [0, d_0]$, using the expression for α_0, the contribution to λ_ζ is

$$\begin{aligned}\mu_f([k/\nu, (k+1)/\nu]) &= \mu_f([0, (k+1)/\nu]) - \mu_f([0, k/\nu]) \\ &\approx \alpha_0 \left\{\left(\frac{k+1}{\nu}\right)^{1/2} - \left(\frac{k}{\nu}\right)^{1/2}\right\} \\ &\approx \frac{\alpha_0}{2k^{1/2}\nu^{1/2}}.\end{aligned}$$

Similarly, for $x \in [d_1, 1]$, the expression for α_1 gives contribution

$$\begin{aligned}\mu_f\left(\left[\frac{\nu - k}{\nu}, \frac{\nu - k + 1}{\nu}\right]\right) &= \mu_f\left(\left[\frac{\nu - k}{\nu}, 1\right]\right) - \mu_f\left(\left[\frac{\nu - k + 1}{\nu}, 1\right]\right) \\ &\approx \alpha_1 \left\{\left(1 - \frac{\nu - k}{\nu}\right)^{1/2} - \left(1 - \frac{\nu - k + 1}{\nu}\right)^{1/2}\right\} \\ &\approx \frac{\alpha_1}{2k^{1/2}\nu^{1/2}}.\end{aligned}$$

So, for preimages $x \in [0, d_0] \bigcup [d_1, 1]$, that is $\zeta \in [0, z] \bigcap \mathbf{L}_\nu$,

$$\lambda_\zeta \approx \frac{\alpha_0 + \alpha_1}{2k^{1/2}\nu^{1/2}}.$$

On the other hand, if $\zeta \in [z, 1] \bigcap \mathbf{L}_\nu$ and is the preimage $x \in (d_0, d_1)$, it is itself a preimage of some point in $(d_1, 1]$. Since the measure μ is invariant under f, $\lambda_\zeta = \mu_f(W_{\nu,f}(\zeta))$ may be measured as if x were a preimage in $(d_1, 1]$. That is

$$\lambda_\zeta \approx \frac{\alpha_1}{2k^{1/2}\nu^{1/2}}$$

for ζ corresponding to preimages $x \in (d_0, d_1)$. Hence, asymptotically,

$$H(T(\,\cdot\,|\,\Lambda_{\nu,f})) \approx \frac{\alpha_1^2 + (\alpha^0 + \alpha_1)^2}{4\nu} \ln \nu. \tag{7}$$

In the context of this model, the analogue of the set of collapsing points is the collapsing component of $T_\nu = T(\,\cdot\,|\,\Lambda_{\nu,f})$,

$$Q = \{\eta \in \mathbf{V}_{\nu,f} : T_\nu^n(\eta) = 0 \text{ for some } n\}.$$

It should be emphasised that T_ν is *not* a random mapping with single attracting centre.

Step 2. Now replace the random mapping $T(\,\cdot\,|\,\Lambda_{\nu,f})$ constructed in the previous step by a random mapping $T_{\Delta(\nu,f),K(\nu,f)}$ with a single attracting centre with asymptotically the same 0-weight as λ_0 and, asymptotically, the same mathematical expectation H as $T(\,\cdot\,|\,\Lambda_{\nu,f})$. It is sufficient to define the corresponding parameters $K(\nu;f)$ and $\Delta(\nu;f)$ of this induced random mapping. From (2) and (7), define $K(\nu;f)$ by

$$K(\nu;f) = \frac{4}{\alpha_1^2 + (\alpha_0 + \alpha_1)^2} \frac{\nu}{\ln \nu}. \tag{8}$$

Then (2), (6) and (8) give

$$\Delta(\nu, f) = \frac{2\sqrt{2}(\alpha_0 + \alpha_1)}{\alpha_1^2 + (\alpha_0 + \alpha_1)^2} \frac{\nu^{1/2}}{\ln \nu}. \tag{9}$$

From (8) and (9), reasonable estimates for $a(f)$ and $b(f)$ are thus

$$a(f) = \frac{2\sqrt{2}(\alpha_0(f) + \alpha_1(f))}{\alpha_1^2 + (\alpha_0 + \alpha_1)^2}, \tag{10}$$

$$b(f) = \frac{4}{\alpha_1^2 + (\alpha_0 + \alpha_1)^2}. \tag{11}$$

These heuristics are similar to some physical reasoning that is commonly used, for example in the construction of phenomenological models of hysteresis, see the bibliography in [17].

3 Theoretical properties of the model

Proposition 1. *The sequence of random variables*

$$C_{\nu;f} = Q_{\lambda(\nu;f),K(\nu;f)} \ln \nu, \qquad \nu = 1,2,\ldots \tag{12}$$

converges in distribution to a random variable with density $D(x;c_f)(x) = c_f d(c_f x)$, *where*

$$d(x) = \frac{1}{\sqrt{2\pi} x^{3/2}} e^{-1/(2x)}, \qquad 0 < x < \infty, \tag{13}$$

and

$$c_f = \frac{a(f)^2}{b(f)}. \tag{14}$$

Proof: The *central component* of the mapping $T_{\Delta(\nu,f),K(\nu,f)}$ [4] is a (random) subset A of $\{0,1,\ldots,K(\nu,f)\}$ defined by

$$A = \bigcup_{i=0}^{\infty} T^{-i}_{\Delta(\nu,f),K(\nu,f)} \left(\bigcup_{j=0}^{\infty} T^{j}_{\Delta(\nu,f),K(\nu,f)}(0) \right).$$

Clearly, A contains a unique cycle C of the mapping $T_{\Delta(\nu,f),K(\nu,f)}$. Adapting the argument of Burtin ([4], p.409), group the elements comprising the central component in a pairwise disjoint union $A = \bigcup_{i=1}^{4} A_i$, where

(1) A_1 is the set of ancestors of 0, that is

$$A_1 = A = \left(\bigcup_{i=0}^{\infty} T^{-i}_{\Delta(\nu,f),K(\nu,f)}(0) \right) \setminus C;$$

(2) A_2 is the set of elements of $\{0,1,\ldots,K(\nu,f)\}$ which belong to an oriented path between the centre and the cycle, excluding the last vertex of this path;

(3) $A_3 = C$;

(4) A_4 is the set of all the other elements of the central component.

From (4),

$$\lim_{\nu \to \infty} K(\nu;f) = \infty, \qquad \lim_{\nu \to \infty} \Delta(\nu;f) = \infty, \qquad \lim_{\nu \to \infty} \frac{\Delta(\nu;f)}{\sqrt{K(\nu;f)}} = 0.$$

Now Burtin's statement (II), page 411, [4], can be reformulated as

Lemma 2. $\#(A_1)/\Delta(\nu;f)^2$ *and the random vector*

$$\left(\frac{\#(A_2)}{\sqrt{K(\nu;f)}}, \frac{\#(A_3)}{\sqrt{K(\nu;f)}}, \frac{\#(A_4)}{K(\nu;f)} \right) \tag{15}$$

are asymptotically independent as ν goes to ∞. Moreover, $\#(A_1)/(\Delta(\nu;f)^2)$ converges in distribution to a random variable with density (13), *while the vector* (15) *converges in distribution to a random variable with density h given by*

$$h(x,y,z) = \frac{1}{\sqrt{2\pi}} \frac{(x+y)\exp(-(x+y)^2/2z)}{z^{3/2}(1-z)^{1/2}} \qquad 0 < x,y < \infty,\ 0 < z < 1.$$

Clearly, the collapsing component $Z_{\Delta(\nu,f),K(\nu,f)}$ satisfies

$$A_1 \subseteq Z_{\Delta(\nu,f),K(\nu,f)} \subseteq A_1 \bigcup A_3$$

and, furthermore,

$$\#(A_1) \leq \#(Z_{\Delta(\nu,f),K(\nu,f)}) \leq \#(A_1) + \#(A_3). \tag{16}$$

Therefore, from Lemma 2 and (16) it follows that the the sequence of random variables

$$Q^*_{\Delta(\nu,f),K(\nu,f)} = \frac{\#(Z_{\Delta(\nu,f),K(\nu,f)})}{\Delta(\nu,f)^2}$$

converges in distribution to a random variable with density (13) as $\nu \to \infty$. On the other hand, by (4)

$$Q^*_{\Delta(\nu,f),K(\nu,f)} = Q_{\Delta(\nu,f),K(\nu,f)} \frac{K(\nu,f)}{\Delta(\nu,f)^2} = \frac{b(f)}{a(f)^2} Q_{\Delta(\nu,f),K(\nu,f)} \ln \nu.$$

So $(b(f)/(a(f)^2)) Q_{\Delta(\nu,f),K(\nu,f)} \ln \nu$ converges in distribution to a random variable with density (13) as $\nu \to \infty$. This is equivalent to the statement of the lemma. □

An obvious corollary of this proposition is more convenient to use in numerical experiments.

Corollary 1. *Distributions of random variables* (12) *converge to the function* $erfc(1/\sqrt{2c_f x})$ *where* $erfc(y)$ *is the complementary error function* [20]

$$erfc(x) = \frac{2}{\sqrt{\pi}} \int_x^\infty e^{-t^2} dt, \quad x \geq 0. \tag{17}$$

Denote by $q(\nu;f)$ the probability that the relation

$$Q_{\Delta(\nu;f),K(\nu;f)} = K(\nu;f) + 1$$

holds. That is, that the collapsing component is the whole set $\{0,1,\ldots,K\}$.

Proposition 2. $$q(\nu;f) = \frac{\Delta(\nu,f)}{K(\nu,f)+1} \approx \frac{a(f)}{b(f)} \frac{1}{\sqrt{\nu}}.$$

Proof: Clearly, $q(\nu;f)$ coincides with the probability P_0 that none of the cycles of $T_{\Delta,K}$ is contained in the set $\{1,2,\ldots,K(\nu,f)\}$. Then, by Proposition 1 ([4], p. 404),

$$q(\nu;f) = P_0 = \lambda_0 = \frac{\Delta}{K+\Delta}. \quad \square$$

4 Interpretation and experimental results

Let **S** a finite set of non-negative real numbers. Define the *distribution function* of the set **S**

$$D(x;\mathbf{S}) = \frac{\#(\{s \in \mathbf{S} : s \leq x\})}{\#(\mathbf{S})}, \quad x \geq 0.$$

The hypothesis linking collapse to the collapsing component and Proposition 1 imply that for large N and $1 \ll n \ll N$,

$$D(x; (\ln N)\mathbf{P}(N, n; f))$$

is close to the probability distribution with the density function $c_f d(c_f x)$ where $d(\cdot)$ and c_f are defined by (13), (14). Here, for brevity we use the same symbol $\mathbf{P}(N, n; f)$ to denote the unordered set of elements of the sequence of proportions of collapsing points defined by (3).

To test this implication, computational experiments were performed with the logistic mapping F. Recall that $F(x) = 4x(1-x)$ has a unique invariant measure μ_F defined by

$$\mu_F([0,\gamma]) = \frac{2}{\pi}\arcsin(\sqrt{\gamma}).$$

Hence, $\alpha_0(F) = \alpha_1(f) = 2/\pi$, while the formulas (10), (11) give the estimates

$$a(F) = \frac{2\sqrt{2}\pi}{5}, \qquad b(F) = \frac{\pi^2}{5}.$$

In particular, $c_F = a(F)^2/b(F) = 8/5$.

Figure 3 shows $D(x; (\ln N)\widetilde{\mathbf{P}}(N, n; F))$ for $N = 2^{27}$, $n = 500$ as a step function, compared against the smooth curve of the distribution function with the density $\frac{8}{5}d(8x/5)$. Here,

$$\widetilde{\mathbf{P}}(N, n; F) = \widetilde{p}_N(F), \widetilde{p}_{N+1}(F), \ldots, \widetilde{p}_{N+n}(F)$$

and $\widetilde{p}_\nu$ were calculated as described in the introduction.

Since each $\widetilde{p}_{N+m}(F) \leq 1$, $m = 0, 1, \ldots, n$, and $\ln N \approx 18.7$,

$$D(x; (\ln N)\widetilde{\mathbf{P}}(N, n; F)) := D_{27}$$

is a truncated distribution as shown, with domain $[0, 18.7]$. On the other hand, $d(x)$ is a distribution with infinite domain $[0, \infty)$. Obviously, the truncated tail of D_{27} is inflated by the tail of d in $[18.7, \infty)$. When this is taken into account, the two distributions are in good agreement.

Say that the lattice $\mathbf{L}_\nu$ is *absolutely collapsing* for f if each point $\xi \in \mathbf{L}_\nu$ is eventually zero. Some general results about absolutely collapsing discretizations $f_\mathbf{L}$ were discussed in [7, 8]. The hypothesis, linking the collapsing component to collapse, and Proposition 2 together imply that the quantity $w(N; f)$ of absolutely collapsing lattices in the sequence

$$\mathbf{L}_1, \mathbf{L}_2, \ldots, \mathbf{L}_N$$

should be of order $(2a(f)/b(f))N^{1/2}$. In other words, $w(N)^2$ should be approximately linear in N with the slope $4(a(f)/b(f))^2$. Another series of experiments was carried out to estimate $w(N; F)$ as a function of N. Figure 4 graphs the observed $w(N; F)^2$ and the agreement with linearity is very close. Slope is approximately 9, giving an estimate of 1.5 for $a(F)/b(F)$. This is not too far away from the heuristic guess of $2\sqrt{2}/\pi \approx 0.9$.

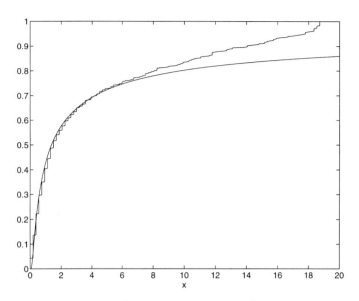

Figure 3: The distribution $D\left(x; (27\ln 2)\widetilde{\mathbf{P}}(2^{27}, 500; F)\right)$ as a step function against the smooth distribution curve with density $d_{8/5}(x) = (8/5)d(8x/5)$.

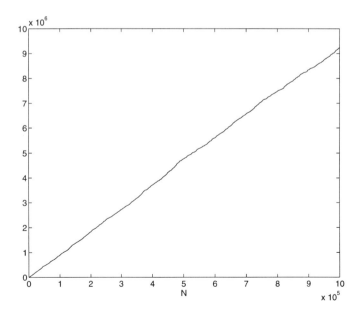

Figure 4: Squared accumulated number, $w(N; F)^2$, of absolutely collapsing lattices, for $0 \leq N \leq 10^6$.

Many simulations of this type were performed, on grids $\mathbf{L}_\nu$ of various sizes and with different nondegenerate unimodal functions satisfying the hypothesis like, for example, $\sin(\pi x)$, $x \in [0,1]$. All experiments were in good agreement with the hypothesis and the corresponding graphs are qualitatively very similar to those shown on Figures 3, 4. However, it must be emphasized that it was more difficult to estimate the quality of quantitative predictions for the parameters Δ and K, given by the heuristics, because of lack of explicit formulas for absolutely continuous measures which are invariant with respect to functions like $\sin(\pi x)$.

5 Other singularities

In this section we briefly consider another class of mappings. Attention will be restricted to unimodal mappings f of $[0,1]$ onto $[0,1]$ which have negative Schwartzian, but the critical point is degenerate and non-flat of order $\ell > 2$. That is, there exist constants L_1, L_2 such that

$$L_1|x-c|^{\ell-1} \leq |Df(x)| \leq L_2|x-c|^{\ell-1}, \qquad l > 2.$$

An appropriate growth condition for $|D^n f(c_1)|$ is also assumed. An example of such a mapping with $\ell = 4$ is

$$G(x) = 1 - (2x-1)^4. \tag{18}$$

Such mappings have absolutely continuous invariant measures μ_f [21]. The essential difference from the nondegenerate case is that now singularities of this measure at the end points of the segment $[0,1]$ are described by relations of the form

$$\mu_f([0,\gamma]) = O(\gamma^{1/l}), \qquad \mu_f([1-\gamma,1]) = O(\gamma^{1/l}) \qquad \text{as} \quad \gamma \to 0.$$

Natural modifications of the heuristic reasons of Section 2 lead to a second hypothesis:

There exist positive constants $a(f)$, $b(f)$ such that for large N and $1 \ll n \ll N$ the statistical properties of the sequence

$$\mathbf{P}(N,n;f) = p_N(f), p_{N+1}(f), \ldots, p_{N+n}(f)$$

are similar to those of the random variable

$$Q_{\Delta_*(\nu;f), K_*(\nu;f)}, \qquad \nu = N, N+1, \ldots, N+n,$$

where

$$\Delta_*(\nu;f) = a(f)\nu^{1/l}, \qquad K_*(\nu;f) = \left[b(f)\nu^{2/l}\right].$$

Adapting results of Burtin [4] give the following analogues of Propositions 1 and 2.

Proposition 3. *The sequence of random variables*

$$C^*_{\nu;f} = Q_{\Delta_*(\nu;f), K_*(\nu;f)}, \qquad \nu = 1, 2, \ldots \tag{19}$$

converges in distributions to a random variable with density

$$D^*(x;c_f)(x) = \frac{\sqrt{c_f}}{\sqrt{2\pi}} x^{-3/2}(1-x)^{-1/2} e^{-c_f(1-x)/(2x)}, \qquad 0 < x < 1$$

where

$$c_f = \frac{a(f)^2}{b(f)}.$$

In numerical experiments it is more convenient to use the following statement instead of the proposition above.

Corollary 2. *Distributions of random variables* (19) *converge to the function*

$$erfc(\sqrt{\frac{c_f}{2}\left(\frac{1}{x}-1\right)})$$

where $erfc(y)$ *is the complementary error function* (17).

Denote by $q_*(\nu; f)$ the probability that the relation

$$Q_{\Delta_*(\nu;f), K_*(\nu;f)} = k_*(\nu; f) + 1$$

holds.

Proposition 4. $\qquad q_*(\nu; f) \approx \dfrac{a(f)}{b(f)} \nu^{-1/l}.$

The qualitative behaviour of experimental calculations with the function (18) are in reasonable agreement with predictions made from the second hypothesis and Propositions 3, 4. Figure 5 is an analogue of Figure 1. Figure 6 shows $D(a; (\ln N)\widetilde{\mathbf{P}}(N, n; G))$ for $N = 2^{27}$, $n = 500$ as a step function, compared against the smooth curve, of the distribution function with the density $d_G^*(x)$).

The second hypothesis and Proposition 4 imply that the function $w(N; G)^{4/3}$ should be approximately linear in N. Figure 7 graphs $w(N; G)^{4/3}$ against N. It is in close agreement with a linear function.

Similar experiments were carried out with other functions from the family $G_\ell(x) = 1 - |2x - 1|^\ell$, for $\ell = 2.5, 3, 3.5, 5, 6, 7$ and a few more. All results are in good agreement with the theoretical predictions.

6 Collapse to cycles

Firstly, consider other logistic mappings of the form

$$F_\eta(x) = (1-\eta) F(x) = 4(1-\eta) x(1-x),$$

where the positive parameter η is sufficiently small for the system to have chaotic behaviour (see [5, 21] for details concerning such values of η). Denote by $F_{\nu,\eta}$ the $\mathbf{L}_\nu$ discretization of F_η and denote by C_ν a cycle of the discretization $F_{\nu,\eta}$ such that the point $\xi_* = [0.5]_\nu$ belongs to its basin of attraction. In other words, C_ν is the cyclic part of the trajectory with the initial point ξ_*. This cycle can be very short for some

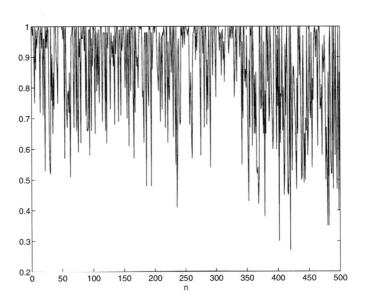

Figure 5: Proportion of collapsing elements $\tilde{p}_{N+n}$ for $N = 2^{27}$, $1 \leq n \leq 500$, for the mapping $G(x) = 1 - (2x - 1)^4$.

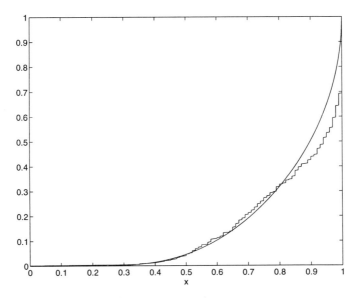

Figure 6: The distribution $D\left(x; \widetilde{\mathbf{P}}(2^{27}, 500; G)\right)$ as a step function against the smooth distribution curve with the density $d_4^*(x)$.

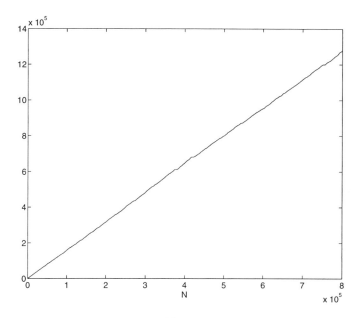

Figure 7: $w(N;G)^{4/3}$, for $0 \leq N \leq 8 \times 10^5$.

values of ν and rather long for other values, with no obvious pattern to or correlation between cycle lengths (see Figure 2). A point $\xi \in \mathbf{L}_\nu$ is called ν-*collapsing* if it is eventually absorbed by the cycle C_ν. Denote by $p_\nu(f)$ the proportion of such points in the lattice. These notations and definitions are consistent with those introduced for the 1-cycle $\{0\}$ in the Introduction. In the same way, for a random mapping $T_{\Delta,K}$ with a single attracting centre, introduce the random variable $Q^*_{\Delta,K}$, the proportion of points belonging to the central component of the mapping, see Section 3. Using much the same reasoning as in the heuristic justification of Hypothesis 1, the asymptotic statistical properties of the sequence $\mathbf{P}(N, n; F_\eta) = p_N(F_\eta), \ldots, p_{N+n}(F_\eta)$ will be similar to those of the random variable $Q^*_{\Delta,K}$ where Δ, K are as in Hypotheses 1 *provided that the mapping F_η has an absolutely continuous invariant measure*. This last assumption concerning invariant measure is valid for very many values of the parameter η but not for all, see the relevant discussion in [21]. The asymptotic distribution of the random variable $Q^*_{\Delta,K}$ is easy to obtain from the results of Burtin [4]. For example, from Lemma 2 it follows that for Δ, K as in (4), $Q^*_{\Delta,K}$ is asymptotically distributed with the density $d(x) = 0.5(1-x)^{-1/2}$. Figure 2 shows the distributions $D(x; \mathbf{P}(10^6, 500; F_{0,01}))$ and $D(x; \mathbf{P}(10^6, 500; F_{0,05}))$ compared with the probability distribution function of the density $d(x)$, that is against graph of the function $1 - \sqrt{1-x}$, $0 \leq x \leq 1$. The three curves are very close, and this strongly supports the hypothesis. Other experimental calculations from this class of functions give similar correspondence with the model. The major distinction between $F(x)$ and these $F_\eta(x)$ is that the collapse to a cycle of F is to the cycle $\{0\}$, while the other logistics have ν−collapse to longer cycles apparently of average length $O(\nu^{1/2})$.

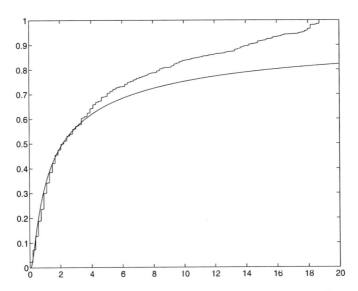

Figure 8: The function $g_1(x) = 4x^3 - 3x$: the distribution $D(a; \ln(N)\widetilde{\mathbf{P}}(N, n; G))$ for $N = 2^{27}$, $n = 500$, as a step function and the distribution function with density given by the model (the smooth curve).

Now consider the question of non unimodal mappings g of an interval $[a, b]$ into itself. Suppose that g has several extreme points, each of which is non-degenerate, and suppose that some iterate of each extreme point is an unstable fixed point of g. Such assumptions are part of the now classical explicit conditions which ensure the existence of an absolutely continuous invariant measure; see, for instance, [16]. p. 211. The same heuristics are valid for collapse of g_ν, iterated on $\mathbf{L}_\nu = [a, b]_\nu$, onto a finite set of such unstable fixed points, *providing that these fixed points belong to the lattice* $\mathbf{L}_\nu$. Experiments were done with the mappings $g_1(x) = 4x^3 - 3x$ and $g_2(x) = 8x^4 - 8x^2 + 1$, $x \in [-1, 1]$, both of which are not unimodal and are well understood ([5], p. 51,52). All results appear to be in very close agreement with the model of random mapping with a single attractive centre, see, for example, Figure 8.

7 Basin of attraction

The previous ideas may also be applied to other characteristics of computed mappings that pertain to computational cycles. Let $f : [0, 1] \to [0, 1]$ as before and consider the dynamical system, generated by $f_\nu : \mathbf{L}_\nu \to \mathbf{L}_\nu$. By $\mathbf{Tr}(\xi_0; f_\nu)$ denote the trajectory of f_ν originating at $\xi_0 \in \mathbf{L}$, that is $\mathbf{Tr}(\xi_0; f_\nu)$ is the sequence $\boldsymbol{\xi} = \xi_0, \xi_1, \ldots, \xi_n, \ldots$ which is defined by $\xi_n = f_\nu(\xi_{n-1})$, $n = 1, 2, \ldots$. For a positive integer m the m-shift of a trajectory $\boldsymbol{\xi}$ is the sequence $\mathbf{S}^m(\boldsymbol{\xi}) = \xi_m, \xi_{m+1}, \ldots$ and this is also a trajectory of f_ν. A trajectory $\mathbf{Tr}(\xi_0; f_\nu)$ is called a *cycle* if there exists a positive integer N with $\xi_N = \xi_0$. Then $\xi_i = \xi_{i+N}$ for all positive integers i. The minimal N satisfying $\xi_N = \xi_0$ is called the *period* of the cycle. Two cycles either do not contain elements in common, or one is a shift of the other. So the totality of cycles is naturally partitioned into a set of equivalence classes. We will call cycles from the same equivalence class *congruent*. Since **L** is finite, every trajectory ξ of the system is eventually cyclic, that is there exists a positive integer m such that the shifted trajectory $S^m(\xi)$ is a cycle. The minimal m such that $S^m(\xi)$ is a cycle is the *length of the transient part* of the trajectory ξ and is denoted by $\mathcal{Q}(\xi)$. The sequence $S^{\mathcal{Q}(\xi)}(\xi)$ is the *cyclic part* of ξ. The set **L** is partitioned into equivalence classes of elements which eventually generate the same cycle. That is, elements ξ_0 and ζ_0 are *equivalent*, if the cyclic parts of trajectories $\mathbf{Tr}(\xi_0; f_\nu)$ and $\mathbf{Tr}(\zeta_0; f_\nu)$ are congruent. Denote by $\mathcal{E}(f_\nu)$ the set of such equivalence classes and by $E(\xi; f_\nu)$ the equivalence class from $\mathcal{E}(f_\nu)$ which contains ξ. Finally, introduce the function $B(x; f_\nu)$ defined by

$$B(\xi; f_\nu) = \frac{\#E(\xi; f_\nu)}{\nu}, \qquad \xi \in \mathbf{L}_\nu. \tag{20}$$

Denote $U(x; f_\nu) = \mathcal{D}(x; \{B(\xi; f_\nu) : \xi \in \mathbf{L}_\nu\})$. The function $U(x; f_\nu)$ can be interpreted as the distribution in the probability theory sense of basin of attraction of the cyclic part of a trajectory $\mathbf{Tr}(\xi_0; f_\nu)$ with the random initial element ξ_0 uniformly distributed in $\mathbf{L}_\nu$. We will be interested in the mean value $V(f_\nu)$ of the function $B(\xi, f_\nu)$ averaged over $\xi \in \mathbf{L}_\nu$. Clearly, $V(f_\nu) = 1 - \int_0^1 U(x; f_\nu)\, dx = \frac{1}{(\nu^2)} \sum_{E \in \mathcal{E}(f_\nu)} (\#(E))^2$. Thus, $V(f_\nu)$ can be interpreted as the probability that two randomly chosen elements of **L** generate the same cycle. We emphasize that the second statistic is a scalar, while the first one is a scalar function $U(\cdot; f_\nu)$ on **L**.

If f_ν is the ν discretization of f, denote $B(\xi; f_\nu)$ by $B_\nu(\xi; f)$, $U(x; f_\nu)$ by $U_\nu(x; f)$ and $V(f_\nu)$ by $V_\nu(f)$.

Consider the sequences

$$\mathbf{U}(f) = U_1(x; f), U_2(x; f), \ldots, U_\nu(x; f), \ldots, \tag{21}$$
$$\mathbf{V}(f) = V_1(f), V_2(f), \ldots, V_\nu(f), \ldots . \tag{22}$$

For large ν, elements of these sequences depend on ν only irregularly when the function f behaves chaotically and the autocorrelation is negligible. For instance, Figure 2 graphs the elements of the finite sequence

$$V_{N+1}(f_{\ell,\varepsilon}), \ldots, V_{N+500}(f_{\ell,\varepsilon})$$

for $\ell = 2$, $\varepsilon = 10^{-3}$, $N = 10^6$. However, in line with previous sections, these sequences do have some asymptotic statistical features which can be described. We will say that the sequence of functions $w_\nu(x)$ is *Cesàro stable with a limit* $D(x)$ if

$$\lim_{N \to \infty} \frac{1}{N} \sum_{\nu=1}^{N} w_\nu(x) = D(x).$$

Cesàro stability of the sequence (21) can be connected with the stable distribution property of a sequence associated with $\{B_\nu(\xi; f)\}$. To this end introduce a random sequence

$$\xi = \xi_1, \xi_2, \ldots, \xi_\nu \ldots$$

where the ξ_ν are independent and each ξ_ν is uniformly distributed on $\mathbf{L}_\nu$ and consider the random sequence

$$\beta(f) = B_1(\xi_1; f), B_2(\xi_2; f), \ldots, B_\nu(\xi_\nu; f), \ldots . \tag{23}$$

Proposition 5. *Let the sequence (23) have with probability 1 (w.p.1), the stable distribution property with a continuous limit $D(x)$. Then the sequence $\mathbf{U}(f)$ is Cesàro stable with the same limit.*

Although the overall approach is the same, because the characteristics that we are considering are more a little intricate than mere collapse, a more detailed phenomenological description is required. Let $f_{\ell,\varepsilon}$ denote the mapping $[0,1] \mapsto [0,1]$ which is defined by

$$f_{\ell,\varepsilon}(x) = (1-\varepsilon)(1 - |1-2x|^\ell) \qquad (24)$$

with parameters $\ell \geq 1$ and $\varepsilon > 0$.

Basin Hypothesis

(a) *Let $1 \leq \ell \leq 2$ and let $f_{\ell,\varepsilon}$ have an absolutely continuous SRB measure with positive density. Then the sequence $\mathbf{U}(f_{\ell,\varepsilon})$ is Cesàro stable with the limit $1 - \sqrt{1-x}$ and the sequence $\mathbf{V}(f_{\ell,\varepsilon})$ has the stable distribution property with a continuous limit $D_V(x)$ which does not depend on ε nor on ℓ.*

(b) *Let $\ell \geq 2$ and let $f_{\ell,\varepsilon}$ have an absolutely continuous SRB measure with positive density. Then the sequence $\mathbf{U}(f_{\ell,\varepsilon})$ is Cesàro stable with a continuous limit $D_U(x;\ell,\varepsilon)$ and the sequence $\mathbf{V}(f_{\ell,\varepsilon})$ has the stable distribution property with a limit $D_V(x;\ell,\varepsilon)$.*

Recall that the set $\tau(\ell) = \{\varepsilon \in (0,1) : f_{\ell,\varepsilon}$ has an SRB measure$\}$ is nonempty and by [16] the mappings $f_{\ell,\varepsilon}$ have absolutely continuous SRB invariant measures $\mu(\ell,\varepsilon)$ for some sets $\tau(\ell) \subset [0,1)$. Moreover, 1 is a density point of each nonempty $\tau(\ell)$, that is the Lebesgue measure of τ, $\mathrm{mes}(\tau(\ell))$ is positive and

$$\lim_{\delta \to 0} \frac{\mathrm{mes}([1-\delta,1] \bigcap \tau(\ell))}{\delta} = 1.$$

Note that the set $\tau(\ell)$ is not generic in a topological sense, but is a set of the first Baire category, that is a closed set without internal points. The Basin Hypothesis implies the following assertion which is more convenient for interpretation and computational verification.

Corollary 3. *For each $\ell \geq 1$ there exist a set $\tau(\ell) \subset [0,1)$ which has the number 1 as its density point, such that the following assertions are true.*

(a) *For $1 \leq \ell \leq 2$ and $\varepsilon \in \tau(\ell)$ the sequence $\mathbf{U}(f_{\ell,\varepsilon})$ is Cesàro stable with the limit $1 - \sqrt{1-x}$ and the sequence $\mathbf{V}(f_{\ell,\varepsilon})$ has the stable distribution property with a continuous limit $D_V(x)$ which does not depend on ε nor on ℓ.*

(b) *For $\ell \geq 2$ and $\varepsilon \in \tau(\ell)$ the sequence $\mathbf{U}(f_{\ell,\varepsilon})$ is Cesàro stable with a continuous limit $D_U(x;\ell,\varepsilon)$ and the sequence $\mathbf{V}(f_{\ell,\varepsilon})$ has the stable distribution property with a continuous limit $D_V(x;\ell,\varepsilon)$.*

This corollary is strongly supported by numerical simulations. We briefly describe the random variables to be considered and a few of the experiments carried out.

Let $T_{\Lambda,K}$ be the random mapping described in Section 2. Any realization T of $T_{\Lambda,K}$ is a dynamical system on $X(K) = \{0, 1, \ldots, K\}$. Therefore the quantity

$V(T)$ and the random variable $\beta(i;T)$, where i is uniformly distributed on $X(K)$, are well defined. For $\nu = 1, 2, \ldots$, let Λ_ν, K_ν be sequences of constants, let T_ν be a sequence of independent realizations of T_{Λ_ν, K_ν} and let i_ν be a sequence of independent random variables uniformly distributed on $X(K_\nu)$. This induces a sequence of random equivalence classes $E(i_\nu; T_\nu)$, each from $X(K_\nu)$ and, associated with this, sequences of random variables

$$\beta(i_\nu; T_\nu) = \frac{\#(E(i_\nu; T_\nu))}{K_\nu + 1},$$
$$v(T_\nu) = \mathbf{E}\left(\beta(i_\nu; T_\nu) | T_\nu\right).$$

Here, $\mathbf{E}(\omega|\zeta)$ denotes the expected value of ω conditioned on ζ.

In much the same way as before, there are constants $K_\nu(f)$, $\Lambda_\nu(f)$ such that the statistical properties of the sequence $\mathbf{V}(f)$ and statistical properties which hold w.p.1 for the random sequence $\beta(f)$ are respectively similar to those which hold w.p.1 for the sequences $V(T_\nu)$, and $\beta(i_\nu; T_\nu)$, $\nu = 1, 2, \ldots$, where T_ν is a sequence of independent realizations of random mappings $T_{\Lambda_\nu(f), K_\nu(f)}$, and i_ν are independent random variables each uniformly distributed in $X(K_\nu)$.

Appropriate $K_\nu(f_{\ell,\varepsilon})$, $\Lambda_\nu(f_{\ell,\varepsilon})$, for $\varepsilon \in \tau(\ell)$ are suggested by arguments in [13]. Take $K_\nu(f_{\ell,\varepsilon}) = [\nu^{\dim_c(\mu_{\ell,\varepsilon})}]$ where $\dim_c(\mu)$ is the correlation dimension of μ ([14]) and $[\alpha]$ denotes the integer part of α. This leads to

$$K_\nu(f_{\ell,\varepsilon}) = \begin{cases} \nu, & \text{if } 1 < \ell < 2, \\ [\nu^{2/\ell}], & \text{if } \ell > 2. \end{cases}$$

It is then natural to define $\Lambda_\nu(f_{\ell,\varepsilon})$ consistent with (1), where $\Delta = \Delta_\nu(f_{\ell,\varepsilon}) = c\nu^{1/\ell}$ and $K = K_\nu(f_{\ell,\varepsilon})$ above.

Define the *central component* ([4]) $A(T_{\Delta,K})$ of $T_{\Delta,K}$ as the random subset of $\{0, 1, \ldots, K\}$,

$$A_{\Delta,K} = \{i \in E(K) : T_{\Delta,K}^n i \in C \text{ for some } n\}.$$

Introduce the family of functions

$$D(x; c) = \text{erfc}\left(\frac{c\sqrt{1-x}}{\sqrt{2x}}\right) - e^{c^2/2}\left(\sqrt{1-x}\right)\text{erfc}\left(\frac{c}{\sqrt{2x}}\right), \quad 0 \leq x \leq 1. \quad (25)$$

Here $\text{erfc}(x)$ is the complementary error function

$$\text{erfc}(x) = \frac{2}{\sqrt{\pi}} \int_x^\infty e^{-t^2}\, dt, \quad x \geq 0.$$

Clearly, $D(x; c)$ is increasing, with $D(0; c) = 0$, $D(1; c) = 1$ and $D(x; 0) = 1 - \sqrt{1-x}$.

Proposition 6.

(a) Let $1 < \ell \leq 2$. Then the random sequence $\beta(i_\nu, T_\nu)$, where $\{T_\nu\}$ is a sequence of independent realizations of random mappings $\{T_{c\nu^{1/\ell},\nu}\}$ has w.p.1 the stable distribution property with the limit $1 - \sqrt{1-x}$.

(b) Let $\ell \geq 2$. Then the random sequence $\{\beta(i_\nu, T_\nu)\}$, where $\{T_\nu\}$ is the sequence of independent realizations of random mappings $\{T_{c\nu^{1/\ell}, [\nu^{2/\ell}]}\}$ has w.p.1 the stable distribution property with the limit

$$D_\beta(x;c) = xD(x;c) + \int_x^1 D(\theta;c)\,d\theta$$
$$-\frac{1}{2}\int_x^1 D(1-\theta;c)\left(\sqrt{\frac{\theta-x}{\theta}} + \sqrt{\frac{\theta}{\theta-x}}\right) d\theta \ . \quad (26)$$

7.1 Numerical Experiments

Case $\ell \leq 2$. The Basin Hypothesis (a), with respect to the sequence $\mathbf{U}(f_{\ell,\varepsilon})$, implies that for $1 \ll n \ll N$ the function

$$\mathbf{u}(x; f_{\ell,\varepsilon}, N, n) = \frac{1}{n}\sum_{\nu=N+1}^{N+n} U_\nu(f_{\ell,\varepsilon}) \quad (27)$$

is close to $1 - \sqrt{1-x}$ for $\varepsilon \in \tau(\ell)$. By Corollary 3(a), the functions $\mathbf{u}(x; f_{\ell,\varepsilon}, N, n)$ are very nearly $1 - \sqrt{1-x}$ for most sufficiently small $\varepsilon > 0$, in the sense of Lebesgue measure, since 1 is a density point of $\tau(\ell)$. Recall that the set $\tau(\ell)$ where Corollary 3 is applicable is thin in the topological sense, because it it the set of the first Baire category. In the context of numerical experiments below the measure theory properties clearly outweigh the topological properties.

The functions $\mathbf{u}(x; f_{\ell,\varepsilon}, N, n)$ were calculated for $\ell = 1.0, 1.2, 1.4, 1.6, 1.8, 2.0$; and $\varepsilon = 10^{-3}$, $N = 10^6$, $n = 10^3$. Results are shown in Figure 9.

Now consider Basin Hypothesis (a) with respect to the sequence $\mathbf{V}(f_{\ell,\varepsilon})$. For positive integers N and n consider functions

$$\mathbf{v}(x; f_{\ell,\varepsilon}, N, n) = \mathcal{D}(x; \{V_{N+1}(f_{\ell,\varepsilon}), \ldots, V_{N+n}(f_{\ell,\varepsilon})\}).$$

The hypothesis with respect to the sequence $\mathbf{V}(f_{\ell,\varepsilon})$ means that for $1 \ll n \ll N$ and for $\varepsilon \in \tau(\ell)$, functions $\mathbf{v}(x; f_{\ell,\varepsilon}, N, n)$ should all be much alike. Moreover, since $D_V(x) \approx D_K(x)$ for large K, all these functions are close to the distribution of the r.v. $V(T_{1,K})$. It is easy to approximate this distribution numerically using a random generator. For an integer K and for $n = 1, \ldots, N$, consider the mappings $i \to T_K^{(n)}(i) = a[i,n]$ where the array $a[i,n]$ is generated by SUN Pascal strings

```
for n:=1 to N do   for i:=0 to K do
begin a[i,n]:=trunc((K+1)*random(i)); end;
```

The formal algorithm is included to indicate the use of a concrete quasi-random generator which can slightly influence the numerical results. Recall the definition of the equivalence class $\mathcal{E}_\varphi$ and for each $n = 1, \ldots, N$, denote

$$V(T_K^{(n)}) = (K+1)^{-2} \sum_{E \in \mathcal{E}(T_K^{(n)})} (\#(E))^2.$$

The mappings T_K^1, $n = 1, \ldots, N$ can be considered as N independent sample realizations of the completely random mapping $T_{1,K}$. Therefore, the distribution function

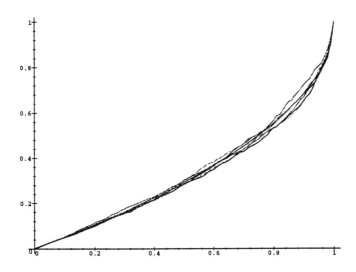

Figure 9: The function $1 - (1-x)^{1/2}$, as a smooth line in the middle, against the functions $u(x; f_{\ell,\varepsilon}, N, n)$ for $\ell = 1.0, 1.2, 1.4, 1.6, 1.8, 2.0$; and $\varepsilon = 10^{-3}$, $N = 10^6$, $n = 10^3$ (jagged lines).

of the set $\{V(T_K^{(1)}), \ldots, V(V_K^{(N)})\}$ should be close to the distribution of $V(T_{1,K})$ for reasonably large K, N. Figure 10 graphs the distribution function

$$\widetilde{\mathcal{D}}^{(0)}(x) = \mathcal{D}(x; \{V(T_K^{(1)}), V(T_K^{(2)}), \ldots, V(T_K^{(10^3)})\}) \qquad (28)$$

for $K = 2^{16} - 1$ and the functions $\mathbf{v}(x; f_{\ell,\varepsilon}, N, n)$ for $\varepsilon = 10^{-3}$, $N = 10^6$, $n = 10^3$ and the same ℓ as in Figure 2. The value $K = 2^{16} - 1$ was chosen simply because this the size of the standard built-in random generator.

Case $\ell > 2$. Let a specific $\ell > 2$ be chosen. The hypothesis with respect to the sequence $\mathbf{U}$ means that for each $1 \ll n \ll N$ and for $\varepsilon \in S(\ell)$ the functions $\mathbf{u}(x; f_{\ell,\varepsilon}, N_1, n)$ and $\mathbf{u}(x; f_{\ell,\varepsilon}, N_2, n)$ should be similar to one another. Further, both functions $\mathbf{u}(x; f_{\ell,\varepsilon}, N_1, n)$ and $\mathbf{u}(x; f_{\ell,\varepsilon}, N_2, n)$ should be close to a function $D(x; c)$ for an appropriate $c = c(\ell, \varepsilon)$ and both functions $\mathbf{v}(x; f_{\ell,\varepsilon}, N_1, n)$ and $\mathbf{v}(x; f_{\ell,\varepsilon}, N_2, n)$ should be close to a function $D_V(x; c)$ for *the same* $c = c(\ell, \varepsilon)$.

Consider, for instance $\ell = 3$, $\varepsilon = 10^{-3}$. Figure 11 graphs functions $\mathbf{u}(x; f_{\ell,\varepsilon}, 10^5, 10^3)$, $\mathbf{u}(x; f_{\ell,\varepsilon}, 10^6, 10^3)$ and $D_W(x; 1.53)$.

To get a similar figure concerning the second statistic V we need to imitate numerically the function $\lim_{K \to \infty} D_K(x; c)$ for $c = 1.53$. To this end for $n = 1, \ldots, N$, consider the mappings $i \mapsto T_K^{(c,n)}(i) = a[i, n]$, where the array $a[i, n]$ was generated by the SUN Pascal strings

```
for n:=1 to N do  for i:=0 to K do
begin a[i,n]:=trunc(random(i)*((K+1)+c*sqrt(K)));
if(a[i,n]>K) then a[i,n]=0; end;
```

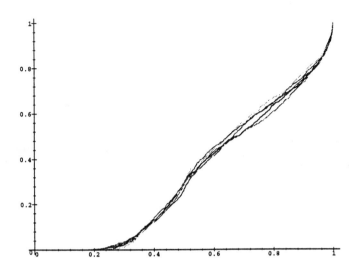

Figure 10: The distribution $\widetilde{\mathcal{D}}^{(0)}(x) = \mathcal{D}(x; \{V(T_K^{(1)}), V(T_K^{(2)}), \ldots, V(T_K^{(10^3)})\})$ for $K = 2^{16} - 1$ and the functions $v(x; f_{\ell,\varepsilon}, N, n)$, $\ell = 1.0, 1.2, 1.4, 1.6, 1.8, 2.0$; and $\varepsilon = 10^{-3}$, $N = 10^6$, $n = 10^3$.

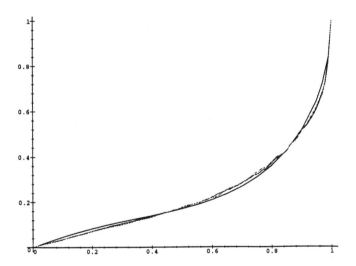

Figure 11: The functions $u(x; f_{\ell,\varepsilon}, 10^5, 10^3)$, $u(x; f_{\ell,\varepsilon}, 10^6, 10^3)$ and $D_W^{(1.53)}(x)$.

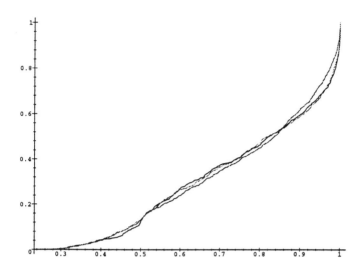

Figure 12: The functions $v(x; f_{\ell,\varepsilon}, 10^5, 10^3)$, $v(x; f_{\ell,\varepsilon}, 10^6, 10^3)$, against the distribution $\widetilde{\mathcal{D}}^{1.53}(x) = \mathcal{D}(x; \{V(T_K^{1.53,1}), \ldots, V(T_K^{1.53,N})\}$ for $K = 2^{16} - 1$.

For each $n = 1, \ldots, K$, denote

$$V(T_K^{c,n}) = (K+1)^{-2} \sum_{E \in \mathcal{E}(T_K^{(n)})} (\#(E))^2.$$

The mappings $T_K^{c,n}$, $n = 1, \ldots, N$, can be considered as N independent sample realizations of the random mapping $T_{c\sqrt{K},K}$. Therefore, the distribution function of the set $\{V(T_K^{c,1}), \ldots, V(T_K^{c,N})\}$ should be close to the distribution of $V(T_{c\sqrt{K},K})$. Figure 12 graphs $\mathbf{v}(x; f_{\ell,\varepsilon}, 10^5, 10^3)$, $\mathbf{v}(x; f_{\ell,\varepsilon}, 10^6, 10^3)$, and the distribution

$$\widetilde{\mathcal{D}}^{(1.53)}(x) = \mathcal{D}(x, \{V(T_K^{1.53,1}), \ldots, V(T_K^{1.53,N})\}$$

for $K = 2^{16} - 1$.

Comparable results were obtained for other values of ℓ, such as $\ell = 2.5, 3.25, 3.5, 4.0$ etc.

7.2 Two-dimensional systems

Consider mappings $f : \mathbb{R}^d \to \mathbb{R}^d$. Let us define the class of computer realizations which we will study. Consider the lattice $\mathbf{L}_\nu^d = \nu^{-1}\mathbf{Z}^d$ where $\mathbf{Z}^d$ is the standard integer lattice in $\mathbb{R}^d$. The $\mathbf{L}_\nu^d$-discretization f_ν of f is defined by $f_\nu(\xi) = ([y^{(1)}]_\nu, \ldots, [y^{(d)}]_\nu)$, where $\xi = (\xi^{(1)}, \ldots \xi^{(d)}) \in \mathbf{L}_\nu^d$, $y = f(\xi) \in \mathbb{R}^d$ and $[\cdot]_\nu$ is the scalar roundoff operator defined above. Write $Q^{z,\rho}$, $z \in \mathbb{R}^d$, $\rho > 0$, for the cube

$$\{x = (x^{(1)}, x^{(2)}, \ldots, x^{(d)}) \in \mathbb{R}^d : |x^{(i)} - z^{(i)}| \leq \rho, \ i = 1, \ldots, d\}.$$

We will consider trajectories of discretizations originating in $Q^{z,\rho}$. Throughout, a fixed vector $z \in \mathbb{R}^d$ and a fixed number $\rho > 0$ will be chosen so that all trajectories ξ satisfying $\xi_0 \in Q^{z,\rho}$ are bounded in $\mathbf{L}_\nu^d$ and therefore eventually periodic. As in Section 1.1, there is a natural partition of the set $Q_\nu^{z,\rho} = Q^{z,\rho} \cap \mathbf{L}_\nu^d$ into the set $\mathcal{E}_\nu^{z,\rho}(f)$ of equivalence classes. Computational statistics of the sequences (21) and (22) are as follows. For $\nu = 1, 2, \ldots$, define

$$U_\nu^{z,\rho}(x; f) = \mathcal{D}\left(x, \left\{ \#(E)\,(\#(Q_\nu^{z,\rho}))^{-1} : E \in \mathcal{E}_\nu^{z,\rho}(f) \right\}\right),$$

$$V_\nu^{z,\rho}(f) = (\#(Q_\nu^{z,\rho}))^{-2} \sum_{E \in \mathcal{E}_\nu^{z,\rho}(f)} (\#(E))^2.$$

Then define the corresponding sequences

$$\mathbf{U}^{z,\rho}(x; f) = \{U_\nu^{z,\rho}(x; f)\}, \qquad \mathbf{V}^{z,\rho}(f) = \{V_\nu^{z,\rho}(f)\}.$$

Suppose that $f \Re^2 \to \Re^2$ has an SRB invariant measure μ. The maximal open set Ω for which μ is a weak limit of the sequence of measures $\mu_n = \frac{1}{n}\sum_{i=0}^{n-1} f_*^i \delta_x$ for almost all initial conditions $x \in \Omega$ with respect to Lebesgue measure, is the basin of attraction of the measure μ.

In particular, recall the Hénon and Lozi mappings which are defined respectively by

$$f_H(\mathbf{x}; \mathbf{a}) = (1 + y - ax^2, bx) \quad \text{and} \quad f_L(\mathbf{x}; \mathbf{a}) = (1 + y - a|x|, bx),$$

where $\mathbf{a} = (a, b)$ is a vector of real parameters. It is thought that the Hénon mapping has an SRB measure μ for some $\mathbf{a}$, while the Lozi mapping certainly has an SRB measure μ for some $\mathbf{a}$ which was rigorously proved in [19].

The earlier hypothesis has to be modified:

For some $\mathbf{a}$, suppose that the Hénon mapping (respectively, the Lozi mapping) has an SRB invariant measure μ and the cube $Q^{z,\rho}$ belong to the basin of attraction of μ. Then the sequence $\mathbf{U}^{z,\rho}(\cdot\,; f_H(\cdot\,; \mathbf{a}))$ (respectively $\mathbf{U}^{z,\rho}(\cdot\,; f_L(\cdot\,; \mathbf{a}))$) is Cesàro stable with the limit $1 - \sqrt{1-x}$ and the sequence $\mathbf{V}^{z,\rho}(\cdot\,; f_H(\cdot\,; \mathbf{a}))$ (respectively $\mathbf{V}^{z,\rho}(\cdot\,; f_L(\cdot\,; \mathbf{a}))$) has the stable distribution property with limit $D_V(x) = \lim_{K \to \infty} D_K$.

Numerical experiments have been carried out to test this hypothesis. Let $z = (-0.05, -0.05)$ and $\rho = 0.2$. Figure 13 compares the functions

$$\mathbf{u}^{z,\rho}(x; f_H(\cdot\,; 1.4, 0.3), 500, 500) \quad \text{and} \quad \mathbf{u}^{z,\rho}(x; f_L(\cdot\,; 1.7, 0.5), 500, 500)$$

against the function $1 - \sqrt{1-x}$. Analogously, Figure 14 graphs the functions

$$\mathbf{v}^{z,\rho}(x; f_H(\cdot\,; 1.4, 0.3), 500, 500) \quad \text{and} \quad \mathbf{v}^{z,\rho}(x; f_L(\cdot\,; 1.7, 0.5), 500, 500)$$

against the function $\widetilde{\mathcal{D}}^{(0)}(x)$, which was defined in (28). The concrete values of parameters a, b were taken from [15], p.269 and from [16], p. 203.

Other combinations of the parameters were tried for the Hénon and Lozi mappings. All experimental results strongly support the hypothesis above. Quite similar results were also obtained for other mappings such as the Belykh mapping defined by

$$f_B(\mathbf{x}; \lambda_1, \lambda_2, k) = \begin{cases} (\lambda_1((x)-1)+1, \lambda_2((y)-1)+1) & \text{if } y > kx, \\ (\lambda_1((x)+1)-1, \lambda_2((y)+1)-1) & \text{if } y < kx. \end{cases}$$

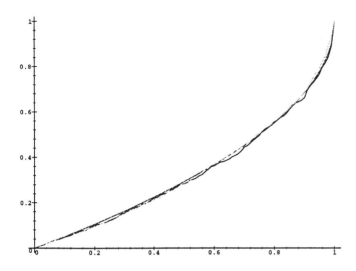

Figure 13: The functions $u^{z,\rho}(x; f_H(\cdot\,; 1.4, 0.3, N, n)$ and $u^{z,\rho}(x; f_L(\cdot\,; 1.7, 0.5, N, n)$ (for the Hénon and Lozi mappings) for $z = (-0.05, -0.05)$, $\rho = 0.2$, $N = n = 500$ against the function $1 - (1-x)^{1/2}$

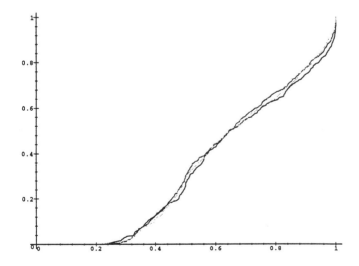

Figure 14: The functions $v^{z,\rho}(x; f_H(\cdot\,; 1.4, 0.3, N, n)$ and $v^{z,\rho}(x; f_L(\cdot\,; 1.7, 0.5, N, n)$ for $z = (-0.05, -0.05)$, $\rho = 0.2$, $N = n = 500$ against the function (refra0E).

Here $\lambda_1 \in (0, 1/2)$, $\lambda_2 \in (1, 2)$, $k \in (-1, 1)$ are parameters and (α) is the fractional part of α. This mapping arises in phase synchronization systems. It is especially interesting because it cannot be reduced in any sense to one-dimensional mappings. See further references in [16], p. 203.

In contrast, consider the shift mapping on a two-dimensional torus defined by $g(x, y) = ((x + a), (y + b))$. This mapping also has SRB invariant measure which coincides with Lebesgue measure, provided that a, b are rationally independent. Nevertheless, the cardinality of the basin of attraction in $\mathbf{L}_\nu^2$ of each cycle of the corresponding ν-discretization clearly does not exceed ν. In particular, the sequence $\mathbf{U}(g)$ is Cesàro stable with a degenerate limit δ_0, the Dirac distribution at the point 0. Similarly, $\mathbf{V}(g)$ has the stable distribution property with the limit δ_0. Much the same behaviour was observed in experiments for discretizations of an algebraic automorphism f_A of the standard 2-torus generated by an integer hyperbolic 2-matrix A with $|\det(A)| = 1$.

Why should circle rotations, toroidal shifts and algebraic automorphisms of the torus behave so differently from the other mappings with an SRB measure considered above? We believe that a reasonable explanation is that for all these mappings the discretizations have their own algebraic structure. This additional structure precludes the possibility of using random mappings as models of discretizations of the original mapping. In particular, the discretizations of these mappings are invertible. and consequently the whole lattice is partitioned into disjoint cycles of the mapping. Hence, such mappings contain many more cycles than do mappings without algebraic structures on discretizations and the basin of attraction of each cycle is much smaller.

8 Summary and Conclusion

This paper demonstrates the existence of a severe pathological artifact of collapse which can occur in computation of systems with an absolutely continuous invariant measure. While it is known that long cycles of computed mappings are endowed with an invariant measure close to the expected result, if significant numbers of initial values collapse either to the zero cycle or to low order cycles, the statistical picture of the attractor which emerges from the simulation will be distorted. Consequently, it is of some importance to evaluate the likely severity of the artifact.

When the artifact is examined at different precisions, the proportion of collapsing points varies erratically, apparently uncorrelated with computer ε. This extremely variable behaviour naturally leads to a probabilistic approach to the modelling of collapse. To increase the size of the sample space, the effect was considered on the sequence of sets $\mathbf{L}_\nu = \{0, 1/\nu, \ldots, (\nu-1)/\nu, 1\}$, which serve, for $\nu = 2^N$, as a model of a fixed point format with N binary digits and radix point in the first position. Using a natural definition of the roundoff operator, many computer experiments were carried out.

The results of these simulations were compared to a theoretical description of collapse by a class of random graphs in which the probabilities of certain edges were weighted. A heuristic argument was given for the weights, in terms of the roundoff operator and the invariant measure of the original mapping. The correspondence between this model and the experiments was very close, over a wide range of ν and all of the mappings studied in one and two dimensions. This gives some confidence that this model (or something very similar), although not rigorously justified by a mathematical argument, describes the actual state of affairs in computation of a large class of chaotic systems.

The presentation at these Proceedings which was perhaps closest to ours is that of Schober. There, in the computation of nonlinear PDEs near a homoclinic point, similar artifacts arose. Apparently, no reliance can be placed on successful computation but, instead, there is collapse to one of several spurious solutions. Sensitivity to initial conditions is such that is not possible even to predict to which artifact a computation will collapse. In some respects this an even more severe problem than we describe, although total collapse can exist in many of the $\mathbf{L}_\nu$.

In our study, the artifact of collapse disappears when very low level noise is injected into the computation, although this again distorts the observed statistics of the computed attractor. Perhaps a similar device would ameliorate the severity of spurious solutions in Schober's computations and provide trajectories which are more valid than those which now arise.

9 Appendix

The heuristics described in Section 2 require more detailed quantitative information concerning the limits α_0, α_1 (5). This is, in some measure, supplied by the lemma below. Let f be a unimodal mapping of $[0, 1]$ onto $[0, 1]$, with negative Schwartzian satisfying the assumptions. Denote by c the unique critical point of f and by m_0, m_1 absolute values of slopes of f at the points 0 and 1.

Lemma 3.

$$\alpha_1 = \lim_{\gamma \to 0} \frac{\mu([0, \gamma])}{\sqrt{\gamma}} = \frac{2\tau(c)}{\sqrt{|f''(c)|}}, \tag{29}$$

$$\alpha_0 = \lim_{\gamma \to 0} \frac{\mu([1-\gamma, 1])}{\sqrt{\gamma}} = \frac{\alpha_1 \sqrt{m_0}}{\sqrt{m_1}(\sqrt{m_0} - 1)} \tag{30}$$

where $\tau(\cdot)$ is the density of the absolutely continuous invariant measure μ_f with respect to Lebesgue measure.

Proof: Since the measure μ_f is invariant,

$$\mu_f([1-\gamma, 1]) = \mu_f(f^{-1}([1-\gamma, 1])).$$

On the other hand, from unimodality of f and $f(c) = 1$, for γ sufficiently close to 1, the set $f^{-1}([1-\gamma, 1]$ belongs to a small neighbourhood of c, where $f(x) \approx 1 + f''(c)(x-c)^2$. Hence,

$$\mu_f([1-\gamma, 1]) \approx \tau(c)|f^{-1}([1-\gamma, 1])| \approx \frac{2\tau(c)\sqrt{\gamma}}{\sqrt{|f''(c)|}}$$

for small γ. The equality (29) follows. Again, by invariance of μ_f we have that

$$\mu_f([0, \gamma]) = \mu_f([0, s_0]) + \mu_f([s_1, 1]),$$

where s_0, s_1 are respectively the lesser and greater of the two preimages of γ. So,

$$\mu_f([0,\gamma]) \approx \mu_f([0,\gamma/m_0]) + \mu_f([1-\gamma/m_1, 1]).$$

Iterating,

$$\mu_f([0,\gamma]) \approx \mu_f([0,\gamma/m_0^2]) + \mu_f([1-\gamma/(m_0 m_1), 1]) + \mu_f([1-\gamma/m_1, 1]).$$

Continuing in this way,

$$\begin{aligned}\mu_f([0,\gamma]) &\approx \sum_{i=0}^{\infty} \mu_f([1-\gamma/(m_0^i m_1), 1]) + \lim_{i\to\infty} \mu_f([0,\gamma/m_0^i]) \\ &= \sum_{i=0}^{\infty} \mu_f([1-\gamma/(m_0^i m_1), 1]).\end{aligned}$$

Now (30) follows from the last relation and (29) which has been just proved. □

Acknowledgements

This research has been supported by the Australian Research Council Grant A 8913 2609.

References

[1] P.M. Binder, Limit cycles in a quadratic discrete iteration. *Phys.-D* **57** (1992), no. 1-2, 31–38.

[2] A. Boyarsky and P. Gora, Why computers like Lebesgue measure, *Comp. Math. Applns*, **16** (1988), 321–329.

[3] B. Bollobas, *"Random graphs"*, Academic Press, London, 1985.

[4] Y.D. Burtin, On a simple formula for random mappings and is applications, *J. Appl. Prob.* **17** (1980), 403–414.

[5] R. Devaney, *"An Introduction to Chaotic Dynamics"*, Addison Wesley, Calif., 1989.

[6] P. Diamond, P. Kloeden and A. Pokrovskii, An invariant measure arising in computer simulation of a chaotic dynamical system, *Journal of Nonlinear Science*, **4**, pp. 59 - 68.

[7] P. Diamond, P. Kloeden and A. Pokrovskii, Weakly chain recurrent points and spatial discretizations of dynamical systems, *Random & Computational Dynamics*, **2**, No 1, pp. 97 - 110.

[8] P. Diamond, P. Kloeden and A. Pokrovskii, Interval stochastic matrices and simulation of chaotic dynamics, in *"Chaotic Numerics"*, P. Kloeden and K.J. Palmer (eds.), Seirie: "Contemporary Mathematics", **172**, 203–217, American Mathematical Society, Providence, 1994.

[9] P. Diamond, A. Klemm, P. Kloeden and A. Pokrovskii, Basin of attraction of cycles of dynamical systems with SRB invariant measures, *J. Statist. Phys.*, to appear.

[10] P. Diamond, P. Kloeden, A. Pokrovskii and A. Vladimirov, Collapsing effects in numerical simulation of a class of chaotic dynamical systems and random mappings with a single attracting centre, *Physica D*, to appear.

[11] P. Diamond P. Kloeden, A. Pokrovskii and M. Suzuki, Collapsing effects in numerical simulation of Chaotic Dynamical Sistems. *Proceedings of 94 Korea Automatic Control Conference. International Session*, 753–757. October 18-20,1994 Taejon, Korea.

[12] I.B. Gerchbakh, Epidemic processes on a random graph: some preliminary results, *J. Appl. Prob.* **14** (1977), 427–438.

[13] C. Grebogi, E. Ott and J.A. Yorke, Roundoff–induced periodicity and the correlation dimension of chaotic attractors, *Phys. Rev. A*, **34** (1988), 3688–3692.

[14] Grassberger P. and Procaccia I., On the characterization of strange attractors, *Phys. Rev. Lett.*, **50** (1983), pp. 346–349.

[15] Guckenheimer J. and Holmes P., *"Nonlinear oscillations, Dynamical Systems and Bifurcations of Vector Fields"*, Springer–Verlag, New York, 1983.

[16] M.V. Jakobson, Ergodic theory of one–dimensional mappings, in: "Sovremennye Problemy Matematiki. Fundamentalnye napravlenija. Tom 2. In Russian [Current problems in mathematics. Fundamental Directions. Vol. 2.] Dinamicheskie Sistemy 2. [Dynamical Systems 2.] AN SSSR, VINITI, Moscow, 1982, 204–232.

[17] M.A. Krasnosel'skii and A.V. Pokrovskii, *"Systems with Hysteresis"*, Springer–Verlag, Berlin, 1989.

[18] Y.E. Levy, Some remarks about computer studies of dynamical systems, *Physics Letters*, **88A** (1982), No 1, 1–3.

[19] Y.E. Levy, Ergodic properties of the Lozi mappings, *Commun. Math. Phys.*, **93** (1984), 461–482.

[20] *"MATLAB. Reference Quide"*, The MathWorks, Inc., Natick, 1992.

[21] W. di Melo and S. van Strien, *"One-dimensional dynamics"*, Springer-Verlag, Berlin, 1993.

[22] V.E. Stepanov, Random mappings with a single attracting centre, *Theory Prob. Appl.*, **16** (1971), 155–161.

[23] H.J. Stetter, *"Analysis of Discretization Methods for Ordinary Differential equations"*, Springer–Verlag, Berlin, 1976.

Commentary by I. Mareels

This paper vividly and candidly reminds us of the fact that chaos does not exist inside a computer based simulation of a dynamical system. In the case of one dimensional interval maps and subject to a classical roundoff operator based on a uniform partion of the interval, the paper illustrates how the behavior of the discretized map may significantly differ from the orginal map. The paper provides an heuristic, well motivated probabilistic model to describe the mechanism through which a supposedly chaotic dynamical behavior may collapse into a fairly mundane one.

Bifurcations in the Falkner-Skan equation

Colin Sparrow
Newton Institute for Mathematical Sciences
20 Clarkson Road, Cambridge CB3 0EH, UK

Abstract

The Falkner-Skan equation is a reversible three dimensional system of ODEs without fixed points. A novel sequence of bifurcations, each of which creates a large set of periodic and other interesting orbits 'from infinity', occurs for each positive integer value of a parameter. Another sequence of bifurcations destroys these orbits as the parameter increases; topological constraints allow us to understand this sequence of bifurcations in considerable detail. While outlining these results, we can also make a number of possibly illuminating remarks connecting parts of the proof, well-known numerical techniques for locating and continuing periodic orbits, and recent ideas in the control of chaos.

1 Introduction

The Falkner-Skan equation is an ordinary differential equation derived originally from a problem in boundary-layer theory [5]. We will not be interested in the derivation of the problem here, and the behaviour we will be discussing is outside the physically interesting region of parameters. Instead we are interested in the bifurcations which create and destroy periodic orbits (and other interesting trajectories) in the flow as a parameter is changed. One sequence of bifurcations, which are the main subject of this paper, is a sequence of 'heteroclinic bifurcations at infinity'. For reasons that will become clear below, we also refer to these as 'gluing' bifurcations. These are analysed in greater detail in Swinnerton-Dyer & Sparrow [13], where the interested reader may look to fill out the details missing from sections 2 – 5 below. Each of the bifurcations in this sequence creates infinitely many periodic orbits and other interesting trajectories as a parameter increases. As far as we are aware, these bifurcations have not been discussed elsewhere in the literature except in the work of Hastings & Troy [8] which is mentioned further below. They would seem, therefore, to be a new mechanism for creating complicated invariant sets in the flows of ordinary differential equations.

There is another sequence of bifurcations in the system that is discussed more briefly in section 6, which will be the subject of another paper [12], and that is still the subject of active research. This second sequence destroys periodic orbits as the parameter increases, and is probably typical of some sequences of bifurcations observed in 3-dimensional *reversible* differential equations (for example, see Roberts, this volume). These two sequences of bifurcations are very different and it is of considerable interest to understand how they are related; periodic orbits created in one sequence are destroyed in the other, and each orbit has various topological invariants associated with it which very much restrict the possibilities for what can occur in between. In particular, we know which orbits are created from the theory, and this gives us detailed insight into the orbits involved in the destruction sequence for orbits of considerably greater complexity than can be examined numerically. This topic is also discussed briefly in section 6.

At first sight this chapter may appear to be a little way away from the main theme of the meeting (and book). However, it is of some interest to note how the proof outlined in section 5 relies on much the same use of hyperbolicity as techniques for the control of chaos described elsewhere in this volume. It is also worth spending a little time thinking about the numerical techniques commonly used to investigate periodic orbits in systems of ordinary differential equations – for example to produce Figs 2 and 3 below. Such thoughts may even suggest refinements of the methods of chaos control currently under investigation, and are discussed further in section 8.

2 The Falkner-Skan Equation; basics

The Falkner-Skan equation [5] is:

$$y''' + y''y + \lambda(1 - y'^2) = 0 \tag{1}$$

with phase space $(y, y', y'') \in \mathbf{R}^3$, and we are concerned with values of the parameter $\lambda > 0$. Differentiation is with respect to an independent variable t that we think of as time-like (though in the derivation of the equation from a model of flow in a boundary layer, the independent variable would be thought of as a space variable). Previous authors who have looked at equation (1) were primarily interested in finding periodic orbits or in finding solutions which exist in $0 < t < \infty$ and satisfy $y(0) = y'(0) = 0$ and $y'(\infty) = 1$. Solutions of the second type are of interest in the physical application, but typically have unbounded behaviour in $t < 0$, and we will have little more to say about them. We do find, however, that it is important for the theory to consider some special unbounded solutions, and so will be interested in a set of solutions to equation (1) which we have called *admissible solutions*. Admissible solutions are those for which $|y'|$ is bounded for all t, and include, in particular, periodic orbits (P-orbits) and orbits satisfying $y' \to 1$ as $t \to \pm\infty$ (which we call Q-orbits). We do not ask that y be bounded on admissible solutions as this would exclude the unbounded solutions $y' = \pm 1$, $y'' = y''' = 0$ which are crucial in the analysis that follows. We call these two straight-line solutions Y_+ and Y_-; the Q-orbits are the solutions biasymptotic to Y_+.

It is possible to determine some properties of admissible solution by standard methods. It follows from equation (1) that every maximum in y' of a trajectory, where $y'' = 0$ and $y''' < 0$, must lie in $|y'| \leq 1$. Similarly, every minimum of y' on a trajectory lies in $|y'| \geq 1$. As illustrated in Fig 1, these facts are consistent with the possibility of recurrent behaviour in the system because solutions may travel in the direction of increasing y close to Y_+ for some time before descending to travel back in the direction of decreasing y close to Y_-, and then back to Y_+ again, etc.

For admissible trajectories standard arguments [13] give the boundedness lemma below, which is illustrated in Fig 1(a). The constants N_i in the lemma depend on the integer n considered, but this is not important; the λ-intervals in the lemma are chosen to ensure that there is an overlap between successive intervals.

Lemma 2.1 *For given n and $\lambda \in [n - \frac{3}{4}, n + \frac{3}{4}]$, there are constants $N_1, \ldots, N_4$ such*

The Falkner-Skan Equation

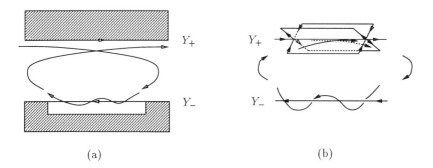

Figure 1: The phase-space of the Falkner-Skan equation projected onto the (y, y') plane. (a) Admissible trajectories cannot enter the shaded region. (b) Behaviour near $Y_{\pm}$.

that any admissible trajectory satisfies

$$1 \geq y' \geq -N_1, \quad \text{with} \quad y' \geq -1 \quad \text{if} \quad |y| > N_2$$

and $|y''| \leq N_3$, $|y'''| \leq N_4$.

Fig 1(b) shows the local behaviour near $Y_{\pm}$. Trajectories near Y_- spiral around it if $|y|$ is not too large. It is relatively easy to see [13] that the maximum number of turns an admissible solution may make about Y_- while staying close to it is bounded; in fact it follows from results later in this paper that the bound is $[\lambda]$, the integer part of λ. The trajectory Y_+ acts as a 'saddle trajectory' and nearby trajectories move with increasing y ($y' \approx 1$) and approach Y_+ as $t \to \infty$ if they lie on a two-dimensional local stable manifold of Y_+ and will approach Y_+ as $t \to -\infty$ if they lie on a two-dimensional local unstable manifold of Y_+.

Note that equation (1) is *reversible* (see Roberts, this volume, and [11]) as it is invariant under a symmetry that is the composition of an involution on phase space and time reversal. The required symmetry for equation (1) is $(t, y, y', y'') \mapsto (t_0-t, -y, y', -y'')$. Trajectories may be invariant under this symmetry, as is the case with each of the trajectories shown in Fig 2, or they may occur in pairs, each of which is taken to the other by the symmetry. Most of our discussion below does not make any essential use of the symmetry, though it simplifies many of the arguments. Of course, Roberts' remarks (this volume) on the stability of invariant sets of reversible systems will apply. We are not aware of parameter ranges in which there are symmetric pairs of attractors and repellors in the flow, but we cannot rule out this possibility; numerical solution of the equations from randomly chosen initial conditions leads to unbounded trajectories in all cases that we have tried.

3 P and Q-orbits and topological invariants

Figure 2 shows four numerically computed admissible solutions projected onto the (y, y') plane. Fig 2(a) shows a periodic trajectory which makes a single pass through

$y' < 0$, during which it makes 3 turns around Y_-; we call the orbit $P3$, following Botta et al [2]. We will see that this orbit is created as λ increases through $\lambda = 3$, though this is not important in this section. Fig 2(b) shows another periodic orbit which makes two passes through the region $y' < 0$ and which makes one and two turns around Y_- during the passages through $y' < 0$; we label it $P12$ (or $P21$ since names of P-orbits can be permuted cyclically). This orbit is created when λ increases through $\lambda = 2$. The total winding number around Y_- (3 in both cases) is an invariant of a periodic orbit, and so cannot change as the orbit is continued for changing λ. Thus, if we continue periodic orbits, the total of the digits in their names cannot change. The names themselves, however, which describe the number of turns around Y_- on successive passes through $y' < 0$, may not be invariant. In fact, the two orbits $P3$ and $P21$ annihilate one another at a larger λ value at which they both make three passes through $y' < 0$ with one turn around Y_- on each pass (and so can legitimately be thought to have both changed into $P111$ orbits) [2, 12].

Periodic orbits are important in the study of equation (1), but an even more central role is played by the Q-orbits. (The importance of the Q-orbits was also noted in [7].) Two of these are shown in Figs 2(c) and (d). We call the orbit shown in Fig 2(d) $Q11$, as it makes two passes into the region $y' < 0$ and on each occasion it turns once around Y_-. This orbit is created as λ increases through $\lambda = 1$. The orbit shown in Fig 2(c) is called $Q2$ and is created as λ increases through $\lambda = 2$.

It is important to note that Q-orbits, like periodic orbits, cannot just disappear as λ varies without their being involved in a bifurcation with one or more other Q-orbits. In fact there is a local bifurcation theory for Q-orbits like the more familiar bifurcation theory for periodic orbits, but simpler because there are no phenomena analogous to period multiplication. We can, in fact, expect saddle-node and pitchfork bifurcations to occur in a generic way between appropriate Q-orbits. As with P-orbits, the total winding number of a Q-orbit around Y_- will be an invariant as the orbit is continued for changing λ. In addition, there is extra information about Q-orbits. Since all Q-orbits approach Y_+ eventually on the same two-dimensional manifold, they can be ordered on this manifold by their closeness to Y_+. We have two such orderings, corresponding to the two limits $t \to \infty$ and $t \to -\infty$ and only symmetric orbits will occupy the same place in both orderings. These orderings will give us rather a lot of information about the order in which Q-orbits can bifurcate [12] and so are important in understanding how orbits disappear from the system (section 6). Yet further information comes from the symmetry. Symmetric P-orbits must each intersect the symmetry line $y = y'' = 0$ in exactly two points; symmetric Q-orbits will intersect the line just once (see Roberts, this volume). The ordering of these intersections on the symmetry line will further restrict the order in which bifurcations involving symmetric P and Q-orbits can occur as λ increases.

These topological considerations are not crucial for our study of gluing bifurcation in sections 4 and 5 below. It is, however, important to note that numerical experiments and theory both show that Q-orbits *are* removed by bifurcation from the set of admissible orbits as λ increases, contrary to the conjecture of Hastings & Troy [8]; this affects the statement of Theorem 5.1 as we cannot in general be precise about the sets of Q-orbits that exist at particular λ values.

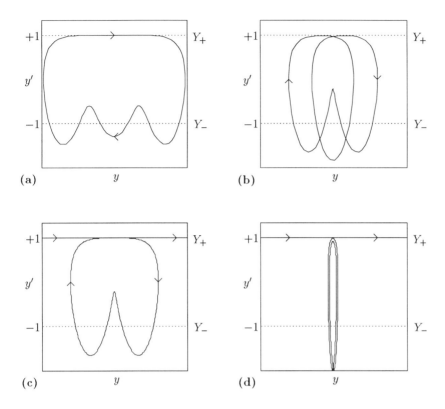

Figure 2: Some admissible solutions for $\lambda = 10$. All orbits are shown at the same scale. (a) Periodic orbit $P3$. (b) Periodic orbit $P21$. (c) The orbit $Q2$. (d) The orbit $Q11$.

4 Creation of Q-orbits

Very crudely, our main result is that large sets of admissible trajectories are created each time λ increases through an integer $n \in \mathbf{Z}$. Slightly more precisely, for integer $\lambda \geq 2$ each of these bifurcations creates an infinity of periodic and other admissible solutions by bifurcation away from the two unbounded trajectories Y_+ and Y_-. When $\lambda = 1$ there is a similar bifurcation that produces many Q-orbits but only one periodic orbit. These bifurcations were discussed by Hastings & Troy [8] who proved some results for the cases $\lambda = 1, 2$ and made some almost correct conjectures about the bifurcations for larger λ values. Our approach is different and more generally applicable than theirs.

The precise statement of the result splits naturally into two parts. The first, Theorem 4.2, states that an orbit Qn is created as λ increases through integer n. Recall that Qn is a Q-orbit which makes n turns around Y_- during a single pass through $y' < 0$. The second, Theorem 5.1, follows from the first, and gives details of the other P and Q-orbits created.

A major step in the proof of Theorem 4.2 is Lemma 4.1 below, which gives a reasonable precise and necessary condition on λ for an admissible trajectory to cross $y' = 0$ with $|y|$ large. This is important, since it will transpire that new admissible solutions created as λ increases through n consist of the union of segments lying very close to Y_+, segments lying very close to Y_-, and nearly vertical segements (on which y is approximately constant) joining the two. In particular, Qn has just two such nearly vertical segments, and as as $\lambda \downarrow n$ the y-values where these segments cross $y' = 0$ satisfy $|y| \to \infty$.

Lemma 4.1 *For λ in a similar interval as in Lemma 2.1, there are constants N_5 and N_6 with the following properties. If (y_0, y'_0, y''_0) is a point of an admissible solution with $|y_0| > N_5$ and $y'_0 = 0$ then $\lambda > n$, and*

$$\left| (\lambda - n)|y_0|^{2n} - \frac{n^2(2n!)^2}{4(n+1)(n!)^3} \right| < N_6 (\lambda - n)^{\frac{1}{(n+1)}} . \tag{2}$$

Moreover, if $y_0 > 0$ the trajectory dips below $y' = -1$ exactly n times before it next crosses $y' = 0$, and it crosses $y' = 0$ upwards through the gateway which satisfies (2) with $y < -N_5$. A similar result holds with time reversed if $y_0 < 0$.

Notice that the gateways are only far out if λ is near an integer. Therefore, the lemma shows that if new admissible orbits are to be created as described above, then this can only occur as λ increases through an integer n.

We omit most of the details of the proof of the lemma, but it is worth mentioning that an important step in the proof involves studying the behaviour of trajectories moving very close to Y_-. For this purpose we take y as the independent variable and $z = 1 + y'$ as the dependent one. Using dots to denote differentiation with respect to y, (1) now becomes

$$\ddot{z} - y\dot{z} + 2\lambda z = Z(y). \tag{3}$$

This equation describes the evolution of the distance from Y_- (given by z) as the trajectory moves along near Y_-, and $Z(y)$ is small compared with z so long as z and its derivatives are all small. The associated linear equation

$$\ddot{z} - y\dot{z} + 2\lambda z = 0 \tag{4}$$

is Weber's equation and the fact that integers are the important parameter values for equation (1) ultimately follows from the fact that the solutions of equation (4) which grow slowly as $t \to \infty$ and as $t \to -\infty$ coincide at these parameter values, but the intervening calculations are non-trivial.

Once the lemma is proved, the remainder of the proof of the creation of Qn is easier and involves matching together several pieces of approximate solution. These include a piece close to Y_-, pieces near to Y_+, and some nearly vertical pieces with large $|y|$. The 'matching' involves showing there is a true trajectory close to the pseudo-trajectory formed of the union of these pieces; we will have more to say about this sort of argument in sections 5 and 8 below. The precise statement of the theorem is:

Theorem 4.2 *In some interval to the right of $\lambda = n$ there is a unique Q-orbit which enters $y' < 0$ just once and does so through the gateway defined by (2). It is symmetric and makes n twists around Y_-.*

5 Gluing Q-orbits

We can next show that the existence of the Qn-orbit in $\lambda > n$ implies the existence of a whole new set of admissible solutions formed by 'gluing' segments of the Qn-orbit to each other, and to segments of other Q-orbits which already exist when $\lambda = n$. A precise statement of the theorem we can prove about the creation of a large set of admissible solutions is:

Theorem 5.1 *Let $Q\alpha_0, Q\alpha_1, \ldots$ be a finite set of Q-orbits which exist and are regular at $\lambda = n$ and let $m_0, m_1, \ldots$ be a finite set of positive integers. Then there exists an $\epsilon > 0$ such that in the interval $n < \lambda < n + \epsilon$ there is a unique Q-orbit $Q\alpha_0 n^{m_0} \alpha_1 n^{m_1} \ldots$ close to the pseudo-orbit obtiained by gluing together in sequence segments of $Q\alpha_0$, m_0 copies of Qn, $Q\alpha_1$ and so on. The name can finish with α_r or n^m and a similar result holds for $Qn^{m_0}\alpha_0 n^{m_1}\ldots$ Furthermore, for any cyclic alternating sequence as above (of minimal period) there exists a unique periodic orbit $P\alpha_0 n^{m_0} \alpha_1 \ldots$.*

[The technical condition that the $Q\alpha$ be regular means, essentially, that $\lambda = n$ cannot be a bifurcation point for $Q\alpha$.]

The theorem and its proof is illustrated in Figs 3 and 4. Given a finite collection of regular $Q\alpha$'s, we first choose an M with the property that every one of our collection remains close to Y_+ in $|y| > M$. For an example, see Fig 3(a) where $n = 2$, the points B and C are at $y = \pm M$, and our finite collection of pre-existing Q-orbits consists of the single orbit $Q1$. We then pick ϵ small enough that in $n < \lambda < n + \epsilon$ the new trajectory Qn that exists in $\lambda > n$ has all its interesting behaviour outside the region $|y| < M$ as illustrated for the orbit $Q2$ at $\lambda = 2.1$ in Fig 3(b). In other words, ϵ is chosen so that Qn is still close to Y_+ at $y = M$ when tending away from Y_+, and is again close to Y_+ at $y = -M$ when tending to Y_+ as $t \to \infty$. We can then create pseudo-trajectories, with small discontinuities at $y = \pm M$, with names like those in the theorem, by gluing together appropriate pieces of Qn and $Q\alpha$ at $y = \pm M$. This process can be used to produce pseudo-Q-orbits or pseudo-P-orbits as illustrated in Figs 3(c) and 3(d). We then only need to show that there will be a unique true P or Q-orbit close to any pseudo-trajectory formed as above. In fact, Figs 3(c) and 3(d) show numerically calculated 'true' trajectories $Q12$ and $P12$ at the same scale and at the same parameter value as the orbits $Q1$ and $Q2$ shown in Figs 3(a) and (b); to the naked eye they are indistinguishable from the pseudo-trajectories created by gluing together pieces of $Q1$ and $Q2$ as described in the caption to Fig 3.

The argument to establish the existence of the required true trajectories is fairly standard in dynamical systems theory, and is illustrated in Fig 4 for the particular case where we wish to prove the existence of a true $Q12$ orbit close to the pseudo-orbit constructed in Fig 3. The figure shows the situation on the plane $y = M$ where we desire to glue together a part of $Q1$ which strikes the return plane at point q_1 ($= C$ on Fig 3(a)) on the local stable manifold of Y_+, and a part of $Q2$ which strikes the return plane at the point q_2 ($= X$ on Fig 3(b)) on the local unstable manifold of Y_+. Since both $Q1$ and $Q2$ lie in the interior of both the global stable and unstable manifolds of Y_+ (as Q-orbits are biasymptotic to Y_+), we can expect that some part of the global stable manifold containing $Q2$ will intersect the plane in a 1-dimensional line segment through q_2, and some part of the global unstable manifold containing $Q1$ will intersect the plane in a 1-dimensional line segment through q_1. The proof then consists of establishing estimates that ensure that these segments exist, are transverse, and intersect in a unique point q_{12}. This point q_{12} will then be a point on

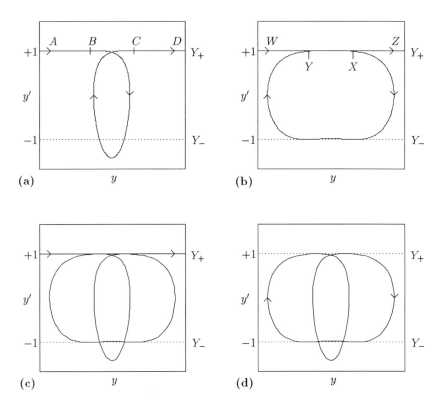

Figure 3: Gluing illustrated at $\lambda = 2.1$. All orbits are shown at the same scale; the y scale is 2.5 times that in Fig 2. (a) A segment ABCD from $Q1$; B and C are at $y = \pm M$. (b) A segment WXYZ from $Q2$; Y and X are at $y = \pm M$. (c) $Q12$ can be formed by 'gluing' ABC from $Q1$ to XYZ from $Q2$. (d) $P21$ can be formed by 'gluing' BC from $Q1$ into XY from $Q2$.

a true $Q12$-orbit with the desired name and properties. For the details of the proof readers should consult [13]. We will have more to say about the relationship between this type of argument and the control of chaos in section 8 below.

Theorem 5.1 does not, unfortunately, give a complete description of the set of admissible solutions created at each bifurcation, and the restriction to a finite set of pre-existing Q-orbits, which is essential for the choice of M in the proof, suggests that it would be difficult to extend this approach as far as we would like. It would be relatively easy to extend the techniques to prove the existence of admissible trajectories which only go to Y_+ at one end, or which are bounded but not periodic; there are no additional complications provided we use only finitely many distinct $Q\alpha$ in constructing each new trajectory.

This means that for $n = 1$, we could extend the theorem to give a complete description. Since we can deduce from earlier work of Coppel [3] that there are no

The Falkner-Skan Equation

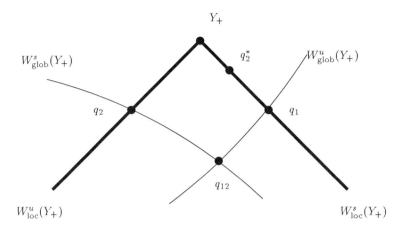

Figure 4: Gluing illustrated on the plane $y = M$ near Y_+. Coordinates are y' vertically and y'' horizontally. Bold lines show parts of the local stable and unstable manifolds of Y_+. Point $q_2 = X$ is on the orbit $Q2$ and $q_1 = C$ is on the orbit $Q1$ (see Figs 2(a) and 2(b) for the points X and C). $Q2$ also intersects the plane again at a point q_2^* (lying between Y and Z on Fig 2(b)), but this is not important here. There is an arc segment through q_2 such that orbits started at these points remain close to $Q2$ as $t \to \infty$, and so make two turns around Y_- and then tend to Y_+. Similarly, there is a segment through q_1 such that orbits through these points remain close to $Q1$ for $t \to -\infty$, and so came from Y_+ and made one turn around Y_- before hitting the return plane. The two segments intersect at a point q_{12} which therefore lies on an orbit $Q12$.

interesting trajectories when $\lambda < 1$, it is only possible to glue $Q1$ to itself. In this case we therefore obtain just one periodic orbit $P1$, an infinity of Q-orbits $Q1^m = Q1\ldots1$, $m \geq 2$ (of which the orbit $Q11$ of Fig 2(d) is the simplest) and exactly two other admissible orbits, each of which is asymptotic to Y_+ at one end and to $P1$ at the other.

If $n > 1$ the set of new admissible trajectories created is large and complicated, and the theorem only gives an exact description of the P and Q-orbits within them, there being infinitely many of each of these. Even if we extend our results as described above, our methods do not allow us to make complete statements about the set of admissible trajectories in these cases. In addition, there is some necessary vagueness in the statement of the theorem because we do not know precisely which Q-orbits exist at $\lambda = n$. As we shall see, with the exception of $Q1$, any Q-orbit eventually disappears as λ increases; however, numerical evidence indicates that most trajectories created at one integer λ value persist over a large range of λ values, and so are available for gluing at many subsequent integer values. For example, it seems that the first Q-orbits to disappear are $Q11$ and $Q2$, which annihilate one another at $\lambda \approx 250$. If this is correct, then the set of admissible solutions for $n < \lambda < n+1$ will contain at least a set equivalent to a full-shift on n symbols for all $n < 250$, as conjectured by

Hastings & Troy [8]. For larger n values some of the trajectories will be missing as Q-orbits disappear from the admissible set.

6 Destruction of P and Q-orbits

There is theoretical and numerical evidence that P and Q-orbits created in the sequence of bifurcations described above are destroyed at larger λ values as $\lambda \to \infty$.

First, we have proved results [12] on the behaviour of the equation (1) for large λ. These results show that $P1$ is the only periodic orbit to exist for all sufficiently large values of λ, and that $Q1$ is similarly unique amongst Q-orbits. This implies that all other P and Q-orbits are ultimately destroyed as λ increases; since we know from the results of the previous sections that infinitely many P and Q orbits are produced as λ increases through each integer $n \geq 2$, this means that the sequence of 'destroying' bifurcations must be relatively complicated and continue for all values of λ however large. It is also relevant here that there are no admissible orbits for $\lambda < 1$ [3]; if we follow orbits created in gluing bifurcations with changing λ we know they cannot just 'turn back' in saddle-node bifurcations and disappear into $\lambda < 0$; something must actually destroy them in the parameter range $1 < \lambda < \infty$.

It is course possible that orbits created as above come together at some λ value larger than the ones at which they were created and annihilate themselves in pairs in saddle-node bifurcations or in larger groupings in more complicated bifurcations. This does actually occur and accounts for all orbit destructions, but there are severe topological restrictions on which orbits can bifurcate with which others, and these can be used to gain a deep understanding of the bifurcation sequences which are possible in the system. For example, two P or two Q orbits can only annihilate one another in a saddle-node bifurcation if they wind the same number of times around Y_-. Furthermore, two Q-orbits can only so annihilate each other if all intermediate Q-orbits in the ordering of Q-orbits on the two-dimensional stable and unstable manifolds of Y_+ (see section 2) have been removed at lower λ values. Similarly, the ordering of intersections of symmetric P and Q orbits with the symmetry line $y = y'' = 0$ places yet further restrictions on the possible bifurcations and their order with λ increasing. Indeed, this last invariant ensures that P and Q-orbit bifurcations cannot happen entirely independently; some P orbits must disappear before certain Q-orbits can bifurcate, and *vice versa*. Further topological invariants concern the number of windings that orbits have around each other, and around $P1$ in particular, but these will not concern us further here.

It would not be useful to describe in great detail all we know about the sequence of bifurcations that destroys P and Q orbits as λ increases. In part this is because the results will appear elsewhere [12], and in part because the research is still in progress and the results are not yet in their final form. It is useful, however, to give some flavour of the argument and to describe the fate of a few example orbits. This is more especially true because the reversibility of the system allows generic period n-tupling bifurcations to occur near symmetric periodic orbits (which must be neutrally stable because of the time reversal symmetry) rather like those that occur in Hamiltonian systems. These bifurcations also play a crucial role in the sequence of bifurcations destroying orbits, and whilst such bifurcations are well-known in Hamiltonian systems, or in reversible systems of even dimension, it is useful to have a further well understood example in an odd dimensional system. Of course, the order in which these period n-tupling bifurcations can occur also suffers some restrictions. For example, orbits undergoing n-tupling bifurcations with the symmetric orbit $P1$ must do so at λ values where the Floquet multipliers of $P1$ have the form $e^{i\theta}$ where $\theta = \pm\frac{2\pi q}{n}$ where q and n are coprime. As the Floquet multipliers of $P1$ (and other symmetric periodic orbits) move continuously round the unit circle with increasing λ, the order in which the rational values of θ are visited orders the infinite sequence of n-tupling bifurcations. [When $n = 2, \theta = \pi$ we expect to see a normal period-doubling bifurcation where one orbit meets up with another of twice the period. For $n > 2$, bifurcations generically involve one orbit of some period T which persists, and two orbits of period nT which we can, in an appropriate orbit-counting interpretation, think of as annihilating each other at the bifurcation.]

That there is some bifurcation scheme which matches the many orbits produced in the gluing bifurcations with the destroying bifurcations in a way that is compatible with all these restrictions is obvious; after all, the Falkner-Skan equation does something that fulfills all the requirements. We could just try to investigate this scheme numerically, though this is difficult for anything other than the shorter periodic orbits, and is in any case not enormously rewarding. It would be much more interesting if we could determine which types of bifurcation scheme are compatible with all the restrictions, and we are certainly close to having results of this kind. It seems likely, however, that in trying to determine what actually does occur, there will always be some assumptions that need to be checked numerically to supplement the rigorous theory.

For example, it seems from numerical experiment that the Floquet multipliers for $P1$ are initially real, but meet at -1 for $\lambda \approx 340$; they then seem to travel round the unit circle towards $+1$ with the complex argument θ decreasing monotonically from π to 0 as $\lambda \to \infty$. If we accept this observation, which is the product of very stable numerical computations and which may be provable, we can argue as follows.

There are only one P and one Q orbit which wind only once around Y_- and these are $P1$ and $Q1$. The large λ theory tells us that orbits with these names continue to exist for all λ. It follows from the 'shooting' results of Hastings & Troy [8], though not directly from our own large λ results, that the orbit $P1$ can be continuously and monotonically followed from its creation at $\lambda = 1$ up to the large λ regime. A similar result is doubtless true for $Q1$. Thus, we have accounted for the creation and ultimate fate of all P and Q-orbits which wind only once around Y_-. The only periodic orbit which winds twice around Y_- is $P2$, and precisely one such orbit is required to be destroyed in a period-doubling bifurcation with $P1$ at $\lambda \approx 340$ when its Floquet multipliers are -1. There are two Q-orbits which wind twice around Y_-, $Q11$ and $Q2$, and these must ultimately disappear; it turns out that the ordering on the stable and unstable manifolds of Y_+, and on the symmetry line, are such as to permit them to destroy each other in a saddle-node bifurcation without requiring any other bifurcation to take place first (i.e. these orbits are 'next to each other' in all these various orderings). Thus we account for all P and Q-orbits which wind twice around Y_-. Continuing, we find that the two periodic orbits $P12$ and $P3$ which wind 3 times round Y_- are just what is needed to meet for mutual annihilation in the vicinity of $P1$ when $\theta = \frac{2\pi}{3}$ at a period-tripling bifurcation. And so it continues. The first element of flexibility in the scheme appears only to occur for Q-orbits winding at least 8 times around Y_-, and even there we believe that slightly more complicated topological argument will reduce all the choices to one. Simple numerical experiments serve to confirm our understanding (at least for the topologically less complicated orbits which are accessible to numerical work), and – in particular – seem to indicate that orbits behave as simply as possible in the intermediate parameter range (between creation and destruction) where we have little theoretical control over their behaviour.

The elements of internal self-consistency in the scheme give us reassurance that our understanding the behaviour of P and Q-orbits in the system is likely to be correct, even where this cannot be checked in detail by numerical experiments or theory. More importantly, in some sense which we have yet to make precise, an accounting scheme of this type comes very close to establishing that any system with the same basic topological features as the Falkner-Skan equation must have the 'same' sequences of bifurcations.

7 Further remarks on the Falkner-Skan equation

It is important to understand how much of the analysis above may apply to perturbations of the Falkner-Skan equation, or to other systems of ordinary differential equations.

First, it is possible to add a further parameter to the Falkner-Skan equation which destroys neither the symmetry, nor the trajectories $Y_\pm$, nor the gluing bifurcations at integer values of λ, nor the large λ analysis. In this case, everything above remains valid and changing the new parameter should only alter the precise λ values at which the destroying bifurcations occur. Of course, this would imply that once the sequence of Q-orbit destructions begins, presumably at a different λ value, the collections of Q-orbits available for gluing at integer λ values will differ, and some details of the bifurcation scheme will change. This implies only that our bifurcation scheme described above cannot be absolutely fixed, and must allow for this kind of flexibility.

Second, we can consider more general perturbations. First let us consider the role of the reversible symmetry. This does not in fact play a crucial role in the analysis of the gluing bifurcations – the essential process in the gluing bifurcation is the creation from two unbounded trajectories of a single new Q-orbit at discrete values of a parameter. Neither this process, nor the gluing from which we then deduce the existence of a much larger set of interesting trajectories, depends on the symmetry. The details of the orbit destroying bifurcations relied more heavily on the symmetry; however, in the usual way we would expect to be able to unfold this scheme and generalise it to non-symmetric but nearby systems.

Another way of looking at the system, pointed out to us by Robert Mackay (personal communication), is illuminating for considering symmetry-breaking and more general perturbations. By doing a standard Poincaré compactification we can convert the Falkner-Skan equation into a form where it is easier to study the 'behaviour at infinity'. In this new form, there are two fixed points (at infinity in the original form), and the transformed trajectories $Y_\pm$ form heteroclinic connections between the two points (as do all the Q-orbits). Analysis of the bifurcations in this system appears to be no easier than the analysis described above (Mackay, personal communication) and the usual theorems about homoclinic and heteroclinic bifurcations in flows do not apply directly because of the very non-generic nature of the compactified fixed points. However, this set-up is more convenient for describing the effects of perturbation. What is important for our analysis is that perturbations should not destroy the fixed points at infinity nor the heteroclinic loops ($Y_\pm$) between them. Providing we keep within the space of systems with these properties our analysis will apply, and the gluing bifurcations will be a co-dimension one phenomenon. We have not yet analysed precisely what this statement means in terms of perturbations to the original Falkner-Skan equation. Certainly it is not clear how, when given an ordinary differential equation to study, one should go about noticing the existence of or locating important unbounded trajectories like $Y_\pm$; in the case of the Falkner-Skan equation their existence is obvious from a cursory examination of the equation, but this is presumably not always going to be the case.

Finally, some similar but essentially different analysis does appear to be useful in the analysis of at least one other set of equations, as noticed by Paul Glendinning. These are the Nosé equations [9, 10] and an analysis of this system is underway [6].

8 Numerical methods, Control of Chaos, and Conclusion

In this section, I would like to make some remarks that arose as a result of hearing other presentations and discussions at the workshop on which this volume is based. There are striking and moderately obvious connections between:

- ideas in the literature about controlling chaotic systems to lock onto unstable periodic orbits;

- the activity of those who use the location and continuation of periodic orbits in flows as tool in understanding the bifurcations and attractors occurring in differential equations and maps;

- the algorithms used in numerical packages to locate and continue periodic solutions

- elements in the proof of the theorems presented here, and in other 'shadowing of pseudo-orbit' type proofs in dynamical systems.

These connections include the following. There are obvious parallels between the way in which chaos-controlling algorithms (such as those described by Glass *et al*, Kostelich & Barreto, and Ott & Hunt, this volume) wait for a nearly periodic orbit to occur before switching on the control, and the type of procedure used by some investigators to find initial conditions for their orbit continuing packages. There is also a close match between the strategy then used to control a chaotic system, and the type of relatively crude (but often effective) numerical continuation algorithm which takes an initial point, computes a numerical trajectory and its Jacobian round to a nearby point, and then uses a Newton method or similar to calculate a new guess for a point on the periodic orbit. In a system like the Falkner-Skan equation, such one-shot numerical methods do not work well, at least for periodic orbits that come close to $Y_\pm$, or for Q-orbits. In this case it is better to use a package that employs a more sophisticated numerical method – in my case I used the package AUTO developed by Doedel [4]. This uses a collocation method and simultaneously manipulates many intermediate points on the orbit instead of merely adjusting an initial point. To put this another way, a whole sequence of points on a pseudo-orbit are adjusted to produce a better pseudo-orbit (closer to a true orbit). This method is actually spectacularly well-adjusted to work on the Falkner-Skan equation, since the easiest way to find more complicated P and Q orbits appears to be to build pseudo-orbits out of pieces from previously located but simpler orbits exactly as suggested by the proof of Theorem 5.1, and to feed these in as initial data for AUTO. The output from AUTO is then a sequence of improving pseudo-orbits, and the algorithm is stable precisely because the hyperbolicity conditions needed in the proof of the theorem hold and cause numerical convergence.

This suggests that in investigating chaos-controllers, there might be much that can be usefully learnt from the numerical procedures and algorithms used and developed over many years by many researchers. In particular, the improved performance obtained from methods like AUTO in some systems, suggests that control of chaos in these systems may be best achieved by locating many short and simple pieces of trajectory and then driving the system to follow desired pseudo-orbits formed from gluing these pieces together. This is a similar idea to that used in targetting as described by Kostelich (this volume), though in his case the short pieces of trajectory were only used to get onto the orbit, rather than while travelling around it. In circumstances where there is a convenient symbolic dynamics, there may be optimal collections of short segments that through gluing provide good pseudo-orbits to shadow all more complicated orbits. For more on this, though it is not what they had in mind, see [1].

Acknowledgements

None of the work described here would have been possible without Peter Swinnerton-Dyer, my collaborator and co-author in references [12] and [13]. I am also grateful to the organisers of the meeting for their efforts and support, and to BRIMS (Hewlett-Packard Basic Research Insitute in Mathematical Sciences, Bristol, UK) and the INI (Isaac Newton Institute in Cambridge, UK).

References

[1] Artuso R, Aurell E & Cvitanović P, 1990. Recycling of strange sets: I. Cycle expansions. *Nonlinearity* **3**:325-359.

[2] Botta E F F, Hut F J & Veldman A E P, 1986. The rôle of periodic solutions in the Falkner-Skan problem for $\lambda > 0$. *J. Eng. Math.*, **20**:81-93.

[3] Coppel W A, 1960. On a differential equation of boundary layer theory. *Philos. Trans. Roy. Soc. London Ser. A* **253**:101-136.

[4] Doedel E, 1986. *AUTO: Software for continuation and bifurcation problems in ordinary differential equations.* C.I.T. Press, Pasadena.

[5] Falkner V M & Skan S W, 1931. Solutions of the boundary layer equations. *Phil. Mag.*, **7**(12):865-896.

[6] Glendinning P & Swinnerton-Dyer H P F, 1996. Bifurcations in the Nosé equation. In preparation.

[7] Hastings S P, 1993. Use of simple shooting to obtain chaos. *Physica D* **62**:87-93.

[8] Hastings S P & Troy W, 1988. Oscillating solutions of the Falkner-Skan equation for positive beta. *J. Diff. Eqn.*, **71**:123-144. [Note: their β is our λ.]

[9] Nosé S, 1984. A unified formulation of the constant temperature molecular-dynamics methods. *J. Chem. Phys.* **81**: 511-519.

[10] Posch H A, Hoover W G & Vesely F J, 1986. Canonical dynamics of the Nosé oscillator: stability, order and chaos. *Phys. Rev. A* **33**: 4253-4265.

[11] Sevryuk M B, 1986. *Reversible systems.* Lect. Notes Math. **1211**, Springer-Verlag, Berlin.

[12] Sparrow C and Swinnerton-Dyer H P F, 1996. The Falkner-Skan Equation II: The bifurcations of P and Q-orbits in the Falkner-Skan equation. *In preparation.*

[13] Swinnerton-Dyer H P F & Sparrow C, 1995. The Falkner-Skan Equation I: The creation of strange invariant sets. *J. Diff. Eqn.*, **119**(2): 336-394.

Commentary by P. Diamond

Colin Sparrow has written an interesting account of a very complicated situation. The paper is not always transparent, but this is inevitable because it is a summary of two lengthy and intricate papers [12,13]. In the situation here, bifurcations occur regularly and indefinitely as the parameter passes through positive integer values. Too, certain significant orbits are destroyed in bifurcations, although portions of them survive glued together and incorporated into new trajectories.

This behaviour is rather more interesting than the structurally stable chaotic flow which prevails between bifurcations. The Falkner-Skan equation is a good example of just how more difficult it is to control a chaotic smooth flow than to control a chaotic discrete dynamical system. Control to a P−orbit might well require an extensive computational description of many P− and Q−orbits. Despite their regularity near Y_-, it is probably even more difficult to control the flow to a Q−orbit, since these do not return like the periodic P−orbits.

Most interestingly for chaotic control, Colin's methods show how some very complicated systems retain some sort of memory through quite drastic bifurcations. In the case of the Falkner-Skan system, new orbits appear which incorporate segments of pre-bifurcatory orbits. Consequently, a controller's experience in the neighbourhood of these portions should possibly be utilisable after a bifurcation occurs. One would like to see how this hysteretic effect showed up in Poincaré sections of the flow, if at all.

Some Characterisations of Low-dimensional Dynamical Systems with Time-reversal Symmetry

John A. G. Roberts
Department of Mathematics
La Trobe University
Bundoora VIC 3083, Australia

Abstract

The structure of the differential or difference equations used to model a complex real-life system should reflect some of the underlying symmetries of the system. Characterising and exploiting this structure can lead to better prediction and explanation of the motion. For example, a well-studied structure is that found in Hamiltonian or conservative dynamical systems. In this paper, we survey our work on dynamical systems with another type of structure, namely a (generalised) time-reversal symmetry. We explain some of the dynamical consequences and structure arising from this property. We explore the question of how systems with this (generalized) time-reversal symmetry are similar to, and how they differ from, Hamiltonian dynamical systems. We pay particular attention to low-dimensional (2D and 3D) systems and use specific examples to illustrate our points.

1 Introduction and Overview

The classical concept of time-reversal symmetry refers to the invariance of equations under the transformation $t \mapsto -t$. Time-reversal symmetry has played, and still plays, an important role in physics and many of the differential equations of physics are time-reversible (for a readable account, see [35]).

Perhaps our first introduction to time-reversal symmetry is via mechanics, e.g. an equation of the form

$$\ddot{q} = W(q), \quad W : \mathbb{R} \mapsto \mathbb{R}, \tag{1}$$

which describes, for example, the motion of a unit-mass particle moving along a line subject to the force $W(q)$ (the dot denotes d/dt). This equation is clearly invariant under $t \mapsto -t$ (with $d/dt \mapsto -d/dt$). As a result, if $q = \gamma(t)$ is a solution of (1), then so is the time-reversed motion $\gamma(-t)$. The latter is what would be seen if a movie was made of $\gamma(t)$ and then played in reverse.

Equation (1) is also clearly derivable from the Hamiltonian

$$H(q,p) = \frac{1}{2}p^2 + V(q), \tag{2}$$

with $p = \dot{q}$ and $dV/dq = -W(q)$. Again, from mechanics, examples abound which are Hamiltonian with time-reversal symmetry, e.g. those of the form 'kinetic energy

+ potential energy' :
$$H(q,p) = \frac{1}{2}p \cdot p + V(q), \tag{3}$$
with $q, p \in \mathbb{R}^m$. Systems described by the Hamiltonian (3) are a subclass of those described by a Hamiltonian which is even in the momenta
$$H(q,p) = H(q,-p). \tag{4}$$
All such Hamiltonian systems have time-reversal symmetry. The $t \mapsto -t$ invariance shows up in their equations of motion
$$\dot{q} = \partial H(q,p)/\partial p \quad \dot{p} = -\partial H(q,p)/\partial q, \tag{5}$$
as invariance under $t \mapsto -t$, together with a change of sign in the momenta (velocities), i.e. $(q,p) \mapsto (q,-p)$ (this is guaranteed by (4)). If $(q(q_0,p_0,t),p(q_0,p_0,t))$ is a solution of the dynamics, then so is $(q(q_0,-p_0,-t),-p(q_0,-p_0,-t))$. In the $2m$-dimensional (q,p) phase space, this corresponds to flipping the original trajectory via $(q,p) \mapsto (q,-p)$ and following the resulting trajectory in the opposite time sense.

The work of Moser [22], Devaney [6] and Arnol'd and Sevryuk [1, 2, 36] has shown that the essence of classical time-reversal symmetry can be usefully generalized to *invariance of a dynamical system by a transformation G of phase space together with reversal of time*. We will call G a reversing symmetry of the system (terminology due to [14]). For autonomous flows,
$$\frac{d}{dt}x = F(x), \quad x \in \mathbb{R}^n, \tag{6}$$
with F a differentiable vector field on $\mathbb{R}^n$, this means that there exists $G: \mathbb{R}^n \mapsto \mathbb{R}^n$ such that
$$\frac{d}{dt}(Gx) = -F(Gx), \tag{7}$$
whence (6) is invariant under the action of G and $t \mapsto -t$. When G is linear, (7) is equivalent to
$$G \circ F = -F \circ G. \tag{8}$$
A mapping (diffeomorphism) $L: \mathbb{R}^n \mapsto \mathbb{R}^n$ has the reversing symmetry $G: \mathbb{R}^n \mapsto \mathbb{R}^n$ if
$$L^{-1} = G^{-1} \circ L \circ G. \tag{9}$$
The two definitions can be linked by, for example, noting that if L is the time-one map of a vector field satisfying (7), then it satisfies (9). [Recall that the time-t map of a flow is the mapping induced by associating to each point in the phase space $x = x(0)$ the (unique) point $x' = x(x(0),t)$ that it evolves to via (6) after time t.] Similarly, the Poincaré map of (6)-(7) satisfies (9). A consequence of (7) is that if $x(x_0,t)$ is a solution of (6) then so is $Gx(Gx_0,-t)$. The latter is precisely the reflection by G of the trajectory $x(x_0,t)$ in phase space followed in the opposite sense. Similarly, the conjugacy between L and L^{-1} expressed by (9) implies that the reflection of the forward (backward) orbit of a point $x_0 \in \mathbb{R}^n$ gives the backward (forward) orbit of the point Gx_0. This is illustrated for $n = 2$ in Figure 1.

We remark that if the $-$ sign is removed from (7), respectively L^{-1} is replaced in (9) by L, we obtain the definition of a *symmetry* of a flow, respectively a diffeomorphism. The set of symmetries of a dynamical system form a group $\mathcal{S}$, and

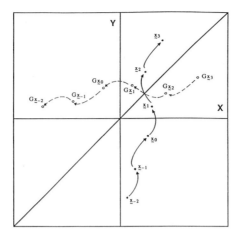

Figure 1: Illustration of the nature of the motion of a 2D mapping L with reversing symmetry $G : x' = y, y' = x$. Shown is part of the trajectory of a point x_0 with unbroken arrows indicating the action of L on each point. The existence of the reversing symmetry implies that the trajectory of Gx_0 is found by reflecting the trajectory of x_0 by G. The forward and backward trajectories are interchanged on reflection as the broken arrows indicate.

symmetries and their consequences are well-studied. It can be shown that if a system has one or more reversing symmetries, then these taken together with $\mathcal{S}$ form a new group $\mathcal{R}$ called the reversing symmetry group [14]. There are definite advantages to studying symmetries and reversing symmetries together. However, in this article we will restrict ourselves to discussion of reversing symmetries.

Some broad classes of dynamical systems which possess reversing symmetries are:
Example 1 All Hamiltonians of the form (4), where $G : (q, p) \mapsto (q, -p)$.
Example 2 All oscillation equations of the form

$$\ddot{q} = W(q), \quad F : \mathbf{R}^m \mapsto \mathbf{R}^m. \tag{10}$$

When an equation of the form (10) is rewritten in the form of (6), where $n = 2m$ and $x = (x_1, \ldots, x_m, x_{m+1}, \ldots, x_{2m}) = (q_1, \ldots, q_m, \dot{q}_1, \ldots, \dot{q}_m)$, it has the reversing symmetry $G : x_j \mapsto x_j (j = 1, \ldots, m), x_j \mapsto -x_j (j = m+1, \ldots, 2m)$. Only in the case $m = 1$ (cf. (1) and (2)) is the resulting flow (6) in general Hamiltonian (i.e. for $m \geq 2$, $W(q)$ is not in general the gradient of some $V(q)$).
Example 3 All autonomous even-order (odd-order) ordinary differential equations in which the odd (even) derivatives occur only in even combinations. When the ODE $D(x^{(n)}, x^{(n-1)}, \ldots, x^{(1)}, x) \equiv 0$ is rewritten in the form (6) using $x = (x_1, \ldots, x_j, \ldots, x_n)$ with $x_j = x^{(j-1)}$ and $n = 2m$ ($n = 2m+1$), the reversing symmetry is $G : x_{2j} \mapsto -x_{2j}, x_{2j-1} \mapsto x_{2j-1}, j = 1, \ldots m.$ ($G : x_{2j} \mapsto x_{2j}, x_{2j+1} \mapsto -x_{2j+1}, j = 0, \ldots m.$) An example with even-order is

$$\dddot{x} + a_2 \ddot{x} + a_1 \dot{x}^2 + a_0 x = 0, \tag{11}$$

which describes various physical situations including evolution of flame fronts [20].

An example with odd-order is

$$\dddot{x} + \ddot{x}x + \lambda(1 - \dot{x}^2) = 0, \tag{12}$$

which is the Falkner-Skan equation studied by Sparrow in this volume.

Example 4 Area-preserving mappings of the form:

$$L: \begin{array}{rcl} x' & = & x + f(y) \\ y' & = & y + g(x'), \end{array} \tag{13}$$

where $f(y)$ and/or $g(x')$ is an odd function. When f is odd, L of (13) satisfies (9) with

$$G: \begin{array}{rcl} x' & = & x + f(y) \quad \text{f odd} \\ y' & = & -y. \end{array} \tag{14}$$

When g is odd, L of (13) satisfies (9) with

$$G: \begin{array}{rcl} x' & = & -x \\ y' & = & y - g(x) \quad \text{g odd}. \end{array} \tag{15}$$

Prominent 1-parameter examples of (13) include:

$$\begin{array}{ll} \text{standard map}: & f(y) = C\sin y \quad g(x') = x' \\ \text{Henon area-preserving map}: & f(y) = -y \quad g(x') = 2x'(1 - C - x'). \end{array} \tag{16}$$

Example 5 The Arnol'd 'cat map' of the 2-torus $\mathbb{T}^2$ [7]. Taking the latter as the square $[0, 1) \times [0, 1)$ in $\mathbf{R}^2$ with edges identified in the usual way, this mapping is

$$\begin{pmatrix} x' \\ y' \end{pmatrix} = \begin{pmatrix} 1 & 1 \\ 1 & 2 \end{pmatrix} \begin{pmatrix} x \\ y \end{pmatrix} \pmod{1}. \tag{17}$$

It has the reversing symmetry G given by [3]

$$\begin{pmatrix} x' \\ y' \end{pmatrix} = \begin{pmatrix} 0 & -1 \\ 1 & 0 \end{pmatrix} \begin{pmatrix} x \\ y \end{pmatrix} \pmod{1}. \tag{18}$$

The above examples motivate various remarks:

Remark 1 A dynamical system having a reversing symmetry need not be Hamiltonian, e.g. cf. Example 2.

Remark 2 Most of the reversing symmetries listed above are linear. The properties (7) and (9) are invariant under coordinate transformation $x \mapsto Tx$ ($T : \mathbb{R}^n \mapsto \mathbb{R}^n$) with $G \mapsto T \circ G \circ T^{-1}$. When G is linear and the transformation T is nonlinear, the transformed reversing symmetry is in general nonlinear.

Remark 3 In many physical applications, the reversing symmetry G in (7) or (9) is an involution (i.e. $G^2 = G \circ G = Id$), and often orientation-reversing as well (i.e. $\det dG(x) < 0$). A dynamical system with an involutory reversing symmetry is called *reversible* [6, 36, 34]. In a neighbourhood of a fixed point, an involution G of $\mathbf{R}^n$ is always conjugate to its linear part, which can be taken as a diagonal matrix comprising k $\{+1\}$'s and $n - k$ $\{-1\}$'s. This shows that locally, up to conjugacy, the fixed points of G:

$$Fix(G) := \{x \mid Gx = x\}, \tag{19}$$

form a k-dimensional plane, and G acts as a reflection in $Fix(G)$. Globally, $Fix(G)$ is useful because it helps to characterise the (symmetric) periodic orbits (cf. Section 3 below). When $n = 2$ and G is orientation-reversing (and C^1), $Fix(G)$ is a non-terminating curve in the plane with no self-intersections. For example, (14) and (15) above are orientation-reversing involutions with fixed sets given by the lines $y = 0$ for (14), and $x = 0$ for (15).

Note, however, that reversing symmetries need not be involutory, e.g. (18) in Example 5 is order 4.

Remark 4 The independent 'time' variable need not be time as such, but rather a spatial coordinate, e.g. this is the case for (11) and (12) of Example 3. This situation arises in reaction-diffusion or other partial differential equations, where the steady states are given by the solutions of a system of reversible ordinary differential equations, with the "time" role being played by the spatial coordinate [20, 9, 13, 37, 41].

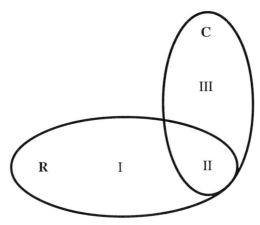

Figure 2: Venn diagram illustrating the relationship between Conservative (Hamiltonian) systems 'C' and dynamical systems with generalized time-reversal symmetry 'R'.

With respect to Remark 1, consider the Venn diagram of Figure 2 showing the set of dynamical systems that are *conservative* or Hamiltonian and the set of dynamical systems with a reversing symmetry. We use conservative/Hamiltonian to mean both Hamiltonian flows and symplectic mappings (in 2D, symplectic is equivalent to area-preserving). Region II represents systems that have both properties. These systems are those which, historically speaking, have received the most attention (in particular, Hamiltonian reversible systems with an involutory reversing symmetry). More recently, there is a growing interest in studying the set of dynamical systems with a reversing symmetry for its own sake (in Region I). It is now known that there exist striking similarities between a reversible (not necessarily conservative) dynamical system and a Hamiltonian one, if one studies the former in the vicinity of its *symmetric* periodic orbits [22, 6, 1, 2, 36, 34]. A symmetric periodic orbit is, by definition, invariant under both the dynamical system *and* the action of the reversing symmetry G (as distinct from an asymmetric periodic orbit, whose image under G produces a different periodic orbit). In recent times, there has also been investigation of Region I where the reversing symmetry is not involutory but of higher order [14, 15] (necessarily even order to avoid trivial dynamics), as in Example 5. Finally, finding elements in Region III of Figure 2 involves identifying Hamiltonian dynamical systems without any reversing symmetry. This can be a subtle mathematical problem in dimension 2, involving checking in some appropriate set for (nonlinear) G's. Identification of candidate irreversible Hamiltonian systems helps in the study of 'seeing' the difference that such a symmetry can make.

In this contribution, we will illustrate via simple examples some of the characteristics of systems with time-reversal symmetry. We also use them to explore the Venn diagram of Figure 2. In Section 2 and Section 3, we highlight some of the general features of, respectively, 2D and 3D reversible mappings in Region I of Figure 2. In Section 4, we consider identification of 2D Hamiltonian flows and area-preserving mappings within Region III. In Section 5, we illustrate a recent generalization of the idea of a reversing symmetry.

2 Reversible dynamical systems that are not conservative

A common textbook division of dynamical systems (i.e. flows or mappings) is into *conservative* systems and *dissipative* systems. It is now known that a mapping or flow with a reversing symmetry can have *both* conservative and dissipative behaviour, associated with, respectively, symmetric and asymmetric (i.e. non-symmetric) periodic orbits.

In [34] we illustrated this by constructing 2-D non-conservative mappings that are *reversible* (cf. Remark 3 of Section 1) and then studying their dynamics. Note from (9) with $G^{-1} = G$ that a reversible mapping L can be written as the composition of two involutions $H := L \circ G$ and G (that $L \circ G$ is an involution follows from applying L to both sides of (9) with $G^{-1} = G$). To construct such mappings, we require at least one of the component involutions, say H, to be non area-preserving (note that although any involution is necessarily measure-preserving, the composition of two measure preserving mappings need not be). To create H, we use conjugation of V, a linear area-preserving involution, by W, an arbitrary (i.e. non area-preserving) invertible coordinate transformation. When H is composed with another involution G we obtain the reversible mapping:

$$L = H \circ G = W \circ V \circ W^{-1} \circ G. \tag{20}$$

If G is area-preserving as well as V and they are both orientation-reversing (the case we usually consider), then the Jacobian determinant of L is

$$J(\boldsymbol{x}) := \det dL(\boldsymbol{x}) = \frac{\det dW^{-1}(G\boldsymbol{x})}{\det dW^{-1}(L\boldsymbol{x})}. \tag{21}$$

We illustrate the construction with an example. Take

$$V : \begin{array}{rcl} x' & = & y \\ y' & = & x, \end{array} \tag{22}$$

and
$$W : \begin{array}{rcl} x' &=& x[1+(y-1)^2] \\ y' &=& y. \end{array} \tag{23}$$

With these choices we obtain the non area-preserving involution $W \circ V \circ W^{-1}$:

$$H_1 : \begin{array}{rcl} x' &=& y[1+(y'-1)^2] \\ y' &=& x/[1+(y-1)^2], \end{array} \tag{24}$$

which we compose with the simple one-parameter area-preserving involution

$$G_1 : \begin{array}{rcl} x' &=& x \\ y' &=& C-y. \end{array} \tag{25}$$

The result $H_1 \circ G_1$ is the reversible mapping

$$L_1 : \begin{array}{rcl} x' &=& (C-y)[1+(y'-1)^2] \\ y' &=& x/[1+(C-y-1)^2], \end{array} \tag{26}$$

whose Jacobian determinant follows from (23), (25) and (21):

$$J_1(x) = \frac{[1+(y'-1)^2]}{[1+(C-y-1)^2]}. \tag{27}$$

For a substantial range of parameter C, the mapping L_1 has both a symmetric fixed point and a pair of asymmetric fixed points. For example, at $C = 3$ these are readily calculated. The symmetric fixed point is at $(15/8, 3/2)$, with $J_1 = 1$ at this point (both G_1 and H_1 map this point to itself). When $C = 3$, the asymmetric pair of fixed points are $(2,1)$ and $(2,2)$ (mapped to one another by both G_1 and H_1), with $J_1 = 0.5$ and $J_1 = 2$ respectively. Further calculation of the linearisation at these points reveals that they are in fact, respectively, attracting and repelling fixed points. In fact, analysis of (26) shows that the attracting and repelling fixed point are born out of the symmetric fixed point in a *pitchfork* bifurcation when $C = 2\sqrt{2}$.

Figure 3 shows part of the phase portrait of the mapping L_1 when $C = 2.87$. The symmetric fixed point at the centre of the picture is a saddle point. It lies at the intersection of the curves $Fix(G_1)$, the line $y = 1.435$, and $Fix(H_1)$ given by $x = y[1+(y-1)^2]$ (cf. Section 3 below for more explanation of this). Below and above the symmetric fixed point are the two asymmmetric fixed points that form the spiral attracting/repelling pair. A trajectory is shown that starts close to the repelling fixed point and moves downwards to spiral into the attracting fixed point. The attractor and repeller and the hyperbolic symmetric fixed point appear to be enclosed by invariant curves. Some elliptic symmetric cycles and their associated island chains are also shown: a symmetric eight-cycle and a symmetric seven-cycle which lie, respectively, inside and outside, one of the curves, and an outer symmetric six-cycle which is surrounded by large islands. The three islands cut off by the border of the picture are part of the island chain surrounding a symmetric five-cycle. Figure 4 shows a plot of the Jacobian determinant $J_1(x,y)$ over a part of the plane.

Figure 5 shows the phase portrait of another reversible non-conservative mapping. Here the origin is an elliptic symmetric fixed point, and again the curves and island chains around it are reminiscent of an area preserving mapping. In the bottom right hand corner, there is a spirally attracting asymmetric fixed point at A surrounded

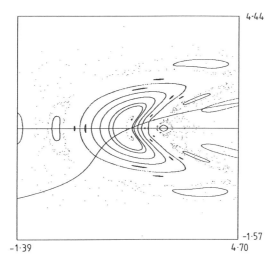

Figure 3: Phase portrait of the 2D non-conservative reversible mapping L_1 of (26). See text for more details.

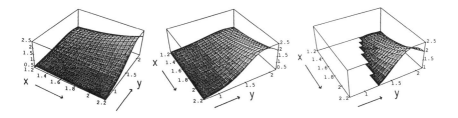

Figure 4: Plot of the Jacobian determinant (27) of L_1 of (26) over part of the (x,y) plane. The far left picture shows only the part of the surface where $J_1(x,y) > 1$.

by an attracting 5-cycle. The reflection of these features by the reversing symmetry, in this case reflection in the line $y = x$, gives a repelling asymmetric fixed point R and a repelling 5-cycle. The mapping corresponding to Figure 5 is a perturbation of the area-preserving Henon mapping in the form $L_h : x' = y, y' = -x + 2Cy + 2y^2$ (this is related to the form (13)-(16) by a simple linear transformation). L_h is the composition of the involutions $H_h : x' = x, y' = -y + 2Cx + 2x^2$, and $G_h : x' = y, y' = x$. Suppose we conjugate H_h with a 1-parameter non area-preserving coordinate transformation T_ϵ with $T_0 \equiv Id$, the identity map. Then, for $\epsilon > 0$, the mapping $L_2 := T_\epsilon \circ H_h \circ T_\epsilon^{-1} \circ G_h$ can be made a reversible non-conservative perturbation of L_h (corresponding in Figure 1 to moving from Region II into Region I). The attracting/repelling pair of fixed points in this example are "born" from infinity as ϵ becomes non-zero. For the explicit form of L_2 and other details, cf. [34, Chapter 4].

In the early 1980's, much numerical and analytical work was done concerning

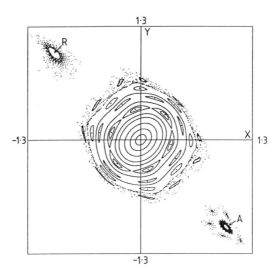

Figure 5: Phase portrait of a 2D non conservative reversible mapping obtained from a reversible perturbation of Hénon's area-preserving reversible mapping. See text for more details.

the universality of several phenomena in 2-D area-preserving mappings [11, 18, 19]. Numerical investigations of 2-D reversible non-conservative mappings have revealed that [34]:

Characterisation 2.1 *In 2-D reversible non-conservative mappings:*

(i) Symmetric fixed points period double symmetrically with the scalings previously found in area-preserving mappings e.g. the parameter rescaling factor $\delta = 8.721\ldots$.

(ii) As a parameter C is varied, the invariant curves with 'noble' winding number surrounding symmetric fixed points break up with the scalings previously found for KAM curve break-up in area-preserving mappings.

(iii) Asymmetric fixed points period double with the scalings previously found in dissipative mappings, e.g. the parameter rescaling factor $\delta = 4.669\ldots$.

(iv) Strange attractors and strange repellors can be found.

In a sense, the results for the asymmetric features are to be expected as the reversibility places no restriction on these features (except for the existence of partnering features under the action of the reversing symmetry). Hence, one expects the generic bifurcations and behaviour of a 1-parameter mapping, which is typically non-conservative. For pictures of strange attractors in reversible mappings, cf. [34, Chapter 6]. [17] shows the phase portrait of a mapping with a 4-fold reversing symmetry, revealing 2 strange attractors and 2 strange repellers.

The more intriguing feature is the great similarities between the symmetric features of a reversible mapping and generic features of area-preserving mappings. To begin with, we have:

Characterisation 2.2 *If $L : \mathbf{R}^n \mapsto \mathbf{R}^n$ satisfies (9) and has a symmetric m-cycle containing the point x, then the eigenvalues of the linearisation $dL^m(x)$ occur in reciprocal pairs (if λ is an eigenvalue, so is λ^{-1}).*

In 2D, Characterisation 2.2 shows that symmetric cycles of an orientation-preserving mapping with a reversing symmetry share the same linearisations as cycles of an area-preserving mapping. This similarity typically extends to all orders of the Taylor expansions around a symmetric cycle of a 2D mapping with a reversing symmetry, compared to around a cycle of an area-preserving mapping. This can be deduced by proving that the *normal forms*, i.e. the systematically-reduced Taylor expansions, are the same in both cases (cf. [15] for a systematic technique of finding the normal forms for 2D time-reversible maps and flows). The one exception is the possibility for the normal form of a symmetric cycle to be non-area preserving by allowing the bifurcation found in L_1 of (26), namely production of an asymmetric attractor/repellor pair from the symmetric cycle as it turns unstable. While the equivalence of the normal forms does not necessarily imply equivalence of the dynamics in a neighbourhood of the cycles, it does form the basis for proofs that linearly-stable symmetric cycles of mappings or flows satisfying (9) or (7) are indeed surrounded by invariant KAM tori [1, 2, 36].

Notice that the example (26) of a non-conservative reversible mapping involves rational functions. Similar examples can be found in [34, Chapter 4], together with an example which involves roots of functions. One might ask if there could be examples nicer than these, namely non-conservative reversible polynomial mappings. For example, in the construction of L_1 in (26), we see that W, V and G_1 are polynomial mappings. The rational form of L_1 arises because the inverse of W is not polynomial. Is it, for instance, possible to choose W polynomial with polynomial inverse *and* not area-preserving. The answer is no, and this essentially denies the possibility of non-conservative reversible polynomial mappings. We can deduce the following:

Characterisation 2.3 *Let L be a polynomial mapping from $\mathbf{R}^n$ to itself. Then:*
  (i) If L has finite order k, then $\det dL(x) = \pm 1$ (-1 only when k even).
  (ii) If L has a polynomial reversing symmetry G of finite order, then $\det dL(x) = \pm 1$.
  (iii) If L has polynomial inverse, a (not necessarily polynomial) reversing symmetry and a fixed point, then $\det dL(x) = \pm 1$.

The important fact underlying Characterisation 2.3 is that a real polynomial mapping $L : \mathbf{R}^n \to \mathbf{R}^n$ with polynomial inverse has a constant non-zero Jacobian determinant. This follows from differentiating $L^{-1} \circ L = Id$ and taking determinants: $\det dL(x)$ and $\det dL^{-1}(Lx)$ must both be real-valued polynomials and inverses of each other (their product is 1). Whence they are real constants and reciprocals of one another. With regard to (i) above, a real polynomial map of finite order obviously has polynomial inverse, and it follows that its Jacobian determinant is a real root of unity. In (ii), if L, its reversing symmetry G, and G^{-1} are real polynomial maps, then so is L^{-1} via (9). Thus $\det dL(x)$ is constant, and differentiating (9) shows that its inverse $\det dL^{-1}(x)$ has the same value. Consequently, $(\det dL(x))^2 = 1$.

Of course, if a mapping L has a reversing symmetry G, then its inverse clearly exists. However, for real polynomial maps that are invertible, the inverse need not be polynomial (2-D counterexample: $x' = x^3, y' = y^3$). When it is, then from above $\det dL(x) = c$, a non-zero constant. Suppose, in addition, that L has a fixed point

x_0. Then differentiating $L^{-1} \circ G = G \circ L$ at $x = x_0$ and taking determinants, shows again that $\det dL(x) = \det dL^{-1}(x) = c$, whence $c^2 = 1$.

We remark that for an invertible polynomial map L of $\mathbb{C}^n$ to itself, the inverse *is* necessarily polynomial [10]. Hence $\det dL(x)$ is a non-zero complex number. When L has a reversing symmetry and a fixed point, again $\det dL(x) = \pm 1$. Note that [5] has studied reversible mappings of the complex plane which contain Hamiltonian-like behaviour together with attracting and repelling asymmetric periodic orbits; the mappings are defined by a root of an implicit equation quadratic in both z and the image z').

Turning to flows, reversible polynomial examples that are non-conservative are readily found - it is easy to create polynomial flows with non-constant divergence [1]. A simple 1D example is

$$\dot{x} = x(1-x). \qquad (28)$$

Comparing to (8), one easily verifies that the involution $G : x \mapsto 1 - x$ is a reversing symmetry. Furthermore, the fixed point $x = 0$, with linearization $\dot{x} = x$, is repelling. Its reflection under G, the fixed point $x = 1$, with linearization $\dot{x} = -x$, is attracting. In 2D, consider the flow

$$\begin{aligned} \dot{x} &= \mu y + f_{11} xy + f_{03} y^3 \\ \dot{y} &= x, \end{aligned} \qquad (29)$$

which is reversible with $G : (x, y) \mapsto (x, -y)$. For $\mu > 0$ and $f_{03} < 0$, the vector field (29) has 3 fixed points : a symmetric saddle point at the origin, an attracting fixed point at $(0, \sqrt{-\mu/f_{03}})$ and a repelling fixed point at $(0, -\sqrt{-\mu/f_{03}})$. In fact, with $f_{03} < 0$, moving μ through 0 from negative to positive corresponds to a generic pitchfork bifurcation in a reversible non-conservative flow in which an asymmetric attractor/repeller pair is produced out of a symmetric fixed point. This is the 2D flow analogue of the bifurcation seen in the 2D mapping L_1 of (26).

Among 3D reversible polynomial flows are examples that have arisen in physical contexts. In their modelling of a particular type of externally-injected laser, [24] arrived at the equations

$$\begin{aligned} \dot{x} &= zx + y + C_1 \\ \dot{y} &= zy - x \\ \dot{z} &= C_2 - x^2 - y^2, \end{aligned} \qquad (30)$$

and noted their reversibility with $G : (x, y, z) \mapsto (-x, y, -z)$, cf. (8). In numerical studies of (30), [24] observed coexistence of KAM tori with attracting and repelling periodic orbits. The latter were born from a symmetric periodic orbit via a symmetry-breaking bifurcation analogous to the one in (29). Attractors and repellers in reversible flows also arise in Hoover's work in molecular dynamics, e.g. cf. [12, Figure 11.4].

---

[1] This is entirely consistent with the above remarks indicating the paucity of polynomial non-conservative maps. The time-one map of a reversible polynomial flow, though also reversible, is rarely polynomial, cf. [4] for a listing of the (polynomial) vector fields of the plane that admit polynomial time-one maps.

3 Universality of period-doubling in 4-D reversible flows

In this section we would like to concentrate on the occurrence in reversible flows of symmetric periodic orbits appearing in 1-parameter families. This is another noted similarity between reversible and Hamiltonian flows, which sets them both apart from generic flows in which a periodic orbit is typically isolated in phase space from other periodic orbits. In the Hamiltonian case, the reason for the appearance of non-isolated families stems from the constant of the motion H. A periodic orbit lies on a particular level set of H, whence the implicit function theorem typically guarantees the existence of a one-parameter family comprising the nearby periodic solutions on adjacent level sets. For reversible flows, the reason has a geometric basis related to the fixed sets (19) of the reversing symmetries (first shown by Devaney in [6]). There are now many examples of reversible flows modelling various physical phenomena, in which the significance of one-parameter families of symmetric periodic orbits has been noted [20, 9, 13, 37, 41]. When there exists 1-parameter families of periodic orbits in a system, one has the opportunity to see if and how different families can *branch* from one another as one moves through a particular family.

To understand how a reversible *flow* can possess 1-parameter families of symmetric periodic orbits, it is useful to note the geometric characterisation of symmetric periodic orbits of a reversible *mapping* in terms of $Fix(G)$ of (19) and its iterates. This is because the latter can occur as the Poincaré map of the former, the reversing symmetry G being inherited from the flow to the map [6, 39, 40].

If a mapping L has a reversing symmetry G, then it has a whole family of reversing symmetries, namely the set $\{L^i \circ G, i \in \mathbb{Z}\}$. An n-cycle of L is called symmetric when its set of n points is left invariant by G, and it follows that it is also invariant under all members of the family.

We have [8]:

Characterisation 3.1 *The symmetric periodic orbits of a reversible mapping L satisfying (9) have the following geometric characterisation:*

$$x \text{ belongs to symmetric } 2j\text{-cycle} \iff \begin{cases} x \text{ or an iterate of } x \\ \quad \in Fix(G) \cap L^j Fix(G) \\ or \\ x \text{ or an iterate of } x \\ \quad \in Fix(L \circ G) \cap L^j Fix(L \circ G) \end{cases} \quad (31)$$

and

$$x \text{ belongs to symmetric } (2j+1)\text{-cycle} \iff \begin{matrix} x \text{ or an iterate of } x \\ \in Fix(G) \cap L^j Fix(L \circ G). \end{matrix} \quad (32)$$

Perhaps the most familiar example of this geometric characterisation is reversible mappings in 2-D in the case that $Fix(G)$ and $Fix(L \circ G)$ are both 1-D (the so-called *symmetry lines*). This is the case for the paradigm area-preserving mappings like the standard mapping, and Henon's area preserving mapping of (16). We also noted in Figure 3(a) that the symmetric fixed point of L_1 of (26) is at the point of intersection of $Fix(G_1)$ and $Fix(H_1) = Fix(L_1 \circ G_1)$.

However, from (31) and (32), it follows that for N-D reversible mappings, $N \geq 3$, and appropriate dimensions of $Fix(G)$ and/or $Fix(L \circ G)$, some symmetric n-cycles may come in families. This follows because in N dimensions when a l-dimensional manifold intersects a k-dimensional manifold, the two surfaces generically meet transversally in a $(l + k - N)$-dimensional manifold (if $l + k < N$, one does not expect intersections to occur).

In particular, in a $(2m-1)$-D reversible mapping with $\dim Fix(G) = \dim Fix(L \circ G) = m$, symmetric cycles generically come in 1-parameter families, i.e. curves. Such a mapping arises naturally when one constructs the Poincaré mapping of a reversible $2m$-D flow with $\dim Fix(G) = m$, cf. [40]. It follows that such reversible flows have 1-parameter families of periodic orbits. For $m = 2$, we can infer something about the arrangement of one-parameter periodic solutions of such 4-D reversible flows through the study of 3-D reversible mappings L with at least one reversing symmetry with $\dim Fix(G) = 2$ (in the same way, the study of the generic arrangement of one-parameter periodic solutions in 4-D Hamiltonian flows has been studied via one-parameter families of 2-D area-preserving mappings [21, 19]).

Consider the non-measure-preserving 3-D reversible mapping (cf. Eq. (12) of [38])

$$L_3: \begin{array}{rcl} x' &=& (k - y)(1 + (y' - 1)^2) \\ y' &=& \frac{x + e(2y-k)(z+e(y-k))}{1+(y+1-k)^2} \\ z' &=& -z + e(k - 2y), \end{array} \tag{33}$$

which has the reversing symmetry

$$G_3: \begin{array}{rcl} x' &=& x + e(2y - k)(z + e(y - k)) \\ y' &=& k - y \\ z' &=& z + e(2y - k). \end{array} \tag{34}$$

One finds $Fix(G_3)$ is the plane $\{(x, k/2, z) \,|\, x, z \in \mathbb{R}\}$, whereas $Fix(L_3 \circ G_3)$ is the curve $\{(y(1 + (y - 1)^2), y, 0) \,|\, y \in \mathbb{R}\}$. Consequently, from (31) we expect symmetric even-cycles to come in curves arising from intersections of $Fix(G_3)$ with its iterates.

This is illustrated in Figure 6 which depicts a portion of the phase portrait of L_3 of (33) for the values $\{e = 2, k = -1\}$. Let us discuss some of the features of this picture.

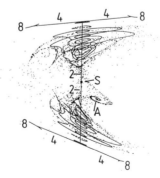

Figure 6: Phase portrait of a 3D non-conservative reversible mapping showing branching 1-parameter families of symmetric $2j$-cycles. See text for more details.

At the point S there is a symmetric fixed point $(-13/8, -1/2, 0)$ of the mapping. This point is at the centre of a 1-parameter family of symmetric 2-cycles with coordinates $(-13/8, -1/2, \pm z)$ — each point on the line above S (with $z > 0$) is mapped to a corresponding point below S (with $z < 0$) and then back onto itself on the second iteration. Also shown in Figure 6 is a 1-parameter family of symmetric 4-cycles connected to the family of 2-cycles and a 1-parameter family of symmetric 8-cycles connected to the 4-cycles etc.

Other features shown are the "rings" encircling the curve of 2-cycles, each of which is invariant under 2 iterations of the mapping. These rings are again explained by KAM theorems for (odd-dimensional) reversible mappings [36]. A curve of symmetric n-cycles in a 3-D reversible mapping with $\dim Fix(G) = 2$ is surrounded by nested L^n- and G-invariant tubes, each tube being fibred into L^n- and G-invariant rings on which the dynamics is conjugate to rotation with a sufficiently irrational winding number. Drifting longitudinal motion can occur parallel to the curve in the 'resonance zones' between the encircling invariant tubes of different rotation number [1, 2]. In the case of the non-measure preserving reversible mapping of Figure 6 (L_3 with $\{e = 2, k = -1\}$), orbits may spiral into the attracting fixed point A at $(-2, 0, -1)$. Because of reversibility $G_3 A$ at $(0, -1, 1)$ is a repelling fixed point. Consequently, L_3 is an example in 3 dimensions of the coexistence of KAM-like behaviour and dissipation/expansion in reversible systems.

In 3D reversible mappings like L_3, we have quantitatively studied what we call the *period-doubling branching tree* made up from connected 1-parameter families of symmetric periodic orbits of successively doubled periods [33]. Firstly, the branching of one family into another can be understood via reversible bifurcation theory. The eigenvalues of the linearization of the mapping around each symmetric even-cycle in the family are $\{1, \lambda, \lambda^{-1}\}$, with λ real, or λ complex on the unit circle. This is a consequence of the general property of Characterisation 2.2. As we move along the family, λ varies. So, for example, in Figure 6 as we move upwards from $z = 0$ along the vertical line of 2-cycles, we have $\lambda = e^{i\theta}$ complex with θ increasing smoothly from 0 when $z = 0$ to π when $z = 7/4$. Thus the 2-cycle $\{(-13/8, -1/2, 7/4), (-13/8, -1/2, -7/4)\}$ has eigenvalue spectrum $\{1, -1, -1\}$ (i.e. $\theta = \pi$). In agreement with the theory of [39, 40], this is precisely where there is an intersection with, or *branching* into, the family of symmetric 4-cycles. Two points of each 4-cycle, halfway around the cycle from one another, lie on the top branch labelled "4" in Figure 6, being symmetrically positioned either side of the intersection with the curve of 2-cycles. Similarly, the remaining two points lie on the bottom branch labelled "4". The value of θ in the spectrum of the 4-cycles varies from $\theta = 0$ at the branching point with the 2-cycles to $\theta = \pi$ at the four points of branching into the family of 8-cycles. Note that in Figure 6 the continuations of the families beyond their point of intersection with one of twice the period are not shown.

Because of reversibility, we can study a subset of the period-doubling branching tree in the 2-D subspace of $Fix(G_3)$. This is because it follows from (31) that a symmetric *even* $2j$-cycle must contain precisely two points $\boldsymbol{y}$ and $L^j\boldsymbol{y}$ in $Fix(G)$, or precisely two such points in $Fix(L \circ G)$. Figure 7(a-b) shows the part of the symmetric period-doubling branching tree of Figure 6 that is seen when looking in this plane (and following the tree to higher period). For each 2^n-cycle, one of the two points lying in $Fix(G_3)$ branches into a curve also lying in $Fix(G_3)$ that comprises two points of 2^{n+1}-cycles. The other point in $Fix(G_3)$ belonging to the 2^n-cycle experiences a simultaneous branching into a curve containing two other points of 2^{n+1}-cycles, this curve cutting across $Fix(G_3)$ and so not shown in Figure 7. The point of the 2^n-cycle on $Fix(G_3)$ at which the period-doubling branching on $Fix(G_3)$ is repeated alternates at each level being on the left or right branch of the new fork.

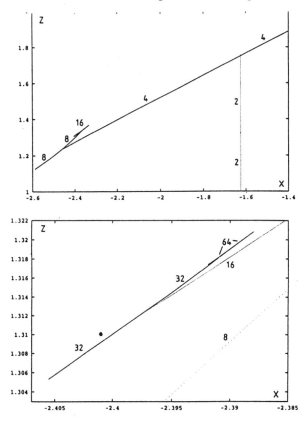

Figure 7: The part of the symmetric period-doubling branching tree of Figure 6 that lies in $Fix(G)$, which in this case is the plane $\{(x, -1/2, z) \,|\, x, z \in \mathbf{R}\}$. We find the top curve of period 4 points from Figure 6 in this plane, and the left-most curve of period 8 points that cuts it, etc. One magnification of this subset of the tree is also shown.

Figure 7(a-b) suggests self-similarity in the period-doubling branching tree. In [33], we showed that this could be quantified for the mapping L_3 and other 3-D reversible mappings with a reversing symmetry G with $\dim Fix(G) = 2$. We found numerically that the period-doubling branching tree exists to quite a high level, and that it possesses geometric self-similarity quantized by the numbers:

$$\begin{cases} \alpha = -4.018\ldots \\ \beta = 16.363\ldots \\ \delta = 8.721\ldots, \end{cases} \tag{35}$$

which are distance rescalings in the relevant directions in which the tree is self-similar. The rescalings can be found by introducing local coordinates for each branching on and across $Fix(G)$ and measuring the distances on and across $Fix(G)$ to the subsequent branching. For Example L_3, considering the part of the tree in $Fix(G_3)$, one could measure in Figure 7 for example, the distance between the point of branching from period 2 to 4 and the point of branching from period 4 to 8, and compare this to the corresponding distance connecting branching points from period 4 to 8 and period 8 to 16. Asymptotically, these distances decrease by $|\alpha|$. However, scaling of the tree in $Fix(G_3)$ is actually governed by *both* α and δ which is seen by resolving the distances between successive branch points into components parallel and perpendicular to the tangent to the curve of cycles at branching. Parallel components scale asymptotically with α, and perpendicular components with δ. The number β characterises rescaling of distances across $Fix(G_3)$.

The scalings (35) were previously found to be those describing period-doubling of symmetric periodic orbits in one-parameter reversible area-preserving 2-D mappings [11, 18], or in one-parameter reversible non-area-preserving 2-D mappings [34] (and cf. Characterisation 2.1 above). The appearance of (35) in period-doubling branching trees in 3-D reversible mappings therefore *extends* the universality class of these numbers beyond 1-parameter 2-D reversible mappings. The latter are special cases of the former if we study them in '*parameter space* $\times$ *phase space*' by incorporating their parameter as an extra dimension:

$$M: \begin{array}{rcl} x' & = & f(x,y,C) \\ y' & = & g(x,y,C) \end{array} \iff L: \begin{array}{rcl} x' & = & f(x,y,z) \\ y' & = & g(x,y,z) \\ z' & = & z \end{array}. \tag{36}$$

In fact, any 3-D reversible mapping with an integral is a special case of those looked at here because it can typically be transformed into the form L of (36) in the neighbourhood of dynamics confined to one set of branches of the level set (and so is effectively a 1-parameter 2-D mapping). This transformation "straightens out" the integral and the foliation of 3-D space by its 2-D level sets (e.g. as occurs in the 3-D Poincare map of a 4-D Hamiltonian flow, also cf. [29]).

In a 3-D reversible mappings with an integral I, the scaling δ is perhaps the most obvious and easily-recovered scaling. The value of the integral is the distinguished variable which "orders" the tree and δ is obtained from the sequence of values of I at which consecutive period-doubling branchings occur. In our study of general 3-D reversible mappings, all coordinates of the phase space are on an equal footing and what should play the role of a parameter is unclear. There seems no obvious way to introduce a global bifurcation parameter when there is no *global* integral.

In [33], it was argued that the normal forms of the 3-D reversible mappings with $\dim Fix(G) = 2$ provide the key to understanding the presence of the scaling factors (35) in the period-doubling branching tree. When the the eigenvalue spectrum of a symmetric cycle in such a mapping is $\{1, -1, -1\}$, the techniques of [15, Proposition 6] yield the following reversible normal form:

$$\begin{array}{rcl} x' & = & -x(1+f(z)) + y - cx^3 \\ y' & = & xf(z) - y + cx^3 \\ z' & = & z \end{array} + o(3), \tag{37}$$

with $f(z) = az + bz^2 + o(3)$. Note that $f(z)$ generically changes sign at $z = 0$.

Significantly, the first two lines of (37) are precisely the unfolding at the period-doubling bifurcation point of a normal form of a 2-D area-preserving or reversible mapping, with z acting as the unfolding parameter. It appears that the normal form also determines the universality class to which the *asymptotic* self-similar structure of the branching tree belongs.

4 Irreversibility in conservative dynamics

In the previous sections, we have looked at consequences of time-reversal symmetry in dynamical systems that possess such a symmetry property. Now we would like to ask, given a dynamical system, how could we identify whether it has a reversing symmetry or not. In the first instance, we can use some of the consequences of reversibility as necessary conditions to help exclude it. For example, a dynamical system which is purely dissipative with all initial conditions evolving to an attractor is clearly irreversible, because of the lack of the time-reversed repellor. Here we will concentrate on identifying 2D flows and mappings that are in Region III of Figure 2, namely Hamiltonian systems without a time-reversal symmetry, in particular irreversible ones with no involutory reversing symmetry (cf. Remark 3 of Section 1). The spectrum of Hamiltonian systems ranges from integrable at one end to fully chaotic at the other. Of course, most 'real-life' Hamiltonian systems fall in between these extremes. They exhibit regular motion via KAM tori *together with* chaotic motion. We will say something about irreversibility across the various parts of this spectrum.

Firstly, restricting to 2-D area-preserving diffeomorphisms, we adopt the definition that a map $L : (x, y) \mapsto (x', y')$ is integrable if there exists a conserved quantity $I(x, y)$ with

$$I(x', y') = I(x, y). \tag{38}$$

In recent years, there has been an explosive interest in 2-D integrable mappings. A large class of such mappings, with 18 free parameters, was introduced in [27]. These mappings were reversible. The component involutions themselves preserved the integral $I(x, y)$, and hence the foliation of the plane provided by its level sets.

Are there 2-D area-preserving integrable mappings that are not reversible? The answer is yes [32], an explicit example being :

$$\begin{array}{rcl} x' & = & \frac{x}{\sqrt{1-2x^2}} \\ y' & = & y(1-2x^2)^{\frac{3}{2}} = x^3 y/x'^3, \end{array} \tag{39}$$

which evidently preserves $I(x, y) = x^3 y$. This irreversible example (and others) can be obtained by writing out explicitly the time-one maps of irreversible 2-D Hamiltonian flows (i.e. 1 degree-of-freedom). In particular, (39) is the time-one map with $H(x, y) = x^3 y$ (see below). Of course all 2-D Hamiltonian systems are integrable (via quadratures), as are their time-t maps, the integral being the Hamiltonian. Incidentally, notice that (39) is a rational map. It can be observed from the work of [4] that all 2-D Hamiltonian flows which have polynomial time-one maps (necessarily area-preserving and integrable) are reversible (and hence so is the map) [4, Theorem 11.8]. Probably then, no example of an integrable area-preserving polynomial mapping which is irreversible exists.

So the above leads to investigation of irreversible 2-D Hamiltonian flows. Let us consider the Hamiltonians defined by binary forms of order n:

$$H_n(x,y) = \sum_{j=0}^{n} C_{n-j,j}\, x^{n-j} y^j \qquad (40)$$

We find [32]:

Characterisation 4.1 *All Hamiltonians of the form (40) with $n = 2$ or $n = 3$ are reversible. For $n = 4$, the Hamiltonian $H = x^3 y$ (and any Hamiltonian transformable to it) is irreversible with respect to both orientation-reversing and orientation-preserving involutory reversing symmetries.*

To see the first part of this result, it suffices to use known results concerning normal forms of binary forms under a non-singular real linear change of variables $M : x' = ax+by,\ y' = cx+dy$ with $\det M = ad-bc \neq 0$. Under such a transformation, the flow arising from (40) remains Hamiltonian with transformed Hamiltonian $H' = \det M \cdot H(x(x',y'), y(x',y'))$. The factor $\det M$ can be absorbed by rescaling time. For the case $n = 2$ in (40), it is well-known that via a suitable linear transformation M, a binary quadratic form can always be reduced to one of the five forms in the following set:

$$\{\pm(x^2 + y^2),\, x^2 - y^2,\, \pm x^2\}. \qquad (41)$$

Note that for convenience we omit the primes in (41). Different forms in the set cannot be transformed to one another, and each of the listed forms is a representative of its conjugacy class. Nevertheless, we observe that all the representative forms in (41) are even in y. Hence, considered now as Hamiltonians, they are reversible with respect to the standard time-reversal symmetry $x' = x,\ y' = -y$ (cf. Example 1 in Section 1). Since reversibility is invariant under transformations, it follows that all Hamiltonians $H_2(x, y)$ of (40) are reversible because they can be transformed to the set (41).

A similar argument shows the reversibility of all Hamiltonians $H_3(x, y)$ of (40). It follows from relatively recent work in catastrophe theory [25] that, via a suitable invertible linear transformation, any cubic binary form can be brought into one of four standard forms:

$$\{y^2 x - x^3,\, y^2 x + x^3,\, y^2 x,\, x^3\}. \qquad (42)$$

Again, considered as the transformed Hamiltonians $H'_3(x', y')$, these forms all possess the standard reversibility.

The theory of characterising the binary quartic forms, i.e. $H_4(x, y)$ of (40), up to linear transformations helps to isolate possible irreversible Hamiltonians of this type. In fact it emerges from a table of standard forms [25] that the only normal form that is not even in the (transformed) variable y, and so not obviously reversible, is $x^3 y$. It could be that $x^3 y$ can be transformed by a linear transformation so as to be even in y (i.e. there is a better representative in the conjugacy class of $x^3 y$ for which to observe reversibility). This can easily be ruled out. Applying $M : x' = ax + by,\ y' = cx + dy$ to $x^3 y$ and demanding that the coefficients of $x'^3 y'$ and $x' y'^3$ be zero leads to

$$a^2(3bc + ad) = 0, \qquad b^2(3ad + bc) = 0. \qquad (43)$$

It is impossible to satisfy (43) and maintain $ad - bc \neq 0$. This shows that the flow generated by the Hamiltonian $H = x^3 y$ is not reversible with any linear orientation-reversing involution, because the latter can always be reduced to $x' = x,\ y' = -y$ using a linear transformation.

We claim in fact that $H = x^3y$ is not reversible with respect to any (nonlinear) orientation-reversing involutory reversing symmetry. This can be seen in two ways. The first is to observe the phase portrait of the resulting flow. This is depicted in Figure 8(a). The y-axis is a line of non-isolated equilibrium points. The origin has two unstable separatrices emanating from it along the x-axis. The x- and y-axes taken together constitute the level set $H = 0$. The level sets $\{H = \mu\}$ with $\mu > 0$ ($\mu < 0$) are the hyperbolic arms in the 1st (2nd) and 3rd (4th) quadrants. The origin is unique among the fixed points because of the invariant separatrices connected to it. If the phase portrait corresponded to a reversible flow, the origin would necessarily be symmetric. If there was an orientation- reversing involutory reversing symmetry G for the flow, it follows that the curve $Fix(G)$ passes through the origin (cf. Remark 3 of Section 1). The separatrices either side of the origin must lie on either side of $Fix(G)$, and the dynamics on one should yield the dynamics on the other via reflection by G and reversal in time. However, the directions of the flow on the separatrices are inconsistent with this. To contrast with Figure 8(a), we show in Figure 8(b) the phase portrait of the flow $\dot{x} = x^2, \dot{y} = -3xy$. This flow is reversible with $G : x' = -x, y' = y$. It also preserves the quantity x^3y, but is not divergence-free and hence not Hamiltonian. The direction of the flow in the left and right halves of the plane are now consistent with reversibility.

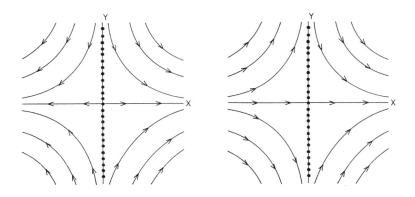

Figure 8: (a) The phase portrait of the irreversible flow $\dot{x} = x^3$, $\dot{y} = -3x^2y$, derived from the Hamiltonian $H = x^3y$. (b) For comparison, the phase portrait of the reversible flow $\dot{x} = x^2$, $\dot{y} = -3xy$ with $G : (x, y) \mapsto (-x, y)$.

A second way to observe the irreversibility of $H = x^3y$ is analytically using the concept of *local reversibility*. This idea has been developed for both flows and mappings [26, 34, 15]. Heuristically, it entails studying the flow/mapping in a neighbourhood of a fixed point which is necessarily also fixed by a reversing symmetry should the system be reversible. Possession of a reversing symmetry in general has implications for the Taylor expansion of the flow/mapping around the fixed point, which result in order-by-order conditions on the expansion coefficients. Satisfying these conditions, even to arbitrarily high order, does not guarantee reversibility in a full neighbourhood of the fixed point because the convergence of the Taylor series for the possible reversing symmetry is not guaranteed. Nevertheless, the conditions act as necessary conditions for global reversibility because global reversibility does imply local reversibility to any order about a symmetric fixed point. For the typical linearisations of a fixed point of a Hamiltonian flow or area-preserving mapping, i.e. hyperbolic, elliptic or parabolic, there exist well-known normal forms about these points that describe the reduced Taylor expansions. These can be observed to be locally reversible to all orders. Only in the atypical case of zero linear part for a Hamiltonian flow, or identity linear part for an area-preserving mapping, do genuine extra conditions for reversibility exist. In summary [31, 32, 34]

Characterisation 4.2 *A Hamiltonian flow (area-preserving mapping) is formally reversible to all orders about a fixed point with linearization not equal to the zero matrix (identity matrix). When the linearization is equal to the zero matrix (identity matrix), the flow (mapping) is generically irreversible.*

For $H = x^3 y$, the origin is necessarily symmetric if H is reversible. The Taylor expansion of the Hamiltonian about the origin, is of course itself, and the linear part of the flow is the zero matrix. An example of an irreversible area-preserving mapping [31] is a mapping of the form (13) where the origin is a fixed point, the expansion of $f(y)$ is $f_n y^n + f_{n+1} y^{n+1} + \ldots$ with $n \geq 2$ and $f_n \neq 0$, and the expansion of $g(x')$ is $g_n x'^n + g_{n+1} x'^{n+1} + \ldots$ with $n \geq 2$ and $g_n \neq 0$. Such a mapping is irreversible when

$$f_n^{n+2} g_{n+1}^{n+1} \neq - g_n^{n+2} f_{n+1}^{n+1}. \qquad (44)$$

An explicit polynomial example is

$$\begin{aligned} x' &= x - y^2(1-y) \\ y' &= y + Cx'^2(1-x'), \end{aligned} \qquad (45)$$

which is irreversible if $C \neq 0$ and, from (44), $C \neq 1$.

The trouble with the above examples of irreversible conservative systems is that they are not persistent in the class of conservative systems, i.e. arbitrarily close to them is a reversible system. For example, even when we just confine ourselves to the equivalence classes of quartic binary forms, almost all arbitrarily small perturbations of the irreversible $H = x^3 y$ form will belong to one of the other reversible equivalence classes ([25] presents a nice geometric interpretation of these equivalence classes showing how $x^3 y$ corresponds to a cusp line separating surfaces corresponding to other (reversible) classes). On the other hand, it is easy to perturb away the singular case of identity (zero) linear part about a fixed point of a mapping (flow).

Very recently, [16] has presented an example of a persistent irreversible area-preserving flow and a persistent irreversible area-preserving mapping. Here, reversibility/irreversibility refer to the existence/non-existence of an involutory orientation-reversing reversing symmetry. The irreversible flow example is constructed by designing a phase portrait of an area-preserving flow (not given explicitly) where the geometry precludes a reversing symmetry. The irreversible mapping example comes from perturbing the time-one mapping of the flow example in such a manner that its obstructions to reversibility are persistent with respect to sufficiently small area-preserving perturbations. Thus this represents an irreversible area-preserving example, whose (open) neighbourhood in the class of area-preserving mappings also comprises irreversible elements. In the same paper, examples are given of persistent reversible area-preserving flows, and it is argued that area-preserving flows are generically reversible inside bounded invariant sets. From these examples, the following conclusions can be drawn [16]

Characterisation 4.3 *Consider the class of C^1 Hamiltonian vector fields of the plane. Reversibility and irreversibility are both nongeneric properties in this class. Consider the class of C^1 area-preserving diffeomorphisms of the plane. Reversibility is a nongeneric property in this class.*

Let us now ask for irreversible examples of 2-D area-preserving mappings at the other end of the spectrum of Hamiltonian systems, namely totally chaotic. More specifically, we consider the hyperbolic toral automorphisms generated by the set $\mathcal{C}$ of hyperbolic unimodular 2×2 integer matrices, i.e.

$$\mathcal{C} := \left\{ \begin{pmatrix} a & b \\ c & d \end{pmatrix} ; \, a,b,c,d \in \mathbb{Z}; \, ad - bc = \pm 1; \, |\lambda_i| \neq 1 \right\}, \tag{46}$$

where λ_i, $i \in \{1, 2\}$, are the eigenvalues of the matrix. When such a matrix acts (mod 1) on the 2-torus $\mathbb{T}^2$, equivalently the square $[0, 1) \times [0, 1)$ in $\mathbb{R}^2$ with edges identified in the usual way, we obtain an example of an *Anosov* diffeomorphism (Anosov diffeomorphisms of a compact space have a more technical definition, but on the torus can be taken to be hyperbolic toral automorphisms without loss of generality). The resulting dynamics is chaotic on the whole torus. More specifically, the periodic orbits are dense in $\mathbb{T}^2$, the dynamics is topologically transitive and has sensitive dependence on initial conditions [7]. The most celebrated element of $\mathcal{C}$ is the cat map of Arnol'd-Avez given by (17). We will use the term 'cat map' to refer more generally to elements of $\mathcal{C}$.

We investigate when a cat map has a reversing symmetry belonging to the group

$$Gl(2, \mathbb{Z}) := \left\{ \begin{pmatrix} a & b \\ c & d \end{pmatrix} ; \, a,b,c,d \in \mathbb{Z}; \, ad - bc = \pm 1 \right\}. \tag{47}$$

This is the natural set in which possible reversing symmetries lie. This is because of the following [3]

Characterisation 4.4 *Suppose two mappings of the torus induced by two hyperbolic elements of $Gl(2, \mathbb{Z})$ are conjugate to one another via a homeomorphism. Then the homeomorphism must correspond to the action of an element of $Gl(2, \mathbb{Z})$, or to the action of an affine extension of an element of $Gl(2, \mathbb{Z})$ constructed by adding to it a rational shift.*

Furthermore, it is easy to show that an affine mapping of the type just described is only a reversing symmetry if the component element of $Gl(2, \mathbb{Z})$ is (cf. Section 5 below). So it turns out that the basic problem is to search for reversing symmetries of cat maps in (47).

Firstly, note that obviously an orientation-reversing element $L \in \mathcal{C}$ of (46) cannot have a reversing symmetry because $\text{tr} L = -\text{tr} L^{-1}$ precluding L and L^{-1} being conjugate. Henceforth we can restrict to the orientation-preserving elements of $\mathcal{C}$, namely the hyperbolic elements of $Sl(2, \mathbb{Z})$:

$$\mathcal{C}^+ := \left\{ \begin{pmatrix} a & b \\ c & d \end{pmatrix} ; \, a,b,c,d \in \mathbb{Z}; \, ad - bc = 1; \, |a + d| \geq 3 \right\}. \tag{48}$$

It turns out that if an element of $\mathcal{C}^+$ has a reversing symmetry G in $Gl(2,\mathbb{Z})$, then G is either (i) an orientation-reversing involution; or (ii) of order 4, equal to v or conjugate to v where

$$v = \begin{pmatrix} 0 & -1 \\ 1 & 0 \end{pmatrix}. \tag{49}$$

Which, if any, possibility holds for a particular $L \in \mathcal{C}^+$ can be resolved by studying related conjugacy problems in the matrix groups $PSl(2,\mathbb{Z})$ and $PGl(2,\mathbb{Z})$ defined by

$$PGl(2,\mathbb{Z}) := Gl(2,\mathbb{Z})/\{\pm\mathbb{1}\}; \quad PSl(2,\mathbb{Z}) := Sl(2,\mathbb{Z})/\{\pm\mathbb{1}\}. \tag{50}$$

That is, each element of $PGl(2,\mathbb{Z})$, for instance, corresponds to the pair of matrices M and $-M$ of $Gl(2,\mathbb{Z})$. To distinguish between the groups, we now use (G) for an element of $Gl(2,\mathbb{Z})$, $[G]$ for an element of $PGl(2,\mathbb{Z})$, etc. So the notation (v) is now used for (49), whereas $[v]$ refers to either $\pm(v)$. We find [3]:

Characterisation 4.5 *Let* $(L) = \begin{pmatrix} a & b \\ c & d \end{pmatrix}$ *be a hyperbolic element of* $Sl(2,\mathbb{Z})$. *There are two possibilities for* (L) *to have a reversing symmetry* $(G) \in Gl(2,\mathbb{Z})$ *(one, both or neither can occur):*

(i) (L) *has a reversing symmetry* $(G) \in Sl(2,\mathbb{Z})$ *of order 4. This occurs if and only if* $[L] \in PSl(2,\mathbb{Z})$ *is reversible with involution* $[G]$, *i.e.* $[G]^{-1}[L][G] = [G][L][G] = [L]^{-1}$ *in* $PSl(2,\mathbb{Z})$. *A necessary and sufficient condition on the entries of* (L) *for this to occur is that* (L) *has* $b = c$, *or is conjugate to such a matrix, whence* (G) *is equal to* (v) *of (49), or is conjugate to* v.

(ii) (L) *is reversible with an orientation-reversing involution* $(G) \in Gl(2,\mathbb{Z}) - Sl(2,\mathbb{Z})$. *If we write* $(G) = (G')(s)$ *with* $(G') \in Sl(2,\mathbb{Z})$ *and* $(s) = \begin{pmatrix} -1 & 0 \\ 0 & 1 \end{pmatrix}$, *this possibility occurs if and only if* $[G']^{-1}[L][G'] = [s][L]^{-1}[s]$ *holds in* $PSl(2,\mathbb{Z})$. *Necessary and sufficient conditions on the entries of* (L) *for this to occur are that* (L) *has* $a = d$, $b = -c$, $b|(d-a)$ *or* $c|(d-a)$, *or is conjugate to such a matrix.*

The advantage of dealing with the group $PSl(2,\mathbb{Z})$ is that conjugacy of any two elements is decideable. This follows because $PSl(2,\mathbb{Z})$ is the free product of a cyclic subgroup of order 2 with one of order 3. Practically speaking, this means that any element of $PSl(2,\mathbb{Z})$ can be written uniquely as a finite word in the order 2 element $[v]$ of $PSl(2,\mathbb{Z})$, an order 3 element $[q]$ defined by

$$[q] = \begin{bmatrix} 0 & 1 \\ -1 & -1 \end{bmatrix}, \tag{51}$$

and its inverse $[q]^{-1}$. Hence

$$[M] \in PSl(2,\mathbb{Z}) \Rightarrow M = [v]^\alpha [q]^{\pm 1} [v] [q]^{\pm 1} \ldots [v] [q]^{\pm 1} [v]^\beta, \tag{52}$$

where α and β are 0 or 1, and the representation is unique. For an explicit algorithm to find this word, see e.g. [3]. Two elements of $PSl(2,\mathbb{Z})$ are conjugate if and only if their words are equal up to cyclic permutation (i.e. equal after each word is wrapped around the circle with the first and last letters adjacent).

For example, consider for $\ell \in \mathbb{Z}$ the element of $\mathcal{C}^+$:

$$\begin{pmatrix} 1 & \ell \\ \ell & 1+\ell^2 \end{pmatrix}, \tag{53}$$

which includes (17) when $\ell = 1$. Consider the corresponding element of $PSl(2,\mathbf{Z})$:

$$\begin{bmatrix} 1 & \ell \\ \ell & 1+\ell^2 \end{bmatrix} = [vq^{-1}]^\ell [vq]^\ell. \tag{54}$$

We have

$$\begin{bmatrix} 1 & \ell \\ \ell & 1+\ell^2 \end{bmatrix}^{-1} = [vq]^{-\ell}[vq^{-1}]^{-\ell} = [q^{-1}v]^\ell [qv]^\ell = [q]^{-1}[vq^{-1}]^{\ell-1}[vq]^\ell [v]. \tag{55}$$

Comparing (55) and (54), we find the words are cyclic permutations of one another. From Characterisation 4.5, the hyperbolic linear automorphisms (53) each have an order 4 reversing symmetry, which can be seen from (55) and (54) to be precisely (v) of (49). Each element of (53) also has an involutory reversing symmetry $(G) = \begin{pmatrix} 1 & 0 \\ \ell & -1 \end{pmatrix}$ which can be deduced from part (ii) of Characterisation 4.5.

A particularly simple reversibility criterion can be developed for cat maps when one uses the structure of $PSl(2,\mathbf{Z})$ [28, 3] to show that, up to possible conjugacy and change of sign, any $L \in \mathcal{C}^+$ can be written as an alternating string in powers of (vq) and (vq^{-1}) where

$$(vq) = \begin{pmatrix} 1 & 1 \\ 0 & 1 \end{pmatrix}, \quad (vq^{-1}) = \begin{pmatrix} 1 & 0 \\ 1 & 1 \end{pmatrix}. \tag{56}$$

This string can be specified by the sequence of powers, i.e.

$$L = (vq^{-1})^{j_1}(vq)^{k_1}\ldots(vq^{-1})^{j_n}(vq)^{k_n} \\ \Rightarrow \{L\} = \{-j_1, k_1, \ldots, -j_n, k_n\}, \; n \geq 1, \; j_i, k_i \in \mathbf{Z}^+. \tag{57}$$

Note that this means that, via a possible conjugacy and change of sign, every cat map can be represented by an element of $\mathcal{C}^+$ whose entries are all positive. The possible conjugacy and sign change will not effect the existence of a reversing symmetry, so it is sufficient to study $L \in \mathcal{C}^+$ of the form (57). The existence of a reversing symmetry of L can now be decided in terms of a symmetry property of the corresponding sequence $\{L\}$ [3]:

Characterisation 4.6 *Regarding reversing symmetries of L of the form (57), possibility (ii) (possibility (i)) of Characterisation 4.5 occurs if and only if the sequence $\{L\}$ is invariant (is invariant with a sign change) up to cyclic permutation when written backwards.*

If we take $n = 1$ in (57) we observe that $(vq^{-1})^{j_1}(vq)^{k_1}$ always has an involutory reversing symmetry because $\{-j_1, k_1\}$ is equivalent to $\{k_1, -j_1\}$. On the other hand, $(vq^{-1})^{j_1}(vq)^{k_1}$ only has an order-4 reversing symmetry if $j_1 = k_1$, because this is needed for $\{-j_1, k_1\}$ to equal $\{-k_1, j_1\}$ (cf. (54)). Once we take $n = 2$ and consider sequences $\{-j_1, k_1, -j_2, k_2\}$, we can avoid *both* possibilities (i) and (ii) and this becomes increasingly easy with higher n (e.g. a necessary condition for possibility (ii) is that the subsequences of j_i's and k_i's are each symmetric). For $n = 2$, we have

$$(vq^{-1})^{j_1}(vq)^{k_1}(vq^{-1})^{j_2}(vq)^{k_2} = \\ \begin{pmatrix} 1 + k_1 j_2 & k_1 + k_2(1 + k_1 j_2) \\ j_2 + j_1(1 + k_1 j_2) & 1 + j_1 k_1 + j_2 k_2 + j_1 k_2(1 + k_1 j_2) \end{pmatrix}. \tag{58}$$

The sequence $\{-j_1, k_1, -j_2, k_2\} = \{-1, 1, -3, 2\}$ corresponds to a cat map which is irreversible in $Gl(2, \mathbb{Z})$. From (58), the trace of this example equals 20. This is the minimum value of tr L with $L \in \mathcal{C}^+$ for which irreversibility can occur. Thus, although reversibility dominates cat maps of low trace, on the whole most cat maps are irreversible (the asymmetric sequences $\{L\}$ far outnumber the symmetric ones). For every value of tr $L \geq 3$, there is a reversible L, namely $\begin{pmatrix} 0 & -1 \\ 1 & t \end{pmatrix}$, which satisfies $b = -c$, cf. Characterisation 4.5(ii). Moreover, there is an infinite subsequence of values of tr L for which all cat maps with that trace are reversible. See [3] for more details.

Recall that an important feature of the hyperbolic toral automorphisms is that they are structurally stable, i.e. all C^1 diffeomorphisms in a neighbourhood of a cat map are conjugate to it. This allows us to make statements for maps of the 2-torus analogous to Characterisation 4.3:

Characterisation 4.7 *Consider the class of C^1 area-preserving diffeomorphisms of the 2-torus. Reversibility and irreversibility are both nongeneric properties in this class.*

This statement follows from our results above on the cat maps. The existence of reversible cat maps shows there exists an open set of reversible area-preserving diffeomorphisms of the torus. This shows that irreversibility is a nongeneric property in the class of C^1 area-preserving diffeomorphisms of the torus. Furthermore, since Characterisation 4.4 forces a reversing symmetry to originate in $Gl(2, \mathbb{Z})$, our irreversible (in $Gl(2, \mathbb{Z})$) examples show that reversibility is also a nongeneric property in the class of C^1 area-preserving diffeomorphisms of the torus.

Of course the above results are essentially to do with *identifying* reversible and irreversible conservative systems. Ideally, one wants to (use these to) get an understanding of how the presence or not of a reversing symmetry influences the dynamics and manifests itself.

5 Further generalizations of time-reversal symmetry

A further generalization of time-reversal symmetry has received some recent attention, one in which some iterate of a mapping L (or the time-k map of a flow) possesses a time-reversal symmetry not possessed by L (or the time-1 map of a flow). This generalization has been observed in a couple of models arising from physical applications. For example [29], consider the so-called *Fibonacci trace map*:

$$F_1 : (x, y, z) \mapsto (y, z, 2yz - x). \tag{59}$$

This is a 3-D polynomial mapping from $\mathbb{C}^3$ to itself (or $\mathbb{R}^3$ to itself) with polynomial inverse and $\det dF_1(x) = -1$ (cf. Characterisation 2.3 and ensuing discussion). It (and similar such 3-D polynomial mappings) are much-studied in the physics literature because they arise from transfer matrix approaches to various physical processes displaying non-periodicity in space or time [29, 30]. Knowledge about the dynamics of such mappings can give corresponding information about the physical model from which they are derived, and vice versa, e.g. [28].

Now consider the 3-D involutory polynomial mapping:

$$G_1 : (x,y,z) \mapsto (z,y,x), \tag{60}$$

and the group

$$\Sigma = \{\sigma_0 := Id, \sigma_1, \sigma_2, \sigma_3\}, \tag{61}$$

where σ_i, $i = 1, 2, 3$, are the pairwise sign changes, e.g.

$$\sigma_1(x,y,z) = (x,-y,-z). \tag{62}$$

It is easy to check that: (i) F_1 is reversible with reversing symmetry G_1; (ii) F_1^3 commutes with each σ_i, $i = 1, 2, 3$, but F_1 and F_1^2 do not commute with any of them; (iii) F_1^3 has reversing symmetries $G_1 \circ \sigma_2$, which is an involution, and $G_1 \circ \sigma_1$ and $G_1 \circ \sigma_3$, which are order 4. However, these mappings are not reversing symmetries for F_1 and F_1^2.

After [17], we call G a reversing k-symmetry, $k \in \mathbf{Z}^+$, of a mapping L if k is the least integer for which G is a reversing symmetry of L^k (so to this point in the article, we have been considering reversing 1-symmetries, cf. (9)). It follows from above that each $G_1 \circ \sigma_i$, $i = 1, 2, 3$, is a reversing 3-symmetry of F_1. Lamb and Quispel have developed a systematic theory for systems with reversing k-symmetries.

Here we will just point out that the cat maps of (48) are simple examples of mappings for which possession of reversing k-symmetries is natural and can be completely determined. Firstly, it can be shown [3] that the cat maps cannot have any reversing k-symmetries for $k > 1$ belonging to $Gl(2, \mathbf{Z})$ of (47). From Characterisation 4.4, it remains to extend the search to the groups $\mathcal{A}_q$, $q \in \mathbf{Z}^+$, of affine transformations of the torus given by

$$\mathcal{A}_q := \{M + \boldsymbol{p}_q; \; M \in Gl(2, \mathbf{Z}); \boldsymbol{p}_q = (\frac{m}{q}, \frac{n}{q})^T \text{ with } 0 \leq m, n < q\}. \tag{63}$$

Recalling that every rational point of the torus is periodic by an element of $Gl(2, \mathbf{Z})$, we have [3]

Characterisation 5.1 *Let* $(L) = \begin{pmatrix} a & b \\ c & d \end{pmatrix}$ *be a hyperbolic element of* $Sl(2, \mathbf{Z})$. *Let* $G \in \mathcal{A}_q$ *of (63) be an affine transformation of the torus. Then G is a reversing k-symmetry of L if and only if M is a reversing symmetry of L. The value of k equals the period of $\boldsymbol{p}_q$ under* (L), *i.e.* $k = j : (L)^j \boldsymbol{p} = \boldsymbol{p} \; (mod \; 1)$.

The periodic orbits of cat maps, and the distribution of the periods, has been studied in [23]. When q in (63) is prime, then for most cat maps, all points on the discrete lattice determined by q have the same period.

Interestingly, the work on reversing k-symmetries of cat maps leads to an explanation of the reversing k-symmetries of polynomial mappings like F_1 above that preserve the quantity $I(x,y,z) = x^2 + y^2 + z^2 - 2xyz$. The set of 3-D polynomial mappings preserving $I(x,y,z)$ forms a group that is simply related to $\mathcal{A}_2$ of (63). The end result is that every question about polynomial reversing symmetries for F_1 and its generalizations can be answered from the study of affine reversing symmetries of cat maps, cf. [3]

6 Concluding Remarks

In this contribution, we have given a survey of some aspects of low-dimensional systems with generalizations of time-reversal symmetry using simple examples. We have emphasized similarities and differences to Hamiltonian dynamics.

Acknowledgements

It is a pleasure, as always, to acknowledge useful discussions with my collaborators M. Baake, H.W. Capel, J. Lamb and G.R.W. Quispel. The author thanks the organizers for a most interesting conference.

References

[1] Arnol'd V I 1984 *Nonlinear and Turbulent Processes in Physics, Vol. 3*, ed. R. Z. Sagdeev (Harwood, Chur) p 1161

[2] V. I. Arnol'd and M. B. Sevryuk, Oscillations and bifurcations in reversible systems, in: *Nonlinear Phenomena in Plasma Physics and Hydrodynamics*, ed. R. Z. Sagdeev (Mir, Moscow,1986) pp. 31-64.

[3] M. Baake and J. A. G. Roberts, Reversing symmetry group of $Gl(2,\mathbb{Z})$ and $PGl(2,\mathbb{Z})$ matrices with connections to cat maps and trace maps, *J. Phys. A* to appear.

[4] H. Bass and G. Meisters, Polynomial flows in the plane, *Adv. Math.* **55** (1985) 173-208.

[5] S. Bullet, Dynamics of quadratic correspondences, *Nonlinearity* **1** 27-50.

[6] R. L. Devaney, Reversible diffeomorphisms and flows, *Trans.Am.Math.Soc.* **218** (1976) 89-112.

[7] R. L. Devaney, *An Introduction to Chaotic Dynamical Systems*, 2nd ed., (Addison-Wesley, Redwood, 1989).

[8] DeVogelaere R 1958 *Contributions to the Theory of Nonlinear Oscillations* Vol 4, ed. S.Lefschetz (Princeton Univ. Press, Princeton) p 53

[9] Eckmann J-P and Procaccia I 1991 *Phys. Rev. Lett.* **66** 891, *Nonlinearity* **4** 567

[10] A. van den Essen, Seven lectures on polynomial automorphisms, in: *Automorphisms of Affine Spaces*, ed. A. van den Essen (Kluwer, Dordrecht, 1995) pp. 3-39.

[11] Greene J M, MacKay R S, Vivaldi F and Feigenbaum M J 1981 *Physica D* **3** 468

[12] W. G. Hoover, *Computational Statistical Mechanics* (Elsevier, Amsterdam, 1992).

[13] Kazarinoff N D and Yan J G G 1991 *Physica D* **48** 147

[14] Lamb J S W 1992 *J. Phys. A* **25** 925

[15] J. S. W. Lamb, J. A. G. Roberts and H. W. Capel, Conditions for local (reversing) symmetries in dynamical systems, *Physica A* **197** (1993) 379-422.

[16] J. S. W. Lamb, Area-preserving dynamics that is not reversible, *Physica A* **228** (1996) 344-365.

[17] J. S. W. Lamb and G. R. W. Quispel, Reversing k-symmetries in dynamical systems, *Physica D* **73** (1994) 277-304.

[18] MacKay R S 1993 *Renormalisation in Area-Preserving Maps* (World Scientific, Singapore) and references herein

[19] MacKay R S and Meiss J D 1987 *Hamiltonian Dynamical Systems : A Reprint Selection* (Hilger, Bristol)

[20] Malomed B A and Tribelsky M I 1984 *Physica D* **14** 67

[21] Meyer K R 1970 *Trans. Am. Math. Soc.* **149** 95

[22] J. Moser, Convergent series expansions for quasi-periodic motions, *Math. Ann.* **169** (1967) 136-176.

[23] I. Percival and F. Vivaldi, *Physica D* **25** (1987) 105-30.

[24] A. Politi, G. L. Oppo and R. Badii, Coexistence of conservative and dissipative behaviour in reversible dynamical systems, *Phys.Rev. A* **33** (1986) 4055-4060.

[25] T. Poston and I. N. Stewart, *Taylor Expansions and Catastrophes* (Pitman, London, 1976)

[26] Quispel G R W and Capel H W 1989 *Phys. Lett. A* **142** 112

[27] G.R.W. Quispel, J.A.G. Roberts and C. J. Thompson, Integrable mappings and soliton equations II, *Physica D* **34** (1989) 183-192.

[28] J. A. G. Roberts, Escaping orbits in trace maps, *Physica A* **228** (1996) 295-325.

[29] J. A. G. Roberts and M. Baake, Trace maps as 3D reversible dynamical systems with an invariant, *J. Stat. Phys.* **74** (1994) 829-888.

[30] Roberts J A G and Baake M 1994, The dynamics of trace maps, in: *Hamiltonian Mechanics: Integrability and Chaotic Behaviour*, ed. J. Seimenis, NATO ASI Series B: Physics (Plenum, New York)

[31] J. A. G. Roberts and H. W. Capel, Area preserving mappings that are not reversible, *Phys. Lett. A* **162** (1992) 243-248.

[32] J. A. G. Roberts and H. W. Capel, Irreversibility in conservative dynamics, to be submitted (1996).

[33] J. A. G. Roberts and J. S. W. Lamb, Self-similarity of period-doubling branching in 3-D reversible mappings, *Physica D* **82** (1995) 317-332.

[34] J. A. G. Roberts and G. R. W. Quispel, Chaos and time-reversal symmetry - order and chaos in reversible dynamical systems, *Phys.Rep.* **216** (1992) 63-177.

[35] R. G. Sachs, *The Physics of Time Reversal*, (Univ. of Chicago Press, Boston, 1987)

[36] Sevryuk M B 1986 *Reversible systems*, Lecture Notes in Mathematics vol. 1211 (Berlin: Springer)

[37] Sevryuk M B 1989 *Funkt. Anal. Prilozh.* **23** 116.

[38] Turner G S and Quispel G R W 1994 *J. Phys. A* **27** 757

[39] Vanderbauwhede A 1990 *SIAM J.Math.Anal.* **21**(4) 954

[40] Vanderbauwhede A 1992 *Geometry and Analysis in Nonlinear Dynamics*, eds. H.Broer and F.Takens (Longman, Harlow) p 97

[41] Yan J G G 1992 *Int. J. Bif. Chaos* **2** 285

Commentary by A. Mazer

John Roberts' paper presents a survey of what is currently known about systems with time reverse symetry. The field is of interest to those in dynamical systems and the paper is a nice introduction to the topic. A sufficeint bibliography is included for those interested in further exploring the field.

As an expert in systems with symetries, John Roberts' views on how symetries might be exploited for control design would be of interest to the controls community. Some comments on this issue as well as references in the bibliography would be welcome. The paper would also be more related to others in the volume.

Commentary by C. Sparrow

This paper, taken together with the many references, gives a fairly comprehensive survey of the current state of knowledge about low-dimensional reversible systems. This theory will develop further and be of increasing importance. In many respects we can think of it as a branch of the study of dynamics with symmetry, though most people working in that field are concerned with symmetries that do not reverse time. It is, however, particularly interesting in the case of reversible symmetry to tease out the relationship with Hamiltonian systems, and this gives it a special flavour.

Control of Chaos by Means of Embedded Unstable Periodic Orbits

Edward Ott and Brian R. Hunt
University of Maryland
College Park, Maryland 20742, USA

Abstract

Large improvements in the performance of a chaotic system can be achieved by use of small controls. This is made possible by the unique properties of chaos. In particular, chaotic attractors typically have an infinite dense set of unstable periodic orbits embedded within them. In this paper we review aspects of controlling chaos by means of feedback stabilizing a chosen embedded unstable periodic orbit. Topics include stabilization methods, flexibility of the method, implementation in the absence of a system model, and the question of whether the optimally performing embedded unstable periodic orbit is expected to be of low period.

1 Controlling Chaos Using Embedded Unstable Periodic Orbits

It is common for dynamical systems to exhibit chaotic time evolution. The presence of chaos in such situations offers attractive possibilities for control strategies. In particular, the following two attributes of chaos are relevant:

(i) An ergodic chaotic region of state space typically has embedded within it an infinite dense set of unstable periodic orbits, and may also have embedded fixed points.

(ii) Orbits in a chaotic region are exponentially sensitive in that a small perturbation to the orbit typically grows exponentially with time.

In (i) the term, fixed point, refers to continuous time systems (flows), since we henceforth regard fixed points of discrete time systems (maps) as periodic orbits of period one. In regard to attribute (ii), given any embedded unstable periodic orbit or fixed point, the ergodic wandering of a chaotic orbit over its strange attractor will eventually (if we wait long enough) bring it arbitrarily close to the given embedded unstable orbit. The chaotic orbit would then appear to closely follow the given orb it for some length of time, after which, because of the instability of the given embedded orbit, the chaotic orbit will veer off to resume its ergodic wandering over its strange attractor. It should be emphasized that (i) and (ii) above are not independe nt, but can be regarded as two sides of the same "chaos coin". The important point is that the attributes of chaos make possible the goal of control using only *small* perturbations. Thus we might hope to make a relatively large improvement in sys tem performance by use of *small* controls. This is typically not possible if the system behavior is nonchaotic. For example, if the system is executing a stable periodic motion, small controls would typically be expected to make only small perturba tions to the uncontrolled periodic motion.

Two general control problems that might be addressed are the targeting problem and the problem of improving long-term system performance.

In the targeting problem one attempts to use the available small control to rapidly move an orbit from some initial condition to some target region in state space. Since, by attribute (ii), chaotic orbits are exponentially sensitive, this appears to be a possible task, and it has been demonstrated that appropriate controls can be found to accomplish this.

The remainder of this paper will focus on the second control problem (i.e., improving long-term system performance). In this case, attribute (i) is most relevant. In particular, proof of principle experiments in a very wide variety of situations have su ccessfully been carried out using the following strategy (originally proposed in Ref. 1). First, one examines the system dynamics to determine several low period unstable periodic orbits and also fixed points, if there are any, embedded in the chaotic at tractor. Next, one evaluates the system performance that would apply if each of the determined unstable orbits were followed. Since the natural measure on a chaotic attractor can often (e.g., for hyperbolic systems) be expressed as an average over the e mbedded periodic orbits, it is to be expected that some of the determined periodic orbits will yield performance better than the uncontrolled performance, and some will yield worse performance. Choosing the embedded unstable orbit that yields the most im proved performance, one then attempts to feedback stabilize that orbit by applying small controls in the neighborhood of that orbit.

In practice, given any initial condition in the chaotic region, one can wait until the ergodic wandering of the naturally chaotic orbit brings it sufficiently close to the desired unstable periodic orbit. When this occurs, it is possible for small contro ls to bring the orbit to the desired unstable periodic (or steady) motion and to maintain it there. In fact, in the ideal case, this can be done by use of *arbitrarily small* controls. The smaller the controls, however, the smaller is the neighborho od in which the control is capable of stabilization, and the longer one must typically wait for the orbit to wander into the stabilizable region (in this regard see the paper in this volume by Kostelich et al. where targeting is used to greatly reduce the time for the orbit to enter the stabilizable region). Furthermore, nonideal effects, such as imperfect knowledge of the periodic orbit and noise, place lower limits on the smallest control size.

It is interesting to note that if the situation is such that control of the type just described is practical, then there is a high degree of flexibility automatically built into the procedure. The point is that any one of a number of different orbits can be stabilized by the small control. If the system is used for different purposes at different times, then, depending on the use desired, the system could be optimized by switching the temporal programming of the small controls to stabilize different orb its. In contrast, in the absence of chaos, completely separate systems might be required for each use. Thus the presence of chaos can conceivably be a benefit when flexibility is required. Several aspects regarding the overall procedure are discussed below.

1.1 Stabilization Methods

In order to briefly illustrate the possibility and effectiveness of control of periodic orbits, we restrict attention to the case of discrete time systems (maps), although continuous time dynamical systems (e.g., ordinary differential equations) are also of great interest. Consider the system,

$$\mathbf{y}_{n+1} = \mathbf{G}(\mathbf{y}_n, c_n), \tag{1}$$

where $\mathbf{y}$ is the d-dimensional system state vector and c_n is a control variable that we assume can be adjusted at each time step n. We only consider the case of a scalar control. We assume that practical considerations constrain c_n to satisfy $|c_n| < c_{\max}$, where $c_{\max}$ is small.

For simplicity, assume that the unstable periodic orbit to be stabilized is a period one orbit of the map. We denote the location of this orbit by $\mathbf{y}_*$, where $\mathbf{y}_* = \mathbf{G}(\mathbf{y}_*, 0)$. Since we are interested in stabilization when $\mathbf{y}_n$ is near $\mathbf{y}_*$, and since c_n is small, we can expand Eq. (1) to linear order in c_n and the deviation of the state from the periodic orbit $\mathbf{Y}_n = \mathbf{y}_n - \mathbf{y}_*$. Equation (1) becomes

$$\mathbf{Y}_{n+1} \cong \mathbf{A}\mathbf{Y}_n + \mathbf{B}c_n, \tag{2}$$

where $\mathbf{A}$ is the $d \times d$ matrix of partial derivatives of $\mathbf{G}$ with respect to $\mathbf{y}$ evaluated at $\mathbf{y} = \mathbf{y}_*$, and $\mathbf{B}$ is the d-vector of partial derivatives of $\mathbf{G}$ with respect to c_n. $\mathbf{Y}$ and $\mathbf{B}$ are column vectors.

The question now is how to program the time variation of the small controls c_n in order to drive $\mathbf{Y}$ to zero (i.e., drive $\mathbf{y}$ to $\mathbf{y}_*$). Since we decide the value to which we set c_n after observing the deviation $\mathbf{Y}_n$ from the periodic orbit, we can regard c_n as a function of $\mathbf{Y}_n$. The choice of this function specifies our chosen control scheme. Consistent with the linearity of (4), we assume a linear dependence of c_n on the deviation, $c_n = \mathbf{K}\mathbf{Y}_n$, where $\mathbf{K}$ is a row vector. Choice of $\mathbf{K}$ now specifies the control scheme. Equation (2) thus becomes $\mathbf{Y}_{n+1} = (\mathbf{A} + \mathbf{B}\mathbf{K})\mathbf{Y}_n$. In the absence of control, we would have $\mathbf{Y}_{n+1} = \mathbf{A}\mathbf{Y}_n$. Since the period one orbit, $\mathbf{y} = \mathbf{y}_*$, is unstable in the absence of control, the matrix $\mathbf{A}$ has at least one eigenvalue whose magnitude is greater than one. Our problem is solved (i.e., stabilization is achieved) if we can find a $\mathbf{K}$ such that the eigenvalues of matrix $(\mathbf{A} + \mathbf{B}\mathbf{K})$ all have magnitudes less than one. One can approach this problem purely empirically: Simply choose a $\mathbf{K}$, and see if it works in the experiment; if it does not, try again. [One could do this without knowledge of $\mathbf{A}$ and $\mathbf{B}$ (only knowledge of $\mathbf{x}_*$ is necessary).] If one knows $\mathbf{A}$ and $\mathbf{B}$ and wants to be more sophisticated, we note that the basic problem of choosing $\mathbf{K}$ is standard in control theory. Note that the method is rather insensitive to error, since if $\mathbf{K}$ makes all the eigenvalues have magnitude sufficiently less than unity, then small changes in $\mathbf{K}$ will not alter this.

To illustrate the stabilization of the period one orbit $\mathbf{y} = \mathbf{y}_*$ in a more geometrical manner, consider the case of a two dimensional map, and assume that $\mathbf{A}$ has one eigenvalue larger than one (the "unstable eigenvalue") and one eigenvalue of magnitude less than one (the "stable eigenvalue"). Correspondingly, as shown in Fig. 1(a), the point $\mathbf{y}_*$ has a *stable manifold* and an *unstable manifold* such that points on the stable manifold approach $\mathbf{y}_*$ as time increases, $n \to +\infty$, while points on the unstable manifold diverge from $\mathbf{y}_*$ (approach $\mathbf{y}_*$ going backwards in time, $n \to -\infty$). The location of an orbit point $\mathbf{y}_n$ at time n is shown in the figure. Without control this point would go to some other point, denoted $\tilde{\mathbf{y}}_{n+1}$ in the figure. If we choose c_n so that $\mathbf{B}c_n$ displaces the orbit to the stable manifold (the point denoted $\mathbf{y}_{n+1}$ in the figure), then the control can subsequently be set to zero, and the orbit will naturally approach $\mathbf{y}_*$. (This works so long as $\mathbf{B}$ is not precisely parallel to the stable manifold.) The analogous method[2] for a period N orbit is illustrated in Fig. 1(b), where th e periodic orbit is $\mathbf{y}_{*1} \to \mathbf{y}_{*2} \to \ldots \to \mathbf{y}_{*p} \to \mathbf{y}_{*p+1} \to \ldots \to \mathbf{y}_{*N} \to \mathbf{y}_{*1}$. Through each of the points of the periodic orbit there are stable and unstable manifold segments. The linearized map at $\mathbf{y}_{*p}$ is $\mathbf{y}_{n+1} - \mathbf{y}_{*p+1} = \mathbf{A}_p(\mathbf{y}_n - \mathbf{y}_{*p}) + \mathbf{B}_p c_n$, and we choose c_n so that the displacement $\mathbf{B}_p c_n$ moves the uncontrolled iter ate $\tilde{\mathbf{y}}_{n+1}$ onto the stable manifold of $\mathbf{y}_{*p+1}$.

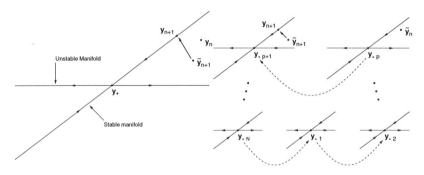

Figure 1: Control puts $\mathbf{y}_{n+1}$ on the stable manifold of (a) $\mathbf{y}_*$ and (b) $\mathbf{y}_{*p+1}$.

We note that the particular stabilization algorithm shown in Fig. 1 is "best" in the sense that, on average, it leads to the shortest transient time before the orbit enters the stabilizable region[1].

1.2 Implementation in the Absence of a System Model

One of the reasons that experimentalists have been able to implement chaos control by means of periodic orbits in such a large number of systems (see review papers in Ref. 3) is that to do this it is not necessary to have a mathematical model of the syste m. One can proceed only using *data* taken from measurement of the system dynamics. In particular, it is possible to use data to "learn" unstable periodic orbits, their stability, and their dependence on the control. For example, in the case of a period one orbit, one would like to find the period one point, $\mathbf{y}_*$ in Fig. 1(a), and possibly the matrix $\mathbf{A}$ and the column vector $\mathbf{B}$ in Eq. (1). Note that such information is local to the region around the periodic orbit; we are no t required to learn the dynamics on the entire attractor (e.g., as one would have to do in order to achieve the harder goals of prediction and targeting).

For example, in the case of period one periodic orbits, one can proceed as follows. Set the control c_n to zero. Accumulate a long orbit string $\{\mathbf{y}_i\}$. Search it for a successive close return pair such that $|\mathbf{y}_j - \mathbf{y}_{j+1}|$ is sm all. Once the first such pair is located, look for more successive close return pairs in the same small region of state space where the first close return occurred. Then, using all these close return pairs, do a least squares fit of the pairs data to de termine a matrix $\tilde{\mathbf{A}}$ and a vector $\mathbf{E}$ satisfying $\mathbf{y}_{j+1} = \tilde{\mathbf{A}}\mathbf{y}_j + \mathbf{E}$. $\tilde{\mathbf{A}}$ is an approximation to the matrix $\mathbf{A}$ in Eq. (1); and $\mathbf{y}_*$ is approximated by $(\mathbf{1} - \tilde{\mathbf{A}})^{-1}\mathbf{E}$. Now setting the small control to some constant value, and repeating the above, one can determine an approximation to the vector $\mathbf{B}$ in Eq. (1).

2 Optimal Periodic Orbits

Imagine that we have determined several low periodic orbits embedded in the strange attractor, e.g., all periodic orbits with period $p \leq 4$. We then examine these orbits to find which of them yields the best performance. A natural question that migh t be asked is whether one can obtain very much better performance by determining and examining many higher period periodic orbits, or by considering stabilization of atypical *nonperiodic* orbits embedded in the strange attractor. We have investigate d this question in a recent paper[4]. The purpose of this section is to present some of our conclusions and to illustrate them with a very simple example [other cases (e.g., the Hénon map) are considered in Ref. 4 and conform to the general conclusion s which we report here]. To begin, we note that long-term performance can often be given as the time average of some function of the system state $F(\mathbf{y})$,

$$\langle F \rangle = \lim_{n \to \infty} \frac{1}{n} \sum_{m=1}^{n} F(\mathbf{y}_m). \tag{3}$$

Thus we seek orbits $\{\mathbf{y}_m\}$ that give values for $\langle F \rangle$ substantially above that for the uncontrolled chaotic orbit.

Our conclusion in Ref. 4 is that the optimal average is typically achieved by a low period unstable periodic orbit embedded in the strange attractor. We now illustrate the sense in which this statement applies by consideration of a simple example.

For our dynamical system we take the doubling map

$$y_{t+1} = 2y_t \text{ modulo} 1, \tag{4}$$

and for F we take

$$F_\gamma(y) = \cos[2\pi(y - \gamma)], \tag{5}$$

where γ is a parameter, $0 \leq \gamma \leq 1$. For each of 10^5 evenly spaced values of γ, we tested the value of $\langle F_\gamma \rangle$ for all periodic orbits of the map (4) with periods 1 to 24. There are on the order of 10^6 such orbits. Figure 2 shows the period of the orbit that maximizes $\langle F_\gamma \rangle$ for Eqs. (4) and (5) as a function of the phase angle γ. The third column of Table I gives the fraction $f(p)$ of phase values γ for which a period p orbit maximizes $\langle F_\gamma \rangle$. For example, if γ is chosen at random in $[0,1]$, then over 92% of the time, the optimal periodic orbit does not exceed 7 in period, and more than half the time the optimal orbit's p eriod is 1, 2, or 3. Though our numerical experiment does not test orbits of period 25 or higher, or nonperiodic orbits, we feel that if these orbits maximized $\langle F_\gamma \rangle$ for a significant proportion of phase values γ, th is would be reflected in our experiment by a large incidence of high period optimal orbits. Further analysis of the data allows us to argue that the set of γ for which a nonperiodic orbit optimizes $\langle F_\gamma \rangle$ is nonempty but has Lebesgue measure zero.

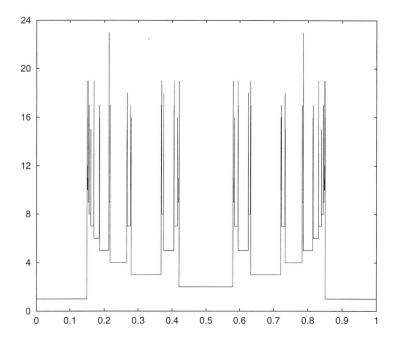

Figure 2: Period that optimizes $\langle F_\gamma \rangle$ as a function of γ.

The second column in Table I gives a conjectured approximate asymptotic prediction of the fraction $f(p)$ of the time a period p orbit maximizes $\langle F_\gamma \rangle$ if γ is chosen at random in $[0, 1]$,

$$f(p) = Kp2^{-p}\phi(p). \qquad (6)$$

Here $\phi(p)$ is the Euler function, which is defined as the number of integers between 1 and p (inclusive) that are relatively prime to p (e.g., the numbers 1, 5, 7, and 11 are relatively prime to 12, and so $\phi(12) = 4$). Thus $\phi(p) \leq p - 1$ for $p \geq 2$, and $\phi(p) = p - 1$ if p is a prime. The factor K is a fitting parameter, which we choose to be $1/6$. We see from Table I that Eq. (6) agrees very well with the numerical results; the agreement is better than 5% for $p > 5$. Note that Eq. (6) apparently has nothing to do with the precise choice of the function F_γ in Eq. (5). We believe that the result (6) is often a good approximation for smooth functions with a single maximum. Tests using other quadratic maximum, single humped functions (e.g., $F_\gamma(x) = -(x - \gamma)^2$) in place of Eq. (5) confirm this[4].

Not only are low period orbits most often optimal, but, even when a somewhat higher period orbit is optimal, it apparently only leads to a relatively small increase in $\langle F_\gamma \rangle$ as compared to a lower period orbit. This point is emphasized by the fourth column in Table I, which gives the fraction of the γ values such that the lowest period orbit that yields a value of $\langle F_\gamma \rangle$ within 90% of the maximum value has period p. Thus, for this example, if one is willing to settle for 90% of optimal, one *never* has to go above period 5. Also for 84% of the γ values it suffices to consider only period 1, 2, and 3.

In all of the other cases examined in Ref. 4, the same qualitative observations hold. The proportion of phase values γ for which a period p orbit maximizes $\langle F_\gamma \rangle$ decays exponentially with p, while nonperiodic maximizing orbits occur only for a Lebesgue measure zero set of γ. Thus optimal performance is usually attained by a low period orbit, and further we find that near-optimal (specifically, 90% of optimal) performance can always be achieved with a low per iod orbit.

The work of E. Ott was partially supported by the Office of Naval Research and that of B. R. Hunt by the National Science Foundation.

References

1. E. Ott, C. Grebogi and J. A. Yorke, Phys. Rev. Lett. **64**, 1996 (1990).

2. E. Ott, C. Grebogi and J. A. Yorke, in Chaos: Soviet-American Perspectives on Nonlinear Science, edited by D. K. Campbell (American Inst. of Physics, New York, 1990) pp. 153–172.

3. E. Ott and M. Spano, Physics Today (May, 1995) p. 34. G. Chen and X. Dong, Int. J. Bif. and Chaos **3**, 1363 (1993); E. Hunt, IEEE Spectrum (1993); T. Shinbrot, C. Grebogi, E. Ott and J. A. Yorke, Nature **363**, 411 (1993).

4. B. R. Hunt and E. Ott, "Optimal Periodic Orbits" (to be published).

Table I. Numerical Results

p	theory f(p)	measured f(p)	$f_{90\%}(p)$
1	0.0833	0.299	0.333
2	0.0833	0.160	0.212
3	0.125	0.176	0.294
4	0.0833	0.0985	0.143
5	0.104	0.116	0.0180
6	0.0313	0.0310	0
7	0.0547	0.0573	0
8	0.0208	0.0211	0
9	0.0176	0.0178	0
10	0.00651	0.00644	0
11	0.00895	0.00918	0
12	0.00195	0.00196	0
13	0.00317	0.00324	0
14	0.00085	0.00084	0
15	0.00061	0.00062	0
16–24	0.00091	0.00092	0

Commentary by K. Glass

The first section of this paper gives a review of a type of control used for chaotic systems. This method of control makes use of the underlying chaotic properties of the system and leads to greater flexibility in control than is available when controlling non-chaotic systems.

A main control problem faced is that of improving the system performance. Unlike the targeting problem (which is discussed in other papers in this volume), it is possible to solve this problem without knowledge of the dynamics over the entire system. Methods for achieving this control are discussed in the first section of the paper.

The second half of the paper examines the achieved performance when orbits of different periods are stabilized. For the performance criterion examined in the paper, the best performance is nearly always achieved by low-period orbits. This is significant as these orbits are, in general the easiest to stabilize. The result indicates that when attempting to improve system performance, it is sufficient to consider low period orbits only.

Notch Filter Feedback Control for k-Period Motion in a Chaotic System

Walter J. Grantham and Amit M. Athalye
School of Mechanical and Materials Engineering
Washington State University
Pullman, WA 99164-2920, USA

Abstract

Chaotic motion can sometimes be desirable or undesirable, and hence control over such a phenomenon has become a topic of considerable interest. Currently available methods involve making systematic time-varying small perturbations in the system parameters. A new method is presented here to achieve control over chaotic motion using notch filter output feedback control. The notch filter controller uses an active negative feedback with fixed controller parameters without affecting the original system parameters. The motivation for using a notch filter in the feedback is to disturb the balance of power at the lower end of the participating frequencies in the power spectrum. This results in a truncation of the period-doubling route to chaos. For low-period motions the harmonic balance method is used to show that a single participating frequency can indeed be eliminated. To deal with relatively complex nonlinear plants, and higher-period motions, a numerical optimal parameter selection scheme is presented to choose the notch filter parameters. The procedures are tested on Duffing's oscillator with a notch filter feedback to achieve desired k-period motion.

1 Introduction

Chaotic behavior in dynamical systems has been observed in a variety of applications in many areas, such as biology, chemistry, ecology, economics, engineering, and physics. Chaotic motion may be desirable in some cases, such as for mixing chemicals effectively. But it is usually undesirable. For example, it can lead to early fatigue failure in some structures due to unevenness of the stress variation with time. Due to the existence of chaotic motion in a variety of systems, it has become important to understand chaotic behaviors in complex nonlinear systems and eventually to control such behaviors.

A recent paper [1] by Ott, Grebogi, and York presented a method to change chaotic motion into a desired attracting periodic motion. The method is based on the observation that a chaotic attractor has embedded within it an infinite number of unstable periodic orbits. In [1] and some subsequent follow-up papers [2, 3], the denseness of such unstable periodic orbits was used to make small time-dependent perturbations in an accessible system parameter and thereby achieve a desired periodic output. This algorithm is commonly known as the "OGY algorithm". It has been extended to higher-dimensional systems and a linear pole placement-type algorithm has been used to achieve feedback stabilization [3] of the periodic orbits. The procedure has been tested in a number of practical applications [4, 5, 6, 7].

In this paper, we consider a method [8] for the control of chaotic systems not through a time-varying system parameter, but by adding a feedback control mechanism consisting of a linear device called a notch filter. The original system parameters are fixed, at their chaotic values. The feedback controller's parameters are also constants, but they can be chosen to change chaotic motion into a desired periodic motion.

The motivation for using a notch filter is to suppress a selected frequency from the power spectrum of the system. Consider a system exhibiting a period-doubling route to chaos as some system parameter is varied. In a typical periodic motion with period T and frequency $\omega = 2\pi/T$, the output of the system will contain higher harmonics, at frequencies 2ω, 3ω, etc. As period-doubling occurs, the output will begin to contain subharmonic terms, at frequencies less than ω. The appearance of such subharmonics is often a precursor to chaotic motion. We use a notch filter controller to suppress the development of subharmonics and in turn avoid chaos. Furthermore, by suitable choice of the notch filter parameters, the period-doubling can be stopped so as to yield a desired k-period motion.

As an example system we will consider the Duffing oscillator

$$\ddot{y} + \delta\dot{y} - \frac{1}{2}y + \frac{1}{2}y^3 = u(t) \tag{1}$$

with $u(t) = F\cos\Omega t$, $\Omega = 1$ rad/sec, $\delta = 0.168$, and $F = 0.21$. For these parameter values the system is chaotic [9]. Figure 1 shows a single-period-sampled Poincaré map for this system as F is varied. Figure 2 shows power spectra [8] at various forcing amplitudes: (a) $F = 0.15$, (b) $F = 0.178$, (c) $F = 0.198$, (d) $F = 0.21$. We will be concerned with converting the chaotic motion in case (d) into a specified k-period motion, but without changing any of the system parameters. We use a notch filter feedback controller to suppress one of the subharmonic frequencies Ω/k in the output feedback of the controlled system.

2 Notch Filter Controller

In general we will be concerned with nonlinear single-input single-output systems of the form

$$\dot{\mathbf{w}} = \mathbf{F}(\mathbf{w}, u) \tag{2}$$
$$y = g(\mathbf{w}, u) \tag{3}$$

where $(\dot{\ }) = d(\ )/dt$, $\mathbf{w}(t) \in R^n$ is the state, $u(t) \in R$ is a control input, and $y(t) \in R$ is the output. We assume that $\mathbf{F}(\cdot) : R^n \times R \to R^n$ is C^1 and dissipative, i.e., $\lim(\mathbf{F}) < 0$. We will be especially concerned with systems that exhibit a period-doubling route to chaotic motion under sinusoidal inputs. For a comparison with results based on the harmonic balance method, we will be particularly concerned with systems of the form

$$\dot{\mathbf{w}} = \mathbf{A}\mathbf{w} - \mathbf{b}\psi(y) + \mathbf{b}u(t) \tag{4}$$
$$y = \mathbf{c}^T\mathbf{w} \tag{5}$$

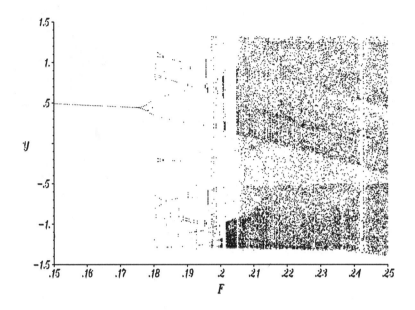

Figure 1: Bifurcation diagram for the Duffing oscillator.

which are linear except for an output nonlinearity $\psi(y)$. It is assumed that the linear part of (4) and (5) constitutes a controllable and observable system.

Suppose that, for a certain initial state $\mathbf{w}(0) = \mathbf{w}_0$ and for a specified sinusoidal reference input $u(t) = r(t) = F\cos\Omega t$ with period $T = 2\pi/\Omega$, the motion $\mathbf{w}(t)$ asymptotically approaches a chaotic strange attractor. We are interested in designing an output feedback control

$$u = r(t) - h(y) \tag{6}$$

so that the resulting output $y(t)$ ceases to be chaotic and becomes periodic with period kT, where k is a positive integer specified by the control system designer. To achieve this, we investigate the use of a notch filter feedback controller. Figure 3 shows a block diagram for notch filter feedback control applied to the Duffing system.

2.1 Notch Filter

A notch filter is a linear device that filters out a particular frequency from an input signal. If the input to the notch filter is denoted by $y(t)$ and the output is denoted by $h(t)$, then the transfer function of a typical notch filter can be written as

$$G(s) = \frac{H(s)}{Y(s)} = K\frac{s^2 + \omega_n^2}{s^2 + 2\zeta\omega_n s + \omega_n^2}, \tag{7}$$

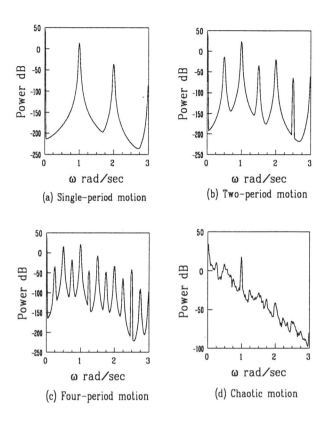

Figure 2: Power spectra for the Duffing oscillator.

where ω_n is the notch frequency, ζ is the notch filter damping coefficient, K is the notch filter gain, and s is the Laplace variable. The Bode plot of the magnitude of $G(i\omega)/K$ is shown in Figure 4. As the name suggests, a "notch" is observed at the notch frequency ω_n in the Bode plot. Away from the notch frequency signals pass through the notch filter essentially unaltered.

In a state space representation, the notch filter equations can be written as

$$\dot{\mathbf{z}} = \mathbf{P}\mathbf{z} + \mathbf{q}\, y \tag{8}$$

with output

$$h = \mathbf{d}\mathbf{z} + K\, y, \tag{9}$$

where

$$\mathbf{P} = \begin{bmatrix} 0 & 1 \\ -\omega_n^2 & -2\zeta\omega_n \end{bmatrix},$$

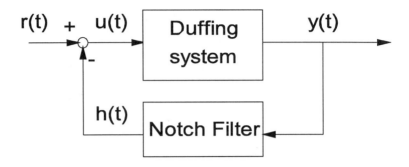

Figure 3: Notch filter feedback control system.

$$\mathbf{q} = \begin{bmatrix} 0 \\ 1 \end{bmatrix}, \quad \mathbf{d} = \begin{bmatrix} 0 & -2\zeta\omega_n K \end{bmatrix}. \tag{10}$$

Defining $\mathbf{x} = (\mathbf{w}, \mathbf{z}) \in R^{n+2}$ and specifying the input as $r(t) = F \cos \Omega t$, the controlled system (4)–(9) is of the form

$$\dot{\mathbf{x}} = \mathbf{f}(\mathbf{x}, \nu, t), \tag{11}$$

where $\nu = [K \ \zeta \ \omega_n]^T$ denotes the parameters of the notch filter.

2.2 Rationale for Notch Filter Feedback

The periodic response of a harmonically excited nonlinear oscillator has a period kT which is an integer multiple $k \geq 1$ of the period $T = 2\pi/\Omega$ of excitation [10]. Consequently, this response can be represented by a Fourier series

$$y(t) = a_0 + \sum_{j=1}^{\infty} \left\{ a_j \cos(j\frac{\Omega}{k}t) + b_j \sin(j\frac{\Omega}{k}t) \right\} \tag{12}$$

consisting of sinusoidal terms at certain frequencies $\omega_j = \frac{j}{k}\Omega$ that are rational fractions or integer multiples of the forcing frequency. The forcing frequency, Ω, is called the primary frequency. The superharmonic frequencies are the frequencies $m\Omega$ and the subharmonic frequencies are Ω/m, $m = 2, 3, \ldots$. Theoretically, these are the only possible frequencies in a periodic response and the power spectrum will consist of discrete peaks. Practically, due to numerical sampling and leakage, the peaks get broadened, as illustrated in Figures 2a–c.

For a nonperiodic response, for example a chaotic response, the power spectrum becomes broad-band and a continuum of frequencies appear in addition to the primary frequency and the super- and subharmonics. For white noise, the power would

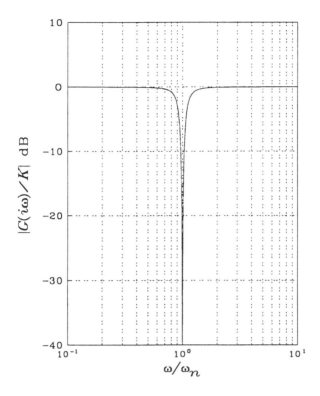

Figure 4: Notch filter Bode magnitude ($\zeta = 0.1$).

be distributed equally at all frequencies, yielding a constant power for all frequencies. In contrast to white noise, the spectrum of a chaotic motion rolls off at higher frequencies, indicating a filtering effect of a dissipative system. However, the power spectrum for chaotic signals retains dominance of certain frequencies. In Figure 2d the dominance of the primary response frequency ($\Omega = 1$ rad/sec) can be clearly seen. Also note the dominance of subharmonics of the primary frequency. The plot indicates that much of the power is indeed concentrated at these subharmonic frequencies, although the powers at lower subharmonics are interacting to generate powers at frequencies other than the subharmonic frequencies. This interaction produces frequencies in between the subharmonic frequencies, making the power spectrum look almost continuous.

For many systems the presence of the subharmonics can be used as a precursor for chaotic motion. For example, the period-doubling route to chaos is characterized by successive formation of subharmonics of periods $2^m T$ and frequency $\Omega/2^m$ as a system parameter varies. Suppression of these subharmonics can disturb the distribution of the power, caused by the interaction with other subharmonics, and can be used to halt the period-doubling sequence that leads to chaotic motion. A notch filter is an ideal candidate for suppressing such subharmonics dynamically, due to its single-frequency filtering characteristic.

A common control systems approach, applicable to eliminating chaos, is to make the system excessively stable, for example by using a negative feedback of the output in (4)-(6), $h(y) = Ky$, with sufficiently large gain K. In reality, such feedback gains are limited by the physical properties of the system or might be too expensive to achieve. Furthermore, such an approach generally will not yield the desired k-period motion, depending on the nonlinearity. Instead, we will show that a notch filter with a relatively small gain can be used in a feedback loop to achieve the goal of filtering a frequency dynamically and producing the desired periodic motion.

3 Harmonic Balance Method

Harmonic balance is a method of finding steady-state periodic solutions of a system using a truncated Fourier series solution. It has also been used to predict the onset of chaos in autonomous systems [11]. Here we will use the method to develop analytic estimates for the notch filter parameters needed to achieve a desired low-period motion.

The state-space system represented by (4)-(9) can be transformed into an $m = n + 2$ order ordinary differential equation, of the form

$$y^{(m)} + c_{m-1} y^{(m-1)} + \ldots + c_0 y + g(y, \dot{y}, \ldots) = r(t), \tag{13}$$

where $y^{(m)} = d^m y/dt^m$ and $g(\cdot)$ contains the nonlinearities of the system. The coefficients c_i are constants and $r(t)$ is the excitation, at frequency Ω.

We assume that the controlled system is exhibiting motion with at least a period of $2\pi k/\Omega$. The most general Nth-order approximation for the k-periodic output can be written as

$$y^*(t) = a_0 + \sum_{j=1}^{N} \left\{ a_j \cos(j\frac{\Omega}{k}t) + b_j \sin(j\frac{\Omega}{k}t) \right\}. \tag{14}$$

Substituting the approximated output (14) into (13) yields a result of the form

$$r(t) = A_0 + \sum_{l=1}^{N} \left\{ A_l \cos(l\frac{\Omega}{k}t) + B_l \sin(l\frac{\Omega}{k}t) \right\}$$

$$+ \text{higher harmonics}. \tag{15}$$

For a periodic input $r(t)$ at the primary frequency Ω, let $r(t) = \alpha_0 + \alpha \cos \Omega t + \beta \sin \Omega t$. Balancing the constants and the first N harmonic terms on both sides of (15), there will be $2N + 1$ algebraic equations with $2N + 4$ unknowns:

$$\begin{aligned}
A_0(a_0, a_1, \ldots, a_N, b_1, \ldots, b_N, K) &= \alpha_0 \\
A_k(a_0, a_1, \ldots, a_N, b_1, \ldots, b_N, K, \zeta, \omega_n) &= \alpha \\
B_k(a_0, a_1, \ldots, a_N, b_1, \ldots, b_N, K, \zeta, \omega_n) &= \beta \\
A_l(a_0, a_1, \ldots, a_N, b_1, \ldots, b_N, K, \zeta, \omega_n) &= 0 \\
B_l(a_0, a_1, \ldots, a_N, b_1, \ldots, b_N, K, \zeta, \omega_n) &= 0
\end{aligned} \tag{16}$$

for $l = 1, \ldots, N$, $l \neq k$.

The describing function method corresponds to $N = 1$ and is frequently used in nonlinear control systems analysis. However, our analysis will require the more general harmonic balance method, since we need to approximate not only the primary response, but also the first k subharmonic responses.

4 Parameter Optimization Method

The harmonic balance method applies to a restricted class of systems and is only practical for low-period motions because of the algebra involved. Furthermore, the truncation errors could yield misleading results. In this section we present an alternate approach based on direct optimization of the notch filter parameters.

Consider the controlled system $\dot{\mathbf{x}} = \mathbf{f}(\mathbf{x}, \nu, t)$, with $\mathbf{f}(\mathbf{x}, \nu, t) = \mathbf{f}(\mathbf{x}, \nu, t + T)$. Let $\mathbf{x}(t) = \mathbf{x}(t; \mathbf{x}_0, \nu)$ denote the solution from $\mathbf{x}(0) = \mathbf{x}_0$. We seek parameter values ν so that $\mathbf{x}(t)$ asymptotically converges to a periodic solution with period kT, where $k \geq 1$ is a specified integer. Numerically, we require that k be the smallest integer for which $||\mathbf{x}(t_0 + kT) - \mathbf{x}(t_0)|| \leq \epsilon$, where ϵ is the machine accuracy, t_0 is the transient time of the system, chosen large enough for transients to die out, and $|| \cdot ||$ denotes the Euclidean norm.

To create k-period motion we will choose optimal parameters ν to minimize a "cost" function $G(\nu)$ corresponding to deviation from k-period motion. For $t_1 = t_0 + kT$ define a terminal time τ and a distance ρ as: $\tau = t$, $\rho = ||\mathbf{x}(t) - \mathbf{x}(t_0)||$ if $\rho \leq \epsilon$ for some $t \in (t_0, t_1)$, otherwise, $\tau = t_1$, $\rho = ||\mathbf{x}(t_1) - \mathbf{x}(t_0)||$. Define a cost function at time τ as

$$G(\nu) = (\tau - [t_0 + kT])^2 + \rho^2. \tag{17}$$

5 Control of Duffing's Oscillator

Notch filter control of the Duffing oscillator (1) yields the following state-space equations for the overall control system:

$$\begin{aligned} \dot{x}_1 &= x_2 \\ \dot{x}_2 &= -\delta x_2 + \tfrac{1}{2}(x_1 - x_1^3) + r(t) - h \\ \dot{x}_3 &= x_4 \\ \dot{x}_4 &= -\omega_n^2 x_3 - 2\zeta \omega_n x_4 + x_1, \end{aligned} \tag{18}$$

where the output of the notch filter is given by

$$h = K x_1 - 2\zeta \omega_n K x_4, \tag{19}$$

the system output is $y = x_1$, and the reference input is $r(t) = F \cos \Omega t$. The parameters $\delta = 0.168$, $F = 0.21$, and $\Omega = 1$ and initial conditions $\mathbf{x}(0) = \mathbf{0}$ are fixed. The notch filter controller parameters K, ζ, and ω_n are also constants, but as yet undetermined.

The unforced system ($F = 0$) described by (18)-(19) has three fixed points: one at the origin and the other two at $\bar{\mathbf{x}} = [\pm\sqrt{1 - 2K}, 0, \pm\sqrt{1 - 2K}/\omega_n^2, 0]^T$. The origin is the only fixed point for $K \geq 0.5$. The two-well potential energy structure of the original system is retained even with notch filter feedback for $K < 0.5$. Detailed results concerning the effects of the notch filter parameters can be found in [8], along with details of the harmonic balance method and the numerical simulations. In most of the simulations, the Lyapunov exponents were also employed to compute the Lyapunov dimension, using the Kaplan-Yorke conjecture, in order to differentiate between chaotic and nonchaotic motion.

5.1 Results of Harmonic Balance

An underlying premise of this study is the idea that one way to attempt a transition from chaotic motion to k-period motion ought to be to tune the notch filter frequency ω_n near the $(k-1)$th subharmonic frequency Ω/k. Because the precise notch frequency required will also depend on the other filter parameters K and ζ and on the system nonlinearities, the harmonic balance method is used to analytically seek k-period motion.

Although the harmonic balance is carried out for general K, only the value $K = 1.0$ is considered for the results presented in this section. For this value of K and for the other two notch filter parameters ω_n and ζ in the range $[0, 1]$, limit cycles of the form (14) are sought by solving the algebraic equations (16) numerically. After the coefficients $a_0, \ldots, a_N$ and $b_1, \ldots, b_N$ are determined, the stability of the approximate solutions is analyzed by considering perturbations about the approximate solutions and using Floquet's theory [8].

The results of the harmonic balance method are presented in Figure 5, which shows predicted one-, two- and three-period motions for $K = 1$. As expected with a forcing frequency $\Omega = 1$ rad/sec, the predicted two-period motion occurs near the notch frequency corresponding to the first subharmonic frequency of 0.5 rad/sec, especially at lower notch filter damping values. Similarly, the predicted three-period motion region occurs near $\omega_n = 1/3$ rad/sec. Single-period motion occurs in a region roughly of the form $\zeta \omega_n < c \approx 0.08$, as well as for small but slightly larger ζ values near $\omega_n = 1$. Some isolated holes appear in the three-period results because the Floquet stability tests generated a run-time error in a LINPACK eigenvalue routine.

5.2 Results of Optimal Parameter Selection

For the numerical experiments two values of notch filter gain K are considered. The value $K = 0.1$ is small enough that it does not affect the two-well potential energy structure of the uncontrolled system, whereas the value $K = 1.0$ corresponds to only one equilibrium point for the unforced system.

For a given value of K the $0 \leq \omega_n \leq 1$ range was divided into 100 parts. At each ω_n value the parameter optimization problem was initiated with $\zeta = 0$ to locate up to ten-period motions. When a particular k-period motion was detected neighboring initial ζ values were used in the optimization process in order to determine the ζ range for which the k-period motion occurs.

5.2.1 High Gain

Figure 6 shows a parametric plot in the (ω_n, ζ)-plane for $K = 1$ and motions up to seven-period motion. Close to the first subharmonic, $\omega_n = 0.53$ rad/sec, two-period motion is observed starting approximately at $\zeta = 0.35$. In general, k-period regions tend to occur near the corresponding harmonic frequency Ω/k. The exceptions to this rule are the four- and six-period motions, which were only found in isolated places

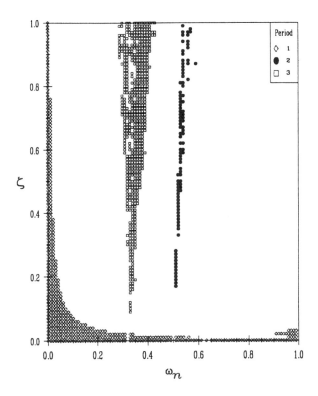

Figure 5: Harmonic balance predictions ($K = 1$).

and along the edge of the one-period region. This result may be due to the cubic nonlinearity in the Duffing system.

For the most part, the white portion of Figure 6 consists of quasiperiodic (or at least greater than ten-period) motions. No chaotic motions were observed, based on Lyapunov exponent and Lyapunov dimension computations [8].

The one-, two-, and three-period optimization results in Figure 6 agree qualitatively with the approximate harmonic balance results in Figure 5. However, harmonic balance consistently over-estimated the corresponding periodic regimes. In particular, the harmonic balance single-period region hides a multitude of k-period motions at the edge of the single-period region in the lower-left corner of Figure 6.

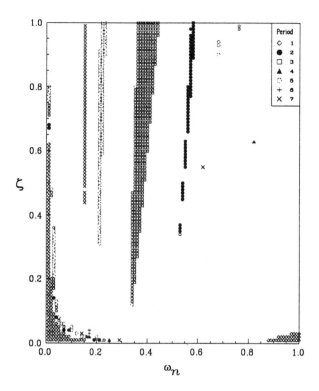

Figure 6: Notch filter periodic regimes ($K = 1$).

5.2.2 Low Gain

Figure 7 shows the parameter optimization results for $K = 0.1$, with up to seven-period motions. In the white portion of the figure, the controller parameter values generally lead to chaotic motion in the controlled plant [8]. Also note that no single-period motions were detected.

6 Discussion

In this section we discuss the original objective of the study, which was to modify the chaotic behavior of a Duffing oscillator. In addition, we revisit the rationale for using notch filter feedback.

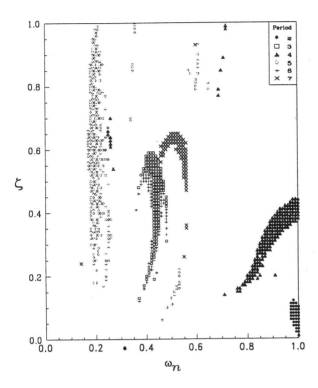

Figure 7: Notch filter periodic regimes ($K = 0.1$).

6.1 Duffing Bifurcation

The original aim of this study was to see what effect notch filter feedback control would have on the bifurcation diagram in Figure 1. In particular, could the period-doubling route to chaos be curtailed? Seeking two-period motion, we tune the notch filter frequency near the first subharmonic $\Omega/2 = 0.5$. Specifically, we take $K = 1$ and, from Figure 6, we select $\omega_n = 0.53$ and $\zeta = 0.36$. Figure 8 shows the resulting single-period-sampled Poincaré map as the amplitude F of the forcing function is varied. This figure shows clearly that not only can chaotic motion at $F = 0.21$ be converted to, in this case, two-period motion, but also that the resulting motion is very robust with respect to variations in F.

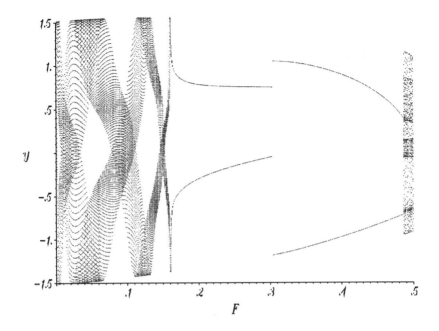

Figure 8: Duffing bifurcation with notch filter control.

6.2 Suppression of a Subharmonic

The use of a notch filter to suppress a frequency has a seemingly strange effect. Suppressing frequency Ω/k from the output feedback produces a k-periodic output, at the frequency Ω/k, that is, we get the frequency that we are filtering out! To understand qualitatively what is happening, we first note that we are suppressing a subharminic frequency from the *feedback* loop, not from the *output*. To see the effect, suppose the notch filter feedback were replaced by a constant feedback with gain $K = 1$. This corresponds, roughly, to $\omega_n = 0$ in (7). Both the harmonic balance results (Figure 5) and the parameter optimization results (Figure 6) indicate single-period motion. In effect this feedback of *all* output frequencies "stabilizes" the system and eliminates the chaos. Now, add in the notch filter, which passes all feedback frequencies *except* Ω/k. All the frequencies that are fedback are "stabilized out" of the output, leaving only the desired output frequency Ω/k.

7 Conclusions

In this paper, a method was presented for control of single-input single-output chaotic systems. The method involved using a notch filter in the feedback loop. The motivation for using this method came from the fact that chaotic motion is accompanied by a range of frequencies that involve the subharmonic frequencies and that there is a pattern of distribution of power in a chaotic power spectrum. It was shown that, by disturbing this distribution by eliminating certain subharmonic frequencies, it is indeed possible to eliminate chaotic motion. In addition, it is possible to achieve a desired k-period motion at the output. The controller parameters for k-period motion were found using both the harmonic balance method and an optimal parameter selection technique. The harmonic balance method is effectively limited to low-period motions and can give erroneous results due to truncation errors. The optimal parameter selection technique is a general numerical method for finding the controller parameters for k-period motion and overcomes the short comings of the harmonic balance method.

References

[1] Ott, E., Grebogi, C., and Yorke, J. A., "Controlling Chaos," *Phys. Rev. Let.*, Vol. 64, No. 11, pp. 1196–1199, 1990.

[2] Ott, E., Grebogi, C., and Yorke, J. A., "Controlling Chaotic Dynamical Systems," *CHAOS/XAOS, Soviet-American Perspectives on Nonlinear Science*, American Institute of Physics, pp. 227–258, 1990.

[3] Romeiras, F.J., Grebogi, C., Ott, E., and Dayawansa W. P., "Controlling Chaotic Dynamical Systems," *Physica D*, Vol. 58, No. 1–4, pp. 165–192, 1992.

[4] Wang, Y., Singer, J., and Bau H. H., "Controlling Chaos in a Thermal Convection Loop," *J. of Fluid Mechanics*, Vol. 237, pp. 479–498, 1992.

[5] Roy, R. T., Murphy Jr., T. W., Maier, T. D., and Gills Z., "Dynamical Control of a Chaotic Laser: Experimental Stabilization of a Globally Coupled System," *Phys. Rev. Let.*, Vol. 68, No. 9, pp. 1259–1262, 1992.

[6] Azevedo, A. and Rezende, S. M., "Controlling Chaos in Spin-Wave Instabilities," *Phys. Rev. Let.*, Vol. 66, No. 10, pp. 1342–1345, 1990.

[7] Hunt, E. R., "Stabilizing High-Period Orbits in a Chaotic System: The Diode Resonator," *Phys. Rev. Let.*, Vol. 67, No. 15, pp. 1953–1955, 1991.

[8] Athalye, A. M., *Notch Filter Control of a Chaotic System*, Ph.D. dissertation, Department of Mechanical and Materials Engineering, Washington State University, Pullman, WA, 1993.

[9] Pezeshki, C., Elgar, S., and Krishna, R. C., "Bispectral Analysis of Systems Possessing Chaotic Motion," *J. of Sound and Vibr.*, Vol. 173, No. 3, pp. 357–368, 1990.

[10] Guckenheimer, J. and Holmes, P., *Nonlinear Oscillations, Dynamical Systems, and Bifurcations of Vector Fields*, Springer-Verlag, New York, 1983.

[11] Genesio, R. and Tesi, A., "Harmonic Balance Methods for Analysis of Chaotic Dynamics in Nonlinear Systems," *Automatica*, Vol. 28, No. 3, pp. 531–548, 1992.

Commentary by A.I.Mees

This is another interesting example of what can be done when control engineering and dynamical systems intuitions are combined. As I'll explain later, it also brings up serendipitous connections with some work that is rather different from what has been mentioned explicitly by the authors.

The rationale for a notch filter feedback is, as I understand it, as follows. If we do the traditional controls trick of putting sufficiently hefty negative feedback around the system, we can probably stabilize it. What would happen if we removed the feedback at one particular frequency? (This is the function of the notch filter.) Intuitively, if the system would have been stabilized by unfiltered feedback then with the filter present it might, in the absence of stabilization at a given frequency, oscillate at a harmonic or a sub-harmonic—or, more generally, at a rational multiple—of the missing frequency.

I don't entirely follow the connection between this and the argument about period-doubling cascades, except insofar as the argument might somehow be developed into an explanation of the resonances in a chaotic or near-chaotic system, but perhaps I am just quibbling. The authors' idea works, and both the harmonic balance and the numerical optimization methods depend only indirectly on the period doubling argument.

It is particularly interesting that Figs 5 and 6 show the presence of Arnold tongues [2] (the stability regions for various driving frequencies). These are well-known to occur in systems that are driven at frequencies near to rational multiples of their resonance frequencies, and the tongues rapidly get both very narrow and very sensitive to small damping or other perturbations as the rational multiples move away from simple fractions like 1/2, 1/3, 2/3 and so on. (This is clear in Fig 6.) A nice physical example, not entirely unconnected with controlling a Duffing oscillator by periodic inputs, is making a child's swing go by standing on it and bobbing up and down; or, equivalently, destabilizing a simple pendulum by periodically changing its length. Arnold [1] calls this *parametric resonance* since at any fixed value of the parameter (the pendulum length) the system is stable, yet the time-varying system is unstable.

There appear to be connections between this work and that of Hubler [3] and others, who also stabilize chaotic systems onto periodic solutions by periodic inputs. One significant difference is that Hubler tends to emphasize control without feedback, since if one can determine the correct excitation frequency it can often be applied as an open-loop control input and the system will phase-lock to it from a wide range of initial states.

References

[1] V. I. Arnold. *Mathematical Methods of Classical Mechanics*, volume 60. Springer, New York, 1978.

[2] E. Ott. *Chaos in Dynamical Systems*. Cambridge University Press, Cambridge, 1993.

[3] A. Hubler, R. Mettin, A. Scheeline, and W. Lauterborn *Parametric Entrainment Control* . Phys. Rev. E. (1995), pp. 4065-4075

Commentary by J. Moore

This research has shown clearly the power of feedback control in a chaotic system to steer the chaotic behaviour. It is perhaps surprising that chaos which is a fundamentally nonlinear system phenomenon is controlled so nicely by linear filters in a feedback loop. In hindsight, of course, a feedback notch filter, which is high gain feedback save in one frequency band, allows the system freedom to resonate (orbit) in this band without modification due to feedback. To converge to orbits rich in harmonic content, it makes sense for the feedback filters to have multiple notches according to the harmonic content. This would be worthy of further investigation.

Targeting and Control of Chaos

Eric J. Kostelich
Department of Mathematics, Box 871804,
Arizona State University, Tempe, Arizona 85287, USA

Ernest Barreto
Department of Physics,
University of Maryland, College Park, Maryland 20742, USA

Abstract

The "control of chaos" refers to a procedure in which a saddle fixed point in a chaotic attractor is stabilized by means of small time dependent perturbations. Control may be switched between different saddle periodic orbits, but it is necessary to wait for the trajectory to enter a small neighborhood of the saddle point before the control algorithm can be applied.

This paper describes an extension of the control idea, called "targeting." By targeting, we mean a process in which a typical initial condition can be steered to a prespecified point on a chaotic attractor using a sequence of small, time dependent changes to a convenient parameter. We show, using a 4-dimensional mapping describing a kicked double rotor, that points on a chaotic attractor with two positive Lyapunov exponents can be steered between typical saddle periodic points extremely rapidly—in as little 12 iterations on the average. Without targeting, typical trajectories require 10,000 or more iterations to reach a small neighborhood of saddle periodic points of interest.

1 Introduction

A chaotic process has sensitive dependence on initial conditions that prevents long-term predictions of the state of the system. Chaotic dynamical processes typically exhibit highly irregular behavior and can be represented mathematically by so-called "strange attractors" whose geometry is very complex. Despite the complexities of chaotic behavior, the sensitive dependence on initial conditions can be exploited to maintain the system about some desired final state (like a saddle periodic orbit embedded in the attractor) by a carefully chosen sequence of small perturbations to a control parameter. This is the basic idea behind the so-called "control of chaos," wherein small perturbations can be used to formulate a feedback stabilization of one of the infinite number of unstable periodic orbits that naturally occur in a chaotic attractor. (See Ref. [1] and the paper by Prof. Ott elsewhere in this volume.) The method relies in part on suitable linear approximations of the stable and unstable manifolds associated with the saddle periodic orbit.

One of the first laboratory experiments to demonstrate the feasibility of this feedback stabilization consisted of a driven, flexible beam whose dynamical behavior was well approximated by a two dimensional map [2]. Although the uncontrolled

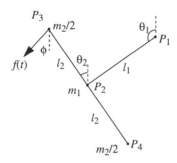

Figure 1: The double rotor.

process was chaotic, the control algorithm maintained the beam about a saddle fixed point that was embedded in the attractor, and only small perturbations were needed to maintain the control. The applicability of the method has since been demonstrated in a variety of laboratory experiments.

A natural extension of this idea is the notion of *targeting*. The problem can be stated simply as follows: Given a typical initial condition on the attractor, determine a sequence of perturbations that directs the resulting trajectory to a small region about some prespecified point on the chaotic attractor as rapidly as possible. Because of the inherent exponential sensitivity of chaotic time evolutions to perturbations, one expects that a suitable alteration of the trajectory can be accomplished using only *small* controlling adjustments of one or more available system parameters.

An initial demonstration of targeting was given by Shinbrot *et al.* [3], who considered some numerical experiments using a two dimensional map. In addition, a targeting procedure was successfully used in a laboratory experiment for which the dynamics were approximately describable by a one dimensional map [4]. In this paper, we describe an alternative approach of the targeting problem that is applicable to systems of higher dimensionality than previously considered.

One potential application of the targeting algorithm is to steer otherwise chaotic trajectories to a neighborhood of a prespecified saddle periodic point. As we will show, the targeting algorithm is particularly effective for systems of moderately high dimension.

We focus attention on the double rotor map [5], which describes the effect of a sequence of impulse kicks on two thin, massless rods connected as illustrated in Fig. 1. This idealized mechanical system exhibits complex dynamical behavior, and its attractor is a subset of $R^2 \times S^2$. For the values of the parameters considered here, the Lyapunov dimension [7] of the attractor is approximately 2.8, and it has two positive Lyapunov exponents.

Romeiras *et al.* [6] have demonstrated a control algorithm that can stabilize some of the saddle periodic points embedded within the double rotor attractor. They showed that control can be achieved by using only one control parameter. However, it often is necessary to wait several thousand iterations before a given trajectory falls in a sufficiently small neighborhood of the desired saddle fixed point so that the required linearizations of the dynamical system are valid.

Figure 2 shows the results of a typical numerical experiment. The plot shows

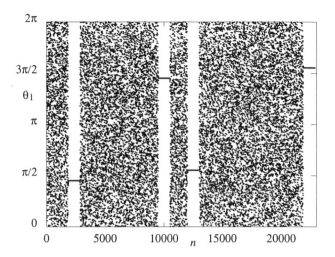

Figure 2: Waiting times for the control of chaos algorithm.

the position θ_1 in radians of one of the rotors as a function of time n. Before the control is applied (for example, during the first 2000 or so iterations of the map), the rotors move chaotically, and the position x_1 assumes values throughout the interval $(0, 2\pi)$. Eventually, one of the iterates lands sufficiently close to a prespecified period-1 saddle point A that the control algorithm can be applied. As long as the control is on, the iterates remain in a small neighborhood of A (for instance, for n approximately between 2000 and 3000). To switch to a different period-1 saddle fixed point B, the control is turned off, allowing the orbit to move away from A and resume its chaotic motion. Eventually, the orbit enters a suitably small neighborhood of B, the control is turned on, the point remains near B until the control is turned off, etc.

This example illustrates an application wherein one wants to switch between different periodic saddle orbits. A difficulty arises in that the waiting times can be quite long. (Although they are not shown in the figure, waiting times of 150,000 or more iterates are not uncommon in this numerical experiment.) In general, one expects that the average waiting time before a typical orbit approaches a given saddle periodic point is proportional to the dimension of the attractor. In this case, the dimension is approximately 2.8, so the average distance between nearest neighbors in a subset of N points on the attractor scales as $N^{-1/2.8}$ [8]. In other words, if the orbit must fall within 10^{-2} of the saddle point for the control algorithm to work, then the waiting time is on the order of $10^{5.6}$ iterations. The observed waiting times in the numerical experiment are consistent with this rough estimate.

The targeting problem has a natural application here, because it can reduce the waiting time by orders of magnitude. In this example, the objective is to choose target points that lie in small neighborhoods of the saddle periodic points of interest. In order to switch between periodic points, one applies the targeting algorithm to steer the trajectory to a small neighborhood of one of the periodic points, then turns on the control algorithm to maintain the orbit near the point for as long as desired.

After the control is turned off, the targeting algorithm can be applied to steer the trajectory to a neighborhood of another prespecified periodic point, and so on.

The targeting problem also is applicable when noise in the system causes occasional loss of control, even when the orbit initially is near the desired fixed point. In such cases, one wants to return to the periodic point as quickly as possible.

2 The double rotor map

In this section, we outline the basic ideas behind the targeting procedure. We begin first with a brief outline of the double rotor map. A derivation can be found in [5]; a slightly different version of the map (that is used here) is described in [6].

The first rod, of length ℓ_1, pivots about P_1 (which is fixed), and the second rod, of length $2\ell_2$, pivots about P_2 (which moves). The angles $\theta_1(t)$, $\theta_2(t)$ measure the position of the two rods at time t. A point mass m_1 is attached at P_2, and point masses $m_2/2$ are attached to each end of the second rod (at P_3 and P_4). Friction at P_1 (with coefficient ν_1) slows the first rod at a rate proportional to its angular velocity $\dot{\theta}_1(t)$; friction at P_2 slows the second rod (and simultaneously accelerates the first rod) at a rate proportional to $\dot{\theta}_2(t) - \dot{\theta}_1(t)$. The end of the second rod marked P_3 receives impulse kicks at times $t = T, 2T, \ldots$, always from the same direction and with strength ρ. Gravity and air resistance are absent.

The double rotor map is the four dimensional map $x_{n+1} = F(x_n)$, defined by

$$x_{n+1} = \begin{pmatrix} \Theta_{n+1} \\ \dot{\Theta}_{n+1} \end{pmatrix} = \begin{pmatrix} (M\dot{\Theta}_n + \Theta_n) \bmod 2\pi \\ L\dot{\Theta}_n + G(\Theta_{n+1}) \end{pmatrix}. \quad (1)$$

Here Θ_n and $\dot{\Theta}_n$ are 2-vectors,

$$\Theta_n = \begin{pmatrix} \theta_1^{(n)} \\ \theta_2^{(n)} \end{pmatrix}, \quad \dot{\Theta}_n = \begin{pmatrix} \dot{\theta}_1^{(n)} \\ \dot{\theta}_2^{(n)} \end{pmatrix},$$

that give the angular positions and velocities of the rods after the nth kick. That is, $\theta_i^{(n)} = \theta_i(nT)$ and $\dot{\theta}_i^{(n)} = \dot{\theta}_i(nT^+)$. The angles θ_1 and θ_2 are taken to lie in $[0, 2\pi)$. Also,

$$G(\Theta) = \begin{pmatrix} c_1 \sin \theta_1 \\ c_2 \sin \theta_2 \end{pmatrix},$$

and L and M are constant 2×2 matrices. For simplicity, we assume $(m_1 + m_2)\ell_1^2 = m_2 \ell_2^2 \equiv I$. Then

$$L = \sum_{i=1}^{2} W_i e^{\lambda_i T}, \quad M = \sum_{i=1}^{2} W_i (e^{\lambda_i T} - 1)/\lambda_i$$

with

$$W_1 = \begin{pmatrix} a & b \\ b & d \end{pmatrix}, \quad W_2 = \begin{pmatrix} d & -b \\ -b & a \end{pmatrix}$$

where $a = \frac{1}{2}(1 + \nu_1/\Delta)$, $d = \frac{1}{2}(1 - \nu_1/\Delta)$, $b = -\nu_2/\Delta$, $\Delta = \sqrt{\nu_1^2 + 4\nu_2^2}$, $\lambda_{1,2} = -\frac{1}{2}(\nu_1 + 2\nu_2 \pm \Delta)$, and $c_{1,2} = \rho \ell_{1,2}/I$. In all the numerical work described in this

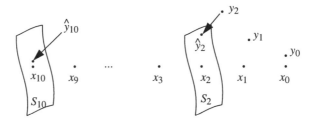

Figure 3: A path of points, consisting of x_{10} and its preimages x_9, x_8,ldots, x_0. The uncontrolled trajectory starting from y_0 rapidly moves away from the target trajectory. The targeting procedure determines perturbations that move the point y_2 to a new point $\hat{y}_2$ on the stable manifold S_2 of x_2. The trajectory starting from $\hat{y}_2$ rapidly approaches the target trajectory, as represented by the point $\hat{y}_{10}$. The surfaces labeled S_2 and S_{10} are representations of the stable manifolds associated with the points x_2 and x_{10}, respectively.

paper, we fix the values of the parameters

$$\nu = T = I = m_1 = m_2 = \ell_2 = 1, \qquad \ell_1 = \frac{1}{\sqrt{2}}$$

and use the force ρ as the control parameter, taking as the nominal value $\rho = \bar{\rho} = 9$.

We write $x_n = F^n(x_0)$ to mean the n times iterated point x_0, i.e., the point obtained by iterating the map n times starting from x_0. The double rotor map is invertible, so $F^{-n}(x_0)$ refers to the nth iterate of x under the inverse map. The notation $F(x)$ means that the map is applied with the kick set to its nominal value (here $\bar{\rho} = 9$); the notation $F(x, \rho)$ means the map applied to x with the kick set to ρ. We let $DF(x, \rho)$ denote the corresponding Jacobian matrix of partial derivatives of F with respect to θ_i and $\dot{\theta}_i$.

3 The targeting procedure

Let T be typical point on the attractor and suppose that T is the target point. A *path* is a set of points consisting of T and a sequence of its preimages. The basic idea is illustrated in Fig. 3. The target point is labeled as x_{10} to emphasize the idea that this path shows the target and ten of its preimages.

Suppose that, as the map is iterated, we find a point y_0 that falls near x_0. We wish to find a sequence of perturbations to some available parameter such that the orbit starting at y_0 approaches the orbit starting at x_0. (As the dynamics are chaotic, the orbits diverge rapidly without targeting.) We now outline the basic targeting algorithm in the case where one parameter (ρ in the case of the double rotor map) is available for control.

If the problem were linear, then in principle we could compute four successive perturbations $\delta\rho_0$, $\delta\rho_1$, $\delta\rho_2$ and $\delta\rho_3$ that allow us to hit the point x_4 starting from y_0, because in general the gradient vectors $\partial F/\partial\rho_0,\ldots, \partial F/\partial\rho_3$ are linearly independent.

In practice, the problem is highly nonlinear, and it is not possible to find a sufficiently accurate linearization so that Newton's method can be applied to determine the perturbations needed to hit x_4 unless y_0 is extremely close to x_0. Instead, we adopt an alternative approach. (For the moment, let us ignore the presence of noise and the effects of numerical roundoff errors.)

For the parameter values given above, numerical results show that the attractor has two positive and two negative Lyapunov exponents. (A numerical estimate of the Lyapunov exponents of the double rotor attractor can be obtained using the algorithm of Benettin et al. [9]) As a result, associated with a typical point on the attractor is a 2-dimensional unstable manifold and a 2-dimensional stable manifold. For example, the stable manifold associated with the point x_2 in Fig. 3 is labeled S_2. If $s \in S_2$, then the orbit starting from s approaches the orbit starting from x_2 as the map is iterated. In general, the rate of approach is rapid because the negative Lyapunov exponents associated with the attractor are fairly large in absolute value. The goal of the targeting algorithm is to try to determine perturbations to the parameters to move the orbit starting from y_0 onto the stable manifold associated with one of the x's. If it is successful, then the dynamics draw the orbits closer together.

In the case where one parameter can be varied, we attempt to find two perturbations $\delta \rho_0$ and $\delta \rho_1$ so that the orbit starting from y_0 is moved onto S_2, the stable manifold of x_2. (Recall that S_2 is a 2-dimensional sheet. The vectors $\partial F(F(x_0, \rho_0), \rho_1)/\partial \rho_0$ and $\partial F(F(x_0, \rho_0), \rho_1)/\partial \rho_1$ typically are linearly independent. Since we are working in R^4, the 2-plane spanned by the gradient vectors intersects S_2 in a unique point, denoted by $\hat{y}_2$ in Fig. 3.)

A basic difficulty arises in approximating S_2. In general, $\hat{y}_2$ is far from x_2, and a linear approximation of S_2 is inadequate. In practice, a better approximation of S_2 can be found by calculating the inverse images of suitable points further down the path. For instance, a reasonable linear approximation of the stable manifold S_{10}, valid in a small neighborhood of x_{10}, can be determined by finding the stable eigenspace associated with the matrix $DF(x_2, \rho)DF(x_3, \rho) \cdots DF(x_{10}, \rho)$. Let s_0 and s_1 denote vectors that span this stable eigenspace at x_{10}. Let z be the point $z = x_{10} + \sigma_0 s_0 + \sigma_1 s_1$, where $|\sigma_i|$ is small. Although z may not lie exactly on S_{10}, the inverse images $F^{-1}(z)$, $F^{-2}(z)$, ... rapidly approach the corresponding sets S_9, S_8, ..., because S_{10} is an expanding set under the inverse map, and components perpendicular to S_{10} contract.

Thus, in the case of one parameter targeting, a successful solution to the steering problem consists of determining two parameter perturbations $\delta \rho_0$ and $\delta \rho_1$, together with values for σ_0 and σ_1, such that

$$F^{-8}(x_{10} + \sigma_0 s_0 + \sigma_1 s_1) = F(F(y_0, \rho_0), \rho_1) = \hat{y}_2. \qquad (2)$$

Equation (2) can be solved numerically using Newton's method. There is no a priori guarantee that Newton's method will converge; however, when it does, we have determined two successive perturbations to the control parameter that steer the orbit of y_0 onto the stable manifold of x_2. Thereafter, in the absence of noise and numerical roundoff, the dynamics bring the two orbits closer together, typically at an exponential rate determined by the negative Lyapunov exponents.

There is nothing special about the choice of x_{10} and the use of eight inverse iterations to estimate S_2. For example, if the target point is x_6, then we can estimate S_2 by the fourth inverse iterate of a point close to x_6. If the target point is farther away, say at x_{12}, then we can look at the inverse images of a point near x_{10} or x_{11} instead. Going further down the path in this manner typically yields a point whose inverse image is a better approximation to the stable manifold S_2. However, the numerical solution of Eq. (2) is more ill-conditioned. For example, if we look at the inverse images of S_8, then it is necessary to evaluate the matrix product $DF^{-1}(x_8)DF^{-1}(x_7)\cdots DF^{-1}(x_3)$. If we look at S_{10} instead, then we must evaluate the matrix product $DF^{-1}(x_{10})DF^{-1}(x_9)\cdots DF^{-1}(x_3)$. These matrix products become more singular as more terms are added. Thus there is a tradeoff between numerical precision and approximation errors arising from the dynamics. For the parameters of the double rotor map used in this investigation, we have found that six inverse iterations is a good compromise. References [10] and [11] discuss some of the numerical issues in more detail.

If two parameters are available for control, then only one perturbation step is necessary. This is because typically there is a 2-plane through y_1, spanned by $g_p = \partial F(y_0, p, q)/\partial p$ and $g_q = \partial F(y_0, p, q)/\partial q$, that intersects the stable manifold S_1 of x_1. The procedure outlined above can be easily extended to other maps in different phase space dimensions and with different numbers of positive Lyapunov exponents. For example, if the attractor sits in a 6-dimensional space and has three positive Lyapunov exponents, then the procedure requires three successive changes to a single parameter to hit the 3-dimensional stable manifold of the appropriate point in the path. If three parameters can be varied independently, then one tries to hit the stable manifold of x_1, and so on.

Because of small errors in the initial approximation of S_8 and numerical roundoff errors, the control described above must be repeated periodically in order to keep the new trajectory close to the path leading to the target. We recalculate the perturbations at each iteration along the path where possible in order to allow for the presence of noise and/or roundoff error. (Recalculation of the control is not possible, for example, at the point just before the end of a path when only one control parameter is being used.)

4 Trees

The procedure described in the previous section works well, but the map must be iterated a large number of times before reaching a neighborhood of one of the points in the path leading to the target. A long path increases the likelihood that a given iterate lies near a point on the path, but the time required to reach the target also increases. Our objective is to steer a typical iterate to the target point in as few steps as possible.

One refinement is to build a hierarchy or "tree" of paths leading to the target. Let a target point x_T be given, together with a "root" path $x_0, x_1, \ldots, x_{T-1}$ leading to it. (In the work described in Refs. [10] and [11], each path typically has about 20 points.) The map is iterated (say from an arbitrary initial condition in the basin of attraction) until a point z_n is found that lies in a suitably small neighborhood of one of the points in the target path. The path leading to z_n (that is, z_n and 20 preimages) is stored in the tree; this path leads to the root path, and forms part of the first level of the tree. A path leading to a neighborhood of one of the preimages of z_n would be in the second level of the tree, and so on. Figure 4 gives a schematic illustration of the procedure.

By storing the paths in a tree, we increase the probability that a given iterate on the attractor lies near a path that can be steered to the target point. For example, if

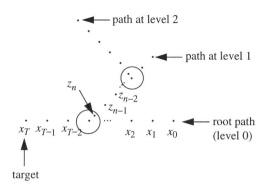

Figure 4: Schematic illustration of the hierarchy of paths leading to the target point.

an iterate lies on the path leading to z_n, the targeting algorithm can be applied with z_n as the target point. Once a small neighborhood of z_n is reached, the targeting algorithm is reapplied to steer the trajectory to a small neighborhood of the point x_T, which is the ultimate target.

A basic advantage of the tree structure is that the maximum amount of time needed to reach x_T grows linearly with the number of levels in the tree, but the number of points that can be stored grows exponentially with the number of levels. It is possible to construct "leafy" trees that reach into many regions of the attractor. With a suitable tree, one does not need to wait very long before some iterate lies near a path in the tree. Similarly, if the targeting algorithm fails at some point (for instance, if Newton's method fails to find a solution of Eq. (2)), then we lose control of the trajectory. However, it is not long before the uncontrolled trajectory again falls near some other point in the tree, whereupon we can attempt the targeting algorithm once again. Reference [11] describes some of the procedures that can be used to construct an "optimal" tree.

In the results described below, we use trees that typically consist of 10,000 points stored in paths to a depth of three levels. (Thus, if each path has length 20, then no more than 60 steps are required to reach the target point.) A separate tree is constructed for each target point of interest.

Once the tree is built, it is possible to steer points to the target very quickly, as follows. Let z_0 be a point on the attractor. If z_0 is not close to any of the points in the path tree, then we create a new set of points A by making n small random perturbations to the kick. Here $A = \{z_1^{(i)} : z_1^{(i)} = F(z_0, \rho_0 + \eta_i), 1 \leq i \leq n\}$ where η_i is a random variable in a small interval around 0. Typically we take η_i from a uniform distribution in the interval $[-0.05, 0.05]$.

We now check whether any of the points in A lies near any of the points in the tree. If so, then the targeting procedure is attempted. If it is successful, then we have steered the point z_0 to the target in no more than 61 steps (the first step consists of the random kick, followed by no more than 60 steps of the control procedure). Each of the points in A can be iterated (using the nominal value of the kick) until one of them can be steered successfully to the target.

Trees can be useful in other contexts. Suppose for instance that we want to avoid certain regions of the attractor. (Perhaps some regions of the phase space correspond

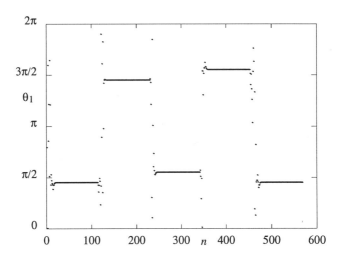

Figure 5: Control of stable periodic points with targeting.

to an undesirable operating regime of the system.) It may be possible to construct trees whose paths lie outside such a region, thus keeping the controlled system within an acceptable operating regime.

Similarly, it is possible in principle to associate a cost function with each of the paths in the tree. In cases where there are two or more sets of paths that lead to the target, one can select the set with the least total cost.

5 Results

Figure 5 shows the results of using the targeting algorithm together with control of saddle periodic points. We have selected four saddle fixed points on the attractor. We then pick four different target points, each of which is in a small neighborhood of the fixed points. A tree of paths leading to each target point is constructed as described above. The uncontrolled process quickly leads to a point that lies near some point in the tree. We apply the targeting algorithm to steer the trajectory to a neighborhood of the saddle fixed point. A control algorithm (see [11]) is then used to stabilize the trajectory near the fixed point. After the control is turned off, the trajectory quickly wanders away from the fixed point, but it approaches another point on the path tree, so the trajectory can be targeted to a neighborhood of the next saddle fixed point. In this way, we can rapidly switch between different saddle fixed points. In contrast to the case described in Fig. 2, where wait times of several thousand iterates are common between controlled states, targeting allows us to reduce the waiting time to as little as 12–16 iterates. Thus, the targeting algorithm can reduce the waiting time by two orders of magnitude.

Of course, there can be significant computational costs associated with building the path trees. In the results described here, 1–10 million iterates are required to build the trees leading to the saddle fixed points. The nature of the application will dictate in practice whether the time savings from repeated use of the targeting procedures outweighs the effort needed to build the trees.

Even so, the necessary computations to build the trees are easily done on a modern workstation (less than five minutes of CPU time on a DEC Alpha or Silicon Graphics Indy). Memory requirements are also modest, generally 4–5 megabytes for the trees. Obviously, the computational and memory requirements depend heavily on the nature of the dynamical system, the dimension of the attractor, and the dimension of the phase space.

The effect of noise on the targeting algorithms remains under study. In the case of the double rotor map, the Lyapunov exponents are large, and small errors in targeting are quickly amplified by the dynamics. Thus the targeting algorithm is somewhat sensitive to noise. However, noise levels on the order of 1 percent or less do not appear to present serious difficulties. Preliminary results suggest that the use of two or more control parameters is more robust in the presence of noise than just one control parameter. See Ref. [11] for more details.

6 Conclusions

We conclude with some brief remarks on the relationship of this work to the other papers presented at the conference. One of the basic themes of the conference is the sensitive dependence on initial conditions of many nonlinear processes. While such dependence makes long-term predictions of such systems difficult or impossible in the presence of noise or measurement uncertainties, it can be made to work in one's favor. Someday, chaos may be seen as a desirable design feature because the behavior of the system can be switched from one state to another with only small perturbations.

Professor Judd has suggested that chaotic dynamics may allow some systems to be operated in a "pseudo-periodic" fashion, wherein trajectories within an attractor are directed from one small region of the attractor to another using only small perturbations. (The trajectories do not necessarily constitute a periodic orbit.) The path trees as described in this article can be adapted for this purpose. All one needs is a collection of targets that forms a pseudo-periodic orbit, and the targeting algorithm can be applied to each target in the sequence.

Acknowledgments

This work is supported by the Department of Energy Office of Scientific Computing. E. K. is supported in part by the Dept. of Energy Program in Applied Mathematics. Many of the computations were done at the Keck Chaos Visualization Laboratory at the University of Maryland.

References

[1] E. Ott, C. Grebogi and J. A. Yorke, *Phys. Rev. Lett.* **64** (1990), 1196.

[2] W. L. Ditto, S. N. Rauseo and M. L. Spano, *Phys. Rev. Lett.* **65** (1990), 3211.

[3] T. Shinbrot, E. Ott, C. Grebogi and J. A. Yorke, *Phys. Rev. Lett.* **65** (1990), 3250.

[4] T. Shinbrot, W. Ditto, C. Grebogi, E. Ott, M. Spano and J. A. Yorke, *Phys. Rev. Lett.* **68** (1992), 2863.

[5] E. J. Kostelich, C. Grebogi, E. Ott and J. A. Yorke, *Physica D* **25** (1987), 347; *Phys. Lett. A* **118** (1986), 448; erratum **120A** (1987), 497.

[6] F. J. Romeiras, C. Grebogi, E. Ott and W. P. Dayawansa, *Physica D* **58** (1992), 165.

[7] J. L. Kaplan and J. A. Yorke, in *Functional Differential Equations and Approximation of Fixed Points*, ed. by H.-O. Peitgen and H.-O. Walter. Berlin: Springer-Verlag, Lecture Notes in Mathematics, Vol. 730, p. 204.

[8] J. D. Farmer, E. Ott and J. A. Yorke, *Physica D* **7** (1983), 153.

[9] G. Benettin, L. Galgani, A. Giogilli, and J.-M. Strelcyn, *Meccanica* **15** (1980), 9, 21.

[10] E. J. Kostelich, C. Grebogi, E. Ott and J. A. Yorke, *Phys. Rev. E* **47** (1993), 305.

[11] E. Barreto, C. Grebogi, E. Ott, and J. A. Yorke, *Phys. Rev. E* **51** (1995), 4169.

Commentary by K. Glass

This paper looks at the control problem of targeting points in a chaotic system. The introduction of the paper outlines the targeting problem and stresses the value of using targeting when stabilizing periodic points. Stabilization algorithms are discussed in the paper in this volume by E. Ott et al.

The third section of the paper outlines in detail the procedure used to target a point in a chaotic system. This procedure involves finding the stable manifold of some point on the trajectory leading to the target point and then attempting to direct the system onto this manifold. A tree is then built up of sections of trajectories leading to these manifolds. Although creating the tree of paths will require some off-line computation, the resulting algorithm will be highly efficient. Not only will the target point be reached after only a few steps, the perturbation required to target the system may be calculated quickly making the method effective for on-line control.

The concept of storing sections of trajectories in a tree of paths would be highly useful in the later sections of the paper: "Creating and Targeting Periodic Orbits". I feel also that it may be possible to use the idea of creating trajectories discussed here to extend the tree of paths used in the above algorithm. By creating trajectories using small perturbations, we might be able to construct paths in less accessible areas of the system.

Commentary by I. Mareels

The contribution by Kostelich and Barreto discusses in the context of a double rotor example a method to achieve fast transitions between orbits on a strange attractor by the aid of small amplitude control.

Under the generic condition that the map is controllable at each point on the attractor the minimal transition time control problem is well posed. A repeated dead beat control action (to use control engineering lingo) with amplitude bounded control is proposed. (The minimum time optimal control problem is considered a difficult optimal control problem.)

The authors propose a numerical method to generate the appropriate control actions. Essentially a tree structure of paths which covers adequately the attractor is oncstructed off line.

Questions to authors

1. could you do a small calculation as in the intro to why 10,000 points in the tree would roughly give you a transition time of 20 (I think it is illuminating) Using the fractal dimension it seems to come out quite nicely.

2. Why are most paths without control, and only the root path contains control action? CPU time savings? Could you comment on this?

Adaptive Nonlinear Control: A Lyapunov Approach

Petar V. Kokotović
Dept. Electrical and Computer Engineering
University of California
Santa Barbara, CA 93106, USA

Miroslav Krstić
Dept. Mechanical Engineering
University of Maryland
College Park, MD 20742, USA

1 Introduction

Realistic models of physical systems are nonlinear and usually contain parameters (masses, inductances, aerodynamic coefficients, etc.) which are either poorly known or dependent on a slowly changing environment. If the parameters vary in a broad range, it is common to employ adaptation: a parameter estimator—**identifier**—continuously acquires knowledge about the plant and uses it to tune the controller "on-line".

Instabilities in nonlinear systems can be more explosive than in linear systems. During the parameter estimation transients, the state can "escape" to infinity in finite time. For this reason, adaptive nonlinear controllers cannot simply be the "adaptive versions" of standard nonlinear controllers.

Here we give a tutorial outline of a new *Lyapunov* methodology for adaptive control design. A complete theory with application examples is presented in [1].

2 Backstepping

Currently the most systematic methodology for adaptive nonlinear control design is **backstepping**. We introduce the idea of backstepping by carrying out a *nonadaptive* design for the system

$$\dot{x}_1 = x_2 + \varphi(x_1)^{\mathrm{T}}\theta, \qquad \varphi(0) = 0 \qquad (1)$$
$$\dot{x}_2 = u \qquad (2)$$

where θ is a *known* parameter vector and $\varphi(x_1)$ is a smooth nonlinear function. Our goal is to stabilize the equilibrium $x_1 = 0$, $x_2 = -\varphi(0)^{\mathrm{T}}\theta = 0$. Backstepping design is recursive. First, the state x_2 is treated as a **virtual control** for the x_1-equation (1), and a **stabilizing function**

$$\alpha_1(x_1) = -c_1 x_1 - \varphi(x_1)^{\mathrm{T}}\theta, \qquad c_1 > 0 \qquad (3)$$

is designed to stabilize (1) assuming that $x_2 = \alpha_1(x_1)$ can be implemented. Since this is not the case, we define

$$z_1 = x_1 \tag{4}$$
$$z_2 = x_2 - \alpha_1(x_1), \tag{5}$$

where z_2 is an error variable expressing the fact that x_2 is not the true control. Differentiating z_1 and z_2 with respect to time, the complete system (1), (2) is expressed in the error coordinates (4), (5):

$$\dot{z}_1 = \dot{x}_1 = x_2 + \varphi^T \theta = z_2 + \alpha_1 + \varphi^T \theta = -c_1 z_1 + z_2 \tag{6}$$
$$\dot{z}_2 = \dot{x}_2 - \dot{\alpha}_1 = u - \frac{\partial \alpha_1}{\partial x_1} \dot{x}_1 = u - \frac{\partial \alpha_1}{\partial x_1}(x_2 + \varphi^T \theta). \tag{7}$$

It is important to observe that the time derivative $\dot{\alpha}_1$ is implemented analytically, without a differentiator. For the system (6)–(7) we now design a control law $u = \alpha_2(x_1, x_2)$ to render the time derivative of a Lyapunov function negative definite. It turns out that the design can be completed with the simplest Lyapunov function

$$V(x_1, x_2) = \frac{1}{2} z_1^2 + \frac{1}{2} z_2^2. \tag{8}$$

Its derivative for (6), (7) is

$$\dot{V} = z_1(-c_1 z_1 + z_2) + z_2 \left[u - \frac{\partial \alpha_1}{\partial x_1}(x_2 + \varphi^T \theta) \right]$$
$$= -c_1 z_1^2 + z_2 \left[u + z_1 - \frac{\partial \alpha_1}{\partial x_1}(x_2 + \varphi^T \theta) \right]. \tag{9}$$

An obvious way to achieve negativity of $\dot{V}$ is to employ u to make the bracketed expression equal to $-c_2 z_2$ with $c_2 > 0$, namely,

$$u = \alpha_2(x_1, x_2) = -c_2 z_2 - z_1 + \frac{\partial \alpha_1}{\partial x_1}(x_2 + \varphi^T \theta). \tag{10}$$

This control may not be the best choice because it cancels some terms which may contribute to the negativity of $\dot{V}$. Backstepping design offers enough flexibility to avoid cancellation. However, for the sake of clarity, we will assume that none of the nonlinearities is useful, so that they all need to be cancelled as in the control law (10). This control law yields

$$\dot{V} = -c_1 z_1^2 - c_2 z_2^2, \tag{11}$$

which means that the equilibrium $z = 0$ is globally asymptotically stable. In view of (4), (5), the same is true about $x = 0$. The resulting closed-loop system in the z-coordinates is linear:

$$\begin{bmatrix} \dot{z}_1 \\ \dot{z}_2 \end{bmatrix} = \begin{bmatrix} -c_1 & 1 \\ -1 & -c_2 \end{bmatrix} \begin{bmatrix} z_1 \\ z_2 \end{bmatrix}. \tag{12}$$

The two main methodologies for adaptive backstepping design are the **tuning functions design** (based on a Lyapunov approach) and the **modular design**. Here we use three examples to illustrate the tuning functions design. A summary of a general design procedure are also provided but without technical details, for which the reader is referred to [1].

3 Introductory Examples

In the tuning functions design both the controller and the parameter update law are designed recursively. At each consecutive step a **tuning function** is designed as a potential update law. The tuning functions are not implemented as update laws. Instead, the stabilizing functions use them to compensate the effects of parameter estimation transients. Only the final tuning function is used as the parameter update law.

The tuning functions design will be introduced through the following three examples in the order of increasing complexity:

A	B	C
$\dot{x}_1 = u + \varphi(x_1)^{\mathrm{T}}\theta$	$\dot{x}_1 = x_2 + \varphi(x_1)^{\mathrm{T}}\theta$	$\dot{x}_1 = x_2 + \varphi(x_1)^{\mathrm{T}}\theta$
	$\dot{x}_2 = u$	$\dot{x}_2 = x_3$
		$\dot{x}_3 = u$

The adaptive problem arises because the parameter vector θ is *unknown*. The nonlinearity $\varphi(x_1)$ is known and for simplicity it is assumed that $\varphi(0) = 0$. The systems A, B, and C differ structurally: the number of integrators between the control u and the unknown parameter θ increases from zero at A, to two at C. Design A will be the simplest because the control u and the uncertainty $\varphi(x_1)^{\mathrm{T}}\theta$ are "matched", that is, the control does not have to overcome integrator transients in order to counteract the effects of the uncertainty. Design C will be the hardest because the control must act through two integrators before it reaches the uncertainty.

Design A. Let $\hat{\theta}$ be an estimate of the unknown parameter θ in the system

$$\dot{x}_1 = u + \varphi^{\mathrm{T}}\theta. \tag{13}$$

If this estimate were correct, $\hat{\theta} = \theta$, then the control law

$$u = -c_1 x_1 - \varphi(x_1)^{\mathrm{T}}\hat{\theta} \tag{14}$$

would achieve global asymptotic stability of $x = 0$. Because $\tilde{\theta} = \theta - \hat{\theta} \neq 0$, we have

$$\dot{x}_1 = -c_1 x_1 + \varphi(x_1)^{\mathrm{T}}\tilde{\theta}, \tag{15}$$

that is, the parameter estimation error $\tilde{\theta}$ continues to act as a disturbance which may destabilize the system. Our task is to find an update law for $\hat{\theta}(t)$ which preserves the boundedness of $x(t)$ and achieves its regulation to zero. To this end, we consider the Lyapunov function

$$V_1(x, \hat{\theta}) = \frac{1}{2}x_1^2 + \frac{1}{2}\tilde{\theta}^{\mathrm{T}}\Gamma^{-1}\tilde{\theta}, \tag{16}$$

where Γ is a positive definite symmetric matrix. The derivative of V_1 is

$$\begin{aligned}\dot{V}_1 &= -c_1 x_1^2 + x_1 \varphi(x_1)^{\mathrm{T}}\tilde{\theta} - \tilde{\theta}^{\mathrm{T}}\Gamma^{-1}\dot{\hat{\theta}} \\ &= -c_1 x_1^2 + \tilde{\theta}^{\mathrm{T}}\Gamma^{-1}\left(\Gamma\varphi(x_1)x_1 - \dot{\hat{\theta}}\right).\end{aligned} \tag{17}$$

Our goal is to select an update law for $\dot{\hat{\theta}}$ to guarantee

$$\dot{V}_1 \leq 0. \tag{18}$$

Adaptive Nonlinear Control: A Lyapunov Approach

The only way this can be achieved for any unknown $\tilde{\theta}$ is to choose

$$\dot{\hat{\theta}} = \Gamma\varphi(x_1)x_1. \qquad (19)$$

This choice yields

$$\dot{V}_1 = -c_1 x_1^2, \qquad (20)$$

which guarantees global stability of the equilibrium $x_1 = 0$, $\hat{\theta} = \theta$, and hence, the boundedness of $x_1(t)$ and $\hat{\theta}(t)$. By LaSalle's invariance theorem, all the trajectories of the closed-loop adaptive system converge to the set where $\dot{V}_1 = 0$, that is, to the set where $c_1 x_1^2 = 0$, which implies that

$$\lim_{t\to\infty} x_1(t) = 0. \qquad (21)$$

△

The update law (19) is driven by the state x_1 and the vector $\varphi(x_1)$, called the **regressor**. This is a typical form of an update law in the tuning functions design: the speed of adaptation is dictated by the nonlinearity $\varphi(x_1)$ and the state x_1.

Design B. For the system

$$\begin{aligned} \dot{x}_1 &= x_2 + \varphi(x_1)^T \theta \\ \dot{x}_2 &= u \end{aligned} \qquad (22)$$

we have already designed a nonadaptive controller in Section 1. To design an adaptive controller, we replace the unknown θ by its estimate $\hat{\theta}$ both in the stabilizing function (3) and in the change of coordinate (5):

$$z_2 = x_2 - \alpha_1(x_1, \hat{\theta}), \qquad \alpha_1(x_1, \hat{\theta}) = -c_1 z_1 - \varphi^T \hat{\theta}. \qquad (23)$$

Because in the system (22) the control input is separated from the unknown parameter by an integrator, the control law (10) will be strengthened by a term $\nu_2(x_1, x_2, \hat{\theta})$ which will compensate for the parameter estimation transients:

$$u = \alpha_2(x_1, x_2, \hat{\theta}) = -c_2 z_2 - z_1 + \frac{\partial \alpha_1}{\partial x_1}\left(x_2 + \varphi^T \hat{\theta}\right) + \nu_2(x_1, x_2, \hat{\theta}). \qquad (24)$$

The resulting system in the z coordinates is

$$\begin{aligned} \dot{z}_1 &= z_2 + \alpha_1 + \varphi^T \theta = -c_1 z_1 + z_2 + \varphi^T \tilde{\theta} \qquad (25) \\ \dot{z}_2 &= \dot{x}_2 - \dot{\alpha}_1 = u - \frac{\partial \alpha_1}{\partial x_1}\left(x_2 + \varphi^T \theta\right) - \frac{\partial \alpha_1}{\partial \hat{\theta}}\dot{\hat{\theta}} \\ &= -z_1 - c_2 z_2 - \frac{\partial \alpha_1}{\partial x_1}\varphi^T \tilde{\theta} - \frac{\partial \alpha_1}{\partial \hat{\theta}}\dot{\hat{\theta}} + \nu_2(x_1, x_2, \hat{\theta}), \qquad (26) \end{aligned}$$

or, in vector form,

$$\begin{bmatrix} \dot{z}_1 \\ \dot{z}_2 \end{bmatrix} = \begin{bmatrix} -c_1 & 1 \\ -1 & -c_2 \end{bmatrix}\begin{bmatrix} z_1 \\ z_2 \end{bmatrix} + \begin{bmatrix} \varphi^T \\ -\frac{\partial \alpha_1}{\partial x_1}\varphi^T \end{bmatrix}\tilde{\theta} + \begin{bmatrix} 0 \\ -\frac{\partial \alpha_1}{\partial \hat{\theta}}\dot{\hat{\theta}} + \nu_2(x_1, x_2, \hat{\theta}) \end{bmatrix}. \qquad (27)$$

The term ν_2 can now be chosen to eliminate the last brackets:

$$\nu_2(x_1, x_2, \hat{\theta}) = \frac{\partial \alpha_1}{\partial \hat{\theta}} \dot{\hat{\theta}}. \tag{28}$$

This expression is implementable because $\dot{\hat{\theta}}$ will be available from the update law. Thus we obtain the **error system**

$$\begin{bmatrix} \dot{z}_1 \\ \dot{z}_2 \end{bmatrix} = \begin{bmatrix} -c_1 & 1 \\ -1 & -c_2 \end{bmatrix} \begin{bmatrix} z_1 \\ z_2 \end{bmatrix} + \begin{bmatrix} \varphi^T \\ -\frac{\partial \alpha_1}{\partial x_1} \varphi^T \end{bmatrix} \tilde{\theta}. \tag{29}$$

When the parameter error $\tilde{\theta}$ is zero, this system becomes the linear asymptotically stable system (12). Our remaining task is to select the update law $\dot{\hat{\theta}} = \Gamma \tau_2(x, \hat{\theta})$. Consider the Lyapunov function

$$V_2(x_1, x_2, \hat{\theta}) = V_1 + \frac{1}{2} z_2^2 = \frac{1}{2} z_1^2 + \frac{1}{2} z_2^2 + \frac{1}{2} \tilde{\theta}^T \Gamma^{-1} \tilde{\theta}. \tag{30}$$

Because $\dot{\tilde{\theta}} = -\dot{\hat{\theta}}$, the derivative of V_2 is

$$\begin{aligned}
\dot{V}_2 &= -c_1 z_1^2 - c_2 z_2^2 + [z_1, \; z_2] \begin{bmatrix} \varphi^T \\ -\frac{\partial \alpha_1}{\partial x_1} \varphi^T \end{bmatrix} \tilde{\theta} - \tilde{\theta}^T \Gamma^{-1} \dot{\hat{\theta}} \\
&= -c_1 z_1^2 - c_2 z_2^2 + \tilde{\theta}^T \Gamma^{-1} \left(\Gamma \left[\varphi, \; -\frac{\partial \alpha_1}{\partial x_1} \varphi \right] \begin{bmatrix} z_1 \\ z_2 \end{bmatrix} - \dot{\hat{\theta}} \right).
\end{aligned} \tag{31}$$

The only way to eliminate the unknown parameter error $\tilde{\theta}$ is to select the update law

$$\dot{\hat{\theta}} = \Gamma \tau_2(x, \hat{\theta}) = \Gamma \left[\varphi, \; -\frac{\partial \alpha_1}{\partial x_1} \varphi \right] \begin{bmatrix} z_1 \\ z_2 \end{bmatrix} = \Gamma \left(\varphi z_1 - \frac{\partial \alpha_1}{\partial x_1} \varphi z_2 \right). \tag{32}$$

Then $\dot{V}_2$ is nonpositive:

$$\dot{V}_2 = -c_1 z_1^2 - c_2 z_2^2, \tag{33}$$

which means that the global stability of $z = 0$, $\tilde{\theta} = 0$ is achieved. Moreover, by applying the LaSalle argument mentioned in Design A, we prove that $z(t) \to 0$ as $t \to \infty$. Finally, from (23), it follows that the equilibrium $x = 0$, $\hat{\theta} = \theta$ is globally stable and $x(t) \to 0$ as $t \to \infty$. △

The crucial property of the control law in Design B is that it incorporates the ν_2-term (28) which is proportional to $\dot{\hat{\theta}}$ and compensates for the effect of parameter estimation transients on the coordinate change (23). It is this term that makes the adaptive stabilization possible for systems with nonlinearities of arbitrary growth.

By comparing (32) with (19), we note that the first term, φz_1 is the potential update law for the z_1-system. The functions

$$\tau_1(x_1) = \varphi z_1 \tag{34}$$

$$\tau_2(x_1, x_2, \hat{\theta}) = \tau_1(x_1) - \frac{\partial \alpha_1}{\partial x_1} \varphi z_2 \tag{35}$$

are referred to as the **tuning functions**, because of their role as potential update laws for intermediate systems in the backstepping procedure.

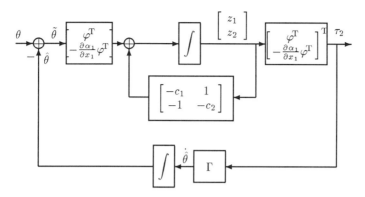

Figure 1: The closed-loop adaptive system (29), (32).

Design C. The system
$$\begin{aligned} \dot{x}_1 &= x_2 + \varphi(x_1)^T \theta \\ \dot{x}_2 &= x_3 \\ \dot{x}_3 &= u \end{aligned} \qquad (36)$$

is obtained by augmenting system (22) with an integrator. The control law $\alpha_2(x_1, x_2, \hat{\theta})$ designed in (24) can no longer be directly applied because x_3 is a state and not a control input. We "step back" through the integrator $\dot{x}_3 = u$ and design the control law for the actual input u. However, we keep the stabilizing function α_2 and use it to define the third error coordinate

$$z_3 = x_3 - \alpha_2(x_1, x_2, \hat{\theta}). \qquad (37)$$

The parameter update law (32) will have to be modified with an additional z_3-term. Instead of $\dot{\hat{\theta}}$ in (28), the compensating term ν_2 will now use the potential update law (35) for the system (29):

$$\nu_2(x_1, x_2, \hat{\theta}) = \frac{\partial \alpha_1}{\partial \hat{\theta}} \Gamma \tau_2(x_1, x_2, \hat{\theta}). \qquad (38)$$

Hence, the role of the tuning function τ_2 is to substitute for the actual update law in the compensation of the effects of parameter estimation transients.

With (23), (23), (37), (35), (38), and the stabilizing function α_2 in (24), we have

$$\begin{bmatrix} \dot{z}_1 \\ \dot{z}_2 \end{bmatrix} = \begin{bmatrix} -c_1 & 1 \\ -1 & -c_2 \end{bmatrix} \begin{bmatrix} z_1 \\ z_2 \end{bmatrix} + \begin{bmatrix} \varphi^T \\ -\frac{\partial \alpha_1}{\partial x_1} \varphi^T \end{bmatrix} \tilde{\theta} + \begin{bmatrix} 0 \\ z_3 + \frac{\partial \alpha_1}{\partial \hat{\theta}} (\Gamma \tau_2 - \dot{\hat{\theta}}) \end{bmatrix}. \qquad (39)$$

This system differs from the error system (29) only in its last term. Likewise, instead of the Lyapunov inequality (33) we have

$$\dot{V}_2 = -c_1 z_1^2 - c_2 z_2^2 + z_2 z_3 + z_2 \frac{\partial \alpha_1}{\partial \hat{\theta}} (\Gamma \tau_2 - \dot{\hat{\theta}}) + \tilde{\theta}^T (\tau_2 - \Gamma^{-1} \dot{\hat{\theta}}). \qquad (40)$$

Differentiating (37), we get

$$\begin{aligned}\dot{z}_3 &= u - \frac{\partial \alpha_2}{\partial x_1}(x_2 + \varphi^{\mathrm{T}}\theta) - \frac{\partial \alpha_2}{\partial x_2}x_3 - \frac{\partial \alpha_2}{\partial \hat{\theta}}\dot{\hat{\theta}} \\ &= u - \frac{\partial \alpha_2}{\partial x_1}\left(x_2 + \varphi^{\mathrm{T}}\hat{\theta}\right) - \frac{\partial \alpha_2}{\partial x_2}x_3 - \frac{\partial \alpha_2}{\partial \hat{\theta}}\dot{\hat{\theta}} - \frac{\partial \alpha_2}{\partial x_1}\varphi^{\mathrm{T}}\tilde{\theta}.\end{aligned} \quad (41)$$

We now stabilize the (z_1, z_2, z_3)-system (39), (41) with respect to the Lyapunov function

$$V_3(x,\hat{\theta}) = V_2 + \frac{1}{2}z_3^2 = \frac{1}{2}z_1^2 + \frac{1}{2}z_2^2 + \frac{1}{2}z_3^2 + \frac{1}{2}\tilde{\theta}^{\mathrm{T}}\Gamma^{-1}\tilde{\theta}. \quad (42)$$

Its derivative along (39) and (41) is

$$\begin{aligned}\dot{V}_3 &= -c_1 z_1^2 - c_2 z_2^2 + z_2\frac{\partial \alpha_1}{\partial \hat{\theta}}(\Gamma\tau_2 - \dot{\hat{\theta}}) \\ &\quad + z_3\left[z_2 + u - \frac{\partial \alpha_2}{\partial x_1}\left(x_2 + \varphi^{\mathrm{T}}\hat{\theta}\right) - \frac{\partial \alpha_2}{\partial x_2}x_3 - \frac{\partial \alpha_2}{\partial \hat{\theta}}\dot{\hat{\theta}}\right] \\ &\quad + \tilde{\theta}^{\mathrm{T}}\left(\tau_2 - \frac{\partial \alpha_2}{\partial x_1}\varphi z_3 - \Gamma^{-1}\dot{\hat{\theta}}\right).\end{aligned} \quad (43)$$

Again we must eliminate the unknown parameter error $\tilde{\theta}$ from $\dot{V}_3$. For this we must choose the update law as

$$\dot{\hat{\theta}} = \Gamma\tau_3(x_1, x_2, x_3, \hat{\theta}) = \Gamma\left(\tau_2 - \frac{\partial \alpha_2}{\partial x_1}\varphi z_3\right) = \Gamma\left[\varphi,\; -\frac{\partial \alpha_1}{\partial x_1}\varphi,\; \frac{\partial \alpha_2}{\partial x_1}\varphi\right]\begin{bmatrix}z_1 \\ z_2 \\ z_3\end{bmatrix}. \quad (44)$$

Upon inspection of the bracketed terms in $\dot{V}_3$, we pick the control law:

$$u = \alpha_3(x_1, x_2, x_3, \hat{\theta}) = -z_2 - c_3 z_3 + \frac{\partial \alpha_2}{\partial x_1}\left(x_2 + \varphi^{\mathrm{T}}\hat{\theta}\right) + \frac{\partial \alpha_2}{\partial x_2}x_3 + \nu_3. \quad (45)$$

The compensation term ν_3 is yet to be chosen. Substituting (45) into (43), we get

$$\dot{V}_3 = -c_1 z_1^2 - c_2 z_2^2 - c_3 z_3^2 + z_2\frac{\partial \alpha_1}{\partial \hat{\theta}}(\Gamma\tau_2 - \dot{\hat{\theta}}) + z_3\left(\nu_3 - \frac{\partial \alpha_2}{\partial \hat{\theta}}\dot{\hat{\theta}}\right). \quad (46)$$

From this expression it is clear that ν_3 should cancel $\frac{\partial \alpha_2}{\partial \hat{\theta}}\dot{\hat{\theta}}$. In order to cancel the cross-term $z_2\frac{\partial \alpha_1}{\partial \hat{\theta}}(\Gamma\tau_2 - \dot{\hat{\theta}})$ with ν_3, it appears that we would need to divide by z_3. However, the variable z_3 might take a zero value during the transient, and should be regulated to zero to accomplish the control objective. We resolve this difficulty by noting that

$$\begin{aligned}\dot{\hat{\theta}} - \Gamma\tau_2 &= \dot{\hat{\theta}} - \Gamma\tau_3 + \Gamma\tau_3 - \Gamma\tau_2 \\ &= \dot{\hat{\theta}} - \Gamma\tau_3 - \Gamma\frac{\partial \alpha_2}{\partial x_1}\varphi z_3,\end{aligned} \quad (47)$$

so that $\dot{V}_3$ in (46) is rewritten as

$$\dot{V}_3 = -c_1 z_1^2 - c_2 z_2^2 - c_3 z_3^2 + z_3\left(\nu_3 - \frac{\partial \alpha_2}{\partial \hat{\theta}}\Gamma\tau_3 + \frac{\partial \alpha_1}{\partial \hat{\theta}}\Gamma\frac{\partial \alpha_2}{\partial x_1}\varphi z_2\right). \quad (48)$$

From (48) the choice of ν_3 is immediate:

$$\nu_3(x_1, x_2, x_3, \hat{\theta}) = \frac{\partial \alpha_2}{\partial \hat{\theta}} \Gamma \tau_3 - \frac{\partial \alpha_1}{\partial \hat{\theta}} \Gamma \frac{\partial \alpha_2}{\partial x_1} \varphi z_2 \,. \tag{49}$$

The resulting $\dot{V}_3$ is

$$\dot{V}_3 = -c_1 z_1^2 - c_2 z_2^2 - c_3 z_3^2 \,, \tag{50}$$

which guarantees that the equilibrium $x = 0$, $\hat{\theta} = \theta$ is globally stable, and $x(t) \to 0$ as $t \to \infty$. The Lyapunov design leading to (48) is effective but does not reveal the stabilization mechanism. To provide further insight we write the (z_1, z_2, z_3)-system (39), (41) with u given in (45) but with ν_3 yet to be selected:

$$\begin{bmatrix} \dot{z}_1 \\ \dot{z}_2 \\ \dot{z}_3 \end{bmatrix} = \begin{bmatrix} -c_1 & 1 & 0 \\ -1 & -c_2 & 1 \\ 0 & -1 & -c_3 \end{bmatrix} \begin{bmatrix} z_1 \\ z_2 \\ z_3 \end{bmatrix} + \begin{bmatrix} \varphi^T \\ -\frac{\partial \alpha_1}{\partial x_1} \varphi^T \\ -\frac{\partial \alpha_2}{\partial x_1} \varphi^T \end{bmatrix} \tilde{\theta} + \begin{bmatrix} 0 \\ \frac{\partial \alpha_1}{\partial \hat{\theta}} (\Gamma \tau_2 - \dot{\hat{\theta}}) \\ \nu_3 - \frac{\partial \alpha_2}{\partial \hat{\theta}} \Gamma \tau_3 \end{bmatrix} \,. \tag{51}$$

While ν_3 can cancel the matched term $\frac{\partial \alpha_2}{\partial \hat{\theta}} \dot{\hat{\theta}}$, it cannot cancel the term $\frac{\partial \alpha_1}{\partial \hat{\theta}} (\Gamma \tau_2 - \dot{\hat{\theta}})$ in the second equation. By substituting (47), we note that $\frac{\partial \alpha_1}{\partial \hat{\theta}} (\Gamma \tau_2 - \dot{\hat{\theta}})$ has z_3 as a factor and absorb it into the "system matrix":

$$\begin{bmatrix} \dot{z}_1 \\ \dot{z}_2 \\ \dot{z}_3 \end{bmatrix} = \begin{bmatrix} -c_1 & 1 & 0 \\ -1 & -c_2 & 1 + \frac{\partial \alpha_1}{\partial \hat{\theta}} \Gamma \frac{\partial \alpha_2}{\partial x_1} \varphi \\ 0 & -1 & -c_3 \end{bmatrix} \begin{bmatrix} z_1 \\ z_2 \\ z_3 \end{bmatrix} + \begin{bmatrix} \varphi^T \\ -\frac{\partial \alpha_1}{\partial x_1} \varphi^T \\ -\frac{\partial \alpha_2}{\partial x_1} \varphi^T \end{bmatrix} \tilde{\theta} + \begin{bmatrix} 0 \\ 0 \\ \nu_3 - \frac{\partial \alpha_2}{\partial \hat{\theta}} \Gamma \tau_3 \end{bmatrix} \,. \tag{52}$$

Now ν_3 in (49) yields

$$\begin{bmatrix} \dot{z}_1 \\ \dot{z}_2 \\ \dot{z}_3 \end{bmatrix} = \begin{bmatrix} -c_1 & 1 & 0 \\ -1 & -c_2 & 1 + \frac{\partial \alpha_1}{\partial \hat{\theta}} \Gamma \frac{\partial \alpha_2}{\partial x_1} \varphi \\ 0 & -1 - \frac{\partial \alpha_1}{\partial \hat{\theta}} \Gamma \frac{\partial \alpha_2}{\partial x_1} \varphi & -c_3 \end{bmatrix} \begin{bmatrix} z_1 \\ z_2 \\ z_3 \end{bmatrix} + \begin{bmatrix} \varphi^T \\ -\frac{\partial \alpha_1}{\partial x_1} \varphi^T \\ -\frac{\partial \alpha_2}{\partial x_1} \varphi^T \end{bmatrix} \tilde{\theta} \,. \tag{53}$$

This choice, which places the term $-\frac{\partial \alpha_1}{\partial \hat{\theta}} \Gamma \frac{\partial \alpha_2}{\partial x_1} \varphi$ at the (2,3) position in the system matrix, achieves skew-symmetry with its positive image above the diagonal. What could not be achieved by pursuing a linear-like form, was achieved by designing a nonlinear system where the nonlinearities are 'balanced' rather than cancelled. △

4 General Recursive Design Procedure

A systematic backstepping design with tuning functions has been developed for the class of nonlinear systems transformable into the **parametric strict-feedback form**:

$$\begin{aligned}
\dot{x}_1 &= x_2 + \varphi_1(x_1)^T \theta \\
\dot{x}_2 &= x_3 + \varphi_2(x_1, x_2)^T \theta \\
&\vdots \\
\dot{x}_{n-1} &= x_n + \varphi_{n-1}(x_1, \ldots, x_{n-1})^T \theta \\
\dot{x}_n &= \beta(x)u + \varphi_n(x)^T \theta \\
y &= x_1
\end{aligned} \qquad (54)$$

where β and

$$F(x) = [\varphi_1(x_1), \varphi_2(x_1, x_2), \cdots, \varphi_n(x)] \qquad (55)$$

are smooth nonlinear functions, and $\beta(x) \neq 0$, $\forall x \in \mathbb{R}^n$.

The general design summarized in Table 1 achieves asymptotic tracking, that is, the output $y = x_1$ of the system (54) is forced to asymptotically track the reference output $y_r(t)$ whose first n derivatives are assumed to be known, bounded and piecewise continuous.

The closed-loop system has the form

$$\dot{z} = A_z(z, \hat{\theta}, t)z + W(z, \hat{\theta}, t)^T \tilde{\theta} \qquad (63)$$

$$\dot{\hat{\theta}} = \Gamma W(z, \hat{\theta}, t) z, \qquad (64)$$

where

$$A_z(z, \hat{\theta}, t) = \begin{bmatrix} -c_1 & 1 & 0 & \cdots & 0 \\ -1 & -c_2 & 1 + \sigma_{23} & \cdots & \sigma_{2n} \\ 0 & -1 - \sigma_{23} & \ddots & \ddots & \vdots \\ \vdots & \vdots & \ddots & \ddots & 1 + \sigma_{n-1,n} \\ 0 & -\sigma_{2n} & \cdots & -1 - \sigma_{n-1,n} & -c_n \end{bmatrix} \qquad (65)$$

and

$$\sigma_{jk}(x, \hat{\theta}) = -\frac{\partial \alpha_{j-1}}{\partial \hat{\theta}} \Gamma w_k . \qquad (66)$$

Because of the skew-symmetry of the off-diagonal part of the matrix A_z, it is easy to see that the Lyapunov function

$$V_n = \frac{1}{2} z^T z + \frac{1}{2} \tilde{\theta}^T \Gamma^{-1} \tilde{\theta} \qquad (67)$$

has the derivative

$$\dot{V}_n = -\sum_{k=1}^n c_k z_k^2 , \qquad (68)$$

which guarantees that the equilibrium $z = 0, \hat{\theta} = \theta$ is globally stable and $z(t) \to 0$ as $t \to \infty$. This means, in particular, that the system state and the control input are bounded and asymptotic tracking is achieved: $\lim_{t \to \infty} [y(t) - y_r(t)] = 0$.

Table 1: Summary of the tuning functions design for tracking. (For notational convenience we define $z_0 \triangleq 0$, $\alpha_0 \triangleq 0$, $\tau_0 \triangleq 0$.)

$$z_i = x_i - y_r^{(i-1)} - \alpha_{i-1} \tag{56}$$

$$\alpha_i(\bar{x}_i, \hat{\theta}, \bar{y}_r^{(i-1)}) = \nu_i - z_{i-1} - c_i z_i - w_i^T \hat{\theta}$$
$$+ \sum_{k=1}^{i-1} \left(\frac{\partial \alpha_{i-1}}{\partial x_k} x_{k+1} + \frac{\partial \alpha_{i-1}}{\partial y_r^{(k-1)}} y_r^{(k)} \right) \tag{57}$$

$$\nu_i(\bar{x}_i, \hat{\theta}, \bar{y}_r^{(i-1)}) = \frac{\partial \alpha_{i-1}}{\partial \hat{\theta}} \Gamma \tau_i + \sum_{k=2}^{i-1} \frac{\partial \alpha_{k-1}}{\partial \hat{\theta}} \Gamma w_i z_k \tag{58}$$

$$\tau_i(\bar{x}_i, \hat{\theta}, \bar{y}_r^{(i-1)}) = \tau_{i-1} + w_i z_i \tag{59}$$

$$w_i(\bar{x}_i, \hat{\theta}, \bar{y}_r^{(i-2)}) = \varphi_i - \sum_{k=1}^{i-1} \frac{\partial \alpha_{i-1}}{\partial x_k} \varphi_k \tag{60}$$

$$i = 1, \ldots, n \qquad \bar{x}_i = (x_1, \ldots, x_i), \qquad \bar{y}_r^{(i)} = (y_r, \dot{y}_r, \ldots, y_r^{(i)})$$

Adaptive control law:

$$u = \frac{1}{\beta(x)} \left[\alpha_n(x, \hat{\theta}, \bar{y}_r^{(n-1)}) + y_r^{(n)} \right] \tag{61}$$

Parameter update law:

$$\dot{\hat{\theta}} = \Gamma \tau_n(x, \hat{\theta}, \bar{y}_r^{(n-1)}) = \Gamma W z \tag{62}$$

To help understand how the control design of Table 1 leads to the closed-loop system (63)–(66), we provide an interpretation of the matrix A_z for $n = 5$:

$$A_z = \begin{bmatrix} -c_1 & 1 & & & \\ -1 & -c_2 & 1 & & \\ & -1 & -c_3 & 1 & \\ & & -1 & -c_4 & 1 \\ & & & -1 & -c_5 \end{bmatrix} + \begin{bmatrix} 0 & 0 & 0 & 0 & 0 \\ 0 & 0 & \sigma_{23} & \sigma_{24} & \sigma_{25} \\ 0 & -\sigma_{23} & 0 & \sigma_{34} & \sigma_{35} \\ 0 & -\sigma_{24} & -\sigma_{34} & 0 & \sigma_{45} \\ 0 & -\sigma_{25} & -\sigma_{35} & -\sigma_{45} & 0 \end{bmatrix}. \tag{69}$$

If the parameters were known, $\hat{\theta} \equiv \theta$, in which case we would not use adaptation, $\Gamma = 0$, the stabilizing functions (57) would be implemented with $\nu_i \equiv 0$, and hence $\sigma_{i,j} = 0$. Then A_z would be just the above constant tri-diagonal asymptotically stable matrix. When the parameters are unknown, we use $\Gamma > 0$ and, due to the change of variable $z_i = x_i - y_r^{(i-1)} - \alpha_{i-1}$, in each of the $\dot{z}_i$-equations a term $-\frac{\partial \alpha_{i-1}}{\partial \hat{\theta}} \dot{\hat{\theta}} = \sum_{k=1}^{n} \sigma_{ik} z_k$ appears. The term $\nu_i = -\sum_{k=1}^{i} \sigma_{ik} z_k - \sum_{k=2}^{i-1} \sigma_{ki} z_k$ in the stabilizing function (57) is crucial in compensating the effect of $\dot{\hat\theta}$. The σ_{ik}-terms above the diagonal in (69) come from $\dot{\hat\theta}$. Their skew-symmetric negative images come from feedback ν_i.

It can be shown that the resulting closed-loop system (63), (64), as well as each intermediate system, has a *strict passivity* property from $\tilde\theta$ as the input to τ_i as the output. The loop around this operator is closed (see Figure 1) with the vector integrator with gain Γ, which is a passive block. It follows from passivity theory that this feedback connection of one strictly passive and one passive block is globally stable.

5 Concluding Remarks

In the tuning functions design the controller and the identifier are derived in an interlaced fashion. This interlacing led to considerable controller complexity and inflexibility in the choice of the update law.

It is desirable to have adaptive designs where the controller can be combined with different identifiers (gradient, least-squares, passivity based, etc.). Such **modular** designs have also been developed [1] and have certain advantages over the tuning functions design.

Judging from the number of publications and significant applications, adaptive nonlinear control is a rapidly growing area of research. In the presence of disturbances and unmodeled dynamics, adaptive nonlinear controllers may lead to complex dynamic phenomena and chaos, which are the topics awaiting further exploration.

Acknowledgements

This work was supported in part by the National Science Foundation under Grant ECS-9203491 and in part by the Air Force Office of Scientific Research under Grant 442530-25335.

References

[1] M. Krstić, I. Kanellakopoulos, and P. V. Kokotović, *Nonlinear and Adaptive Control Design*, New York, NY: Wiley, 1995.

Commentary by W. Grantham

This paper presents a Lyapunov control design method for a special class of nonlinear systems with unknown parameters. The objective is to design a controller to stabilize the origin, despite a vector of unknown system parameters θ. The "backstepping" approach assumes a cascading-state linear portion of the system, with typical equation $\dot x_k = x_{k+1} + \varphi_k(x_1, \ldots, x_k)^T \theta$. The approach is to treat x_{k+1} as a control variable for x_k and then choose x_{k+1} to 1) cancel the parameter terms and 2) add a stabilizing term to the $\dot x_k$ equation. The actual control, which occurs only in the last $\dot x_n$ equation, is used to stabilize the system as well as generate an estimate $\hat\theta$ for the unknown parameters θ.

The stability of the resulting system is established using Lyapunov methods, where $V(x,\theta) > 0$ and $\dot{V} < 0$ for $x, \theta \neq 0$ would imply $V \to 0$. However, since the paper only requires $\dot{V} \leq 0$, there is no assurance that the parameter estimate $\hat{\theta}(t)$ converges to the actual unknown parameter vector θ.

The design approach is quite interesting. In its current state it lacks flexibility in the choice of a Lyapunov function V, and the approach does not survive a linear transformation of coordinates. But these difficulties should not be insurmountable, allowing the method to be extended to more general systems.

Commentary by T. Vincent

The authors combine two topics of considerable interest in dynamical systems, nonlinear control design and parameter identification. The combination is a logical one as the backstepping design method used here assumes knowledge of the system parameters. The basic idea behind the procedure, as outlined in the introductory examples, makes clever use of Lyapunov functions.

Suppose, as an example, under what the authors call design A, we consider the linear system
$$\dot{x}_1 = u + \theta x_1$$
where θ is an unknown parameter.. Following the authors methods, one can arrive at the following nonlinear controller

$$\begin{aligned} u &= -x_1 - x_2 x_1 \\ \dot{x}_2 &= x_1^2 \end{aligned} \quad (70)$$

where x_2 is the estimate of θ. This result is arrived at by using the Lyaponov function

$$V = \frac{1}{2}x_1^2 + \frac{1}{2}(x_2 - \theta)^2$$

so that
$$\dot{V} = -x_1^2 + (x_2 - \theta)(\dot{x}_2 - x_1^2).$$

Clearly the parameter identification system as provided by (70) will guarantee $\dot{V} < 0$. As that authors point out, this will result in asymptotic stability for x_1. However asymptotic stability for x_2 is not guaranteed as $\dot{V}$ may go to zero before $V = 0$. This is a curious result in that we are guaranteed to control the system to the origin, in spite of the fact that the parameter need not be properly identified. The following table illustrates the results of integrating (70) to equilibrium starting the system at $x_1(0) = 0.5$ with $\theta = 3$.

$x_2(0)$	$x_1(t_f)$	$x_2(t_f)$	max$|u|$
-5	0	7.83	31.61
-2	0	6.02	17.17
0	0	4.06	7.20
2	0	2.50	1.52
5	0	5.04	3.00

Thus, as promised, the system goes to equilibrium at $x_1(t_f) = 0$ for every case. However in none of the cases does $x_2(t_f) = 3$. Note that in order to get this stability, a large amount of control may be needed. It is easy to show through simulation that if the control is bounded, say $|u| \leq 10$, then simply using (70) under saturating control will result in an unstable system for the cases where $x_2(0) = -5$ and -2. These observations do not lessen the importance of the method developed by the authors. However it does suggest that using this method with systems operating under bounded control, one can not assume global convergence and the domain of attraction becomes a topic of interest.

If one is interested in identifying the system parameter θ, then, as Petar Kokotovic related to me in a personal communication, one must use persistency of excitation of the input signal. He suggested simply using a nonzero set point in the above example. In this case the control is given by

$$\begin{aligned} u &= 1 - x_1 - x_2 x_1 \\ \dot{x}_2 &= x_1(x_1 - 1) \ . \end{aligned} \qquad (71)$$

Using the Lyaponov function

$$V = \frac{1}{2}(x_1 - 1)^2 + \frac{1}{2}(x_2 - \theta)^2$$

one arrives at

$$\dot{V} = -(x_1 - 1)^2 + (x_2 - \theta)[\dot{x}_2 - x_1(x_1 - 1)] \ .$$

Using the same initial conditions as before one obtains

$x_2(0)$	$x_1(t_f)$	$x_2(t_f)$	$\max|u|$
-5	1	3	46.01
-2	1	3	23.70
0	1	3	12.80
2	1	3	5.40
5	1	3	3.10

In this case the parameter gets identified and the system gets driven to the equilibrium state. However a large amount of control may be needed.

The authors generalize their method for systems far more complex (and nonlinear) systems than suggested by this example. I think it represents a very interesting approach to nonlinear design and I am anxious to examine it use in some nonlinear laboratory systems, which, of course, have bounded control inputs.

Creating and Targeting Periodic Orbits

Kathryn Glass, Michael Renton, Kevin Judd and Alistair Mees
Centre for Applied Dynamics and Optimization
Department of Mathematics
University of Western Australia
Perth, 6907, Australia

Abstract

We look at controlling a two dimensional chaotic map by perturbing the value of a parameter in the system. By extending a simple targeting algorithm, we develop algorithms that will target and stabilize such maps in the presence of a small amount of noise. In particular, we show how to create and stabilize new periodic orbits which are close to existing non-periodic trajectories.

1 Introduction

Many problems encountered in mathematics involve nonlinear dynamical systems that display chaotic properties. One of the properties of chaos is that nearby trajectories diverge exponentially with time. In other words, if we choose any two starting points for the system, the long-term future of these points will appear unrelated however close the starting points are to one another. This property is known as the sensitive dependence on initial conditions. See [6] for a detailed discussion of the use of chaos in control, and also for an extensive bibliography on this area.

Since we can only know the position of the system at any time to some finite degree of accuracy, the property of sensitive dependence on initial conditions makes it impossible to predict the long-term future of the system. Here, the meaning of "long term" will vary for different systems, as is will depend on the rate at which nearby trajectories diverge. However, for all chaotic systems. any small inaccuracy in the specification of a point will lead (eventually) to large inaccuracies in its trajectory. Under these conditions, it is useful to be able to exert control over the system. By applying control to the system, we wish to be able to choose some type of long-term behavior for the system and then force it to display this behavior.

The problem of controlling chaotic systems has been approached in two different ways. Control theorists have looked at applying the methods that have been well developed for controlling linear systems to this class of nonlinear systems. This method of control attempts to remove the nonlinearities from the system. If it can be converted into a linear system we can then apply the types of control used in linear theory. In other words, this approach removes (or attempts to remove) the property of sensitive dependence on initial conditions from the system, and then applies control to the new non-chaotic system.

Researchers in dynamical systems approach the control of chaos from a different viewpoint. Their methods try to exploit the chaotic properties of the system to control its long term future. This type of control, which is generally referred to as OGY control (E. Ott, C. Grebogi, J.A. Yorke), makes use of the variety of behaviors embedded in a chaotic system to control the system. By applying very small perturbations to the system, this form of control uses the sensitive dependence on initial conditions to decide the long term behavior of the system.

We will discuss algorithms presented in [5] that will control chaotic maps of the form

$$x_{n+1} = f(x_n, p_n)$$

where x_n is the state of the system at the nth iteration and p_n is the value of some control parameter at the nth iteration. In general, the control parameter will have an unperturbed or nominal value, $\hat{p}$, and we then define a set P of allowable values for the control parameter, where P is a some neighborhood of $\hat{p}$.

We want to consider the types of trajectories that will result from perturbing the value of p to other values in the set P. This leads us to the following definition:

Definition 1 A sequence $y = \{y_t\}_{t=1}^M$ is a *P-chain* if, for each $t = 1, 2, ..., M-1$, y_{t+1} is in the set $f(y_t, P) = \{f(y_t, p) : p \in P\}$.

That is, a P-chain is a trajectory that may result when perturbations from the set P are made to the state of the system at each iteration. A trivial example of a P-chain will be a trajectory in the unperturbed system. This P-chain is constructed by setting the value of the parameter p_i at all steps to be the nominal value, $\hat{p}$.

We can then extend this idea to consider P-chains with recurrence properties. We define a *periodic P-chain* as follows:

Definition 2 A sequence $y = \{y_t\}_{t=1}^M$ is a *periodic P-chain* of period M if y is a P-chain and $y_1 \in f(y_M, P)$.

In other words, a periodic P-chain is a P-chain where perturbations can be chosen so that the trajectory repeats itself after every M iterations. Although in general it is not a periodic orbit of the original unperturbed system, it behaves like an orbit. We will look at creating these periodic P-chains and discuss methods of targeting and stabilizing them. To do this, we need to review some simple targeting algorithms.

2 Targeting Sets in Chaotic Systems

The ideas behind targeting sets and points in a chaotic system has been well developed in [3, 6, 7, 9]. The basic targeting problem can be stated very simply as follows. We start at some point x_s in state space and we wish to direct the system into some target set T in state space. To express this in terms of P-chains, we are given a start point x_s and a target set T and we want to find a P-chain $y = \{y_t\}_{t=1}^M$ of length M with $y_1 = x_s$ and y_M in T. That is, we choose values for the control parameter at each stage so that if we start at $y_1 = x_s$, after M steps the state of the system, y_M is in our target set T.

This targeting problem may often be solved using very small perturbations to the control parameter p by relying on the property of sensitive dependence on initial conditions. For the purposes of this paper, we will only look at a very simple form of the targeting algorithm. In order to derive this algorithm, we need to consider the sets of points that may be reached from our starting point x_s. If we recall that the equation governing the motion is given by $x_{n+1} = f(x_n, p_n)$, we can define the set S_1

of points that may be reached from x_s under one iterate by

$$S_1 = \{f(x_s, p) : p \in P\}.$$

Similarly, for $n > 1$ we can define the set of points that may be reached from x_s under n iterations by

$$S_n = \{f(x, p) : x \in S_{n-1} \text{ and } p \in P\}.$$

Since these reachable sets contain, by definition, the points that can be targeted from our initial point under a certain number of iterations, as soon as a reachable set intersects the target set, the problem is solved. Any point in this intersection may be reached from the start point and, as it is in the target set, once we reach this point, we have solved the targeting problem.

We find that at each step, the reachable sets are compressed onto and stretched out along the attractor of the unperturbed system. In fact, if n is large enough, the reachable set S_n will contain points arbitrarily close to any point on the attractor (see [8, 7]). Thus, if the target set T intersects the attractor of the unperturbed system, there should be a solution for any perturbation set P containing $\hat{p}$ (if we are willing to wait long enough).

When we come to implement the procedure of recursively calculating reachable sets until a set intersects the target set, we find that it is not practical to evaluate these sets exactly. Due to the chaotic nature of the system, after the first few steps the reachable sets become extremely complicated, and it is preferable to approximate them by finite collections of points. In order to simplify this algorithm, we consider a special case where we allow any perturbation at the first step, but thereafter restrict the control parameter p_i to be its nominal value $\hat{p}$. When we consider the reachable sets, we have S_1 as before, but S_n is given by:

$$S_n = \{f^{n-1}(f(x_s, p), \hat{p}) : p \in P\} = f^{n-1}(S_1).$$

In other words, the recursive procedure of calculating the reachable sets is reduced to that of calculating the first set S_1 and then mapping this set forwards through the system.

We implement the algorithm by choosing a finite collection (k, say) of possible initial perturbations, equally spaced throughout the set P. We then look at the possible images of the initial point x_s under the function, where the control parameter is perturbed initially by some value from this finite set and thereafter it is set to the nominal value $\hat{p}$. In this way, we approximate each reachable set by k points. As we will demonstrate later, these sets become complicated as n increases. In some cases, we will find that k points are not sufficient to approximate reachable sets for large n. We improve the approximation by increasing the number of initial perturbations chosen from the set P (that is, increasing k) until we are satisfied with the approximation. The extra perturbations may be chosen to maintain the even distribution throughout P, or may be selected by a more advanced algorithm, depending on the efficiency required of the method. Once we have chosen the best initial perturbation to target the set T from the set of k possible perturbations, we may improve the choice of perturbation using a similar refining step. We then consider a number of possible initial perturbations in some neighborhood of this first choice, and find the best of these. We may repeat this process as many times as we wish, shrinking the neighborhood of the chosen perturbation each time, until we are satisfied with the degree of accuracy.

To give an example, we introduce a system which is similar to, but more complex than, the Hénon system. The system is:

$$\begin{pmatrix} x_{n+1} \\ y_{n+1} \end{pmatrix} = f \begin{pmatrix} x_n \\ y_n \end{pmatrix} = \begin{pmatrix} p + 0.3 y_n - 0.7 x_n^3 + 2.1 x_n \\ x_n \end{pmatrix} \qquad (1)$$

The control parameter in this system is p and the unperturbed value, $\hat{p}$ is zero. Figure 1 shows reachable sets for this system evaluated by the method outlined above with an arbitrary initial point $(0.1, 0.1)$ and with initial perturbations from the set $P = (-0.01, 0.01)$. The fourth diagram gives the set S_{16}, which is a reasonable approximation to the attractor of this system. So, if the target point lies on, or near the attractor, we should be able to target it to a reasonable degree of accuracy in 16 steps. We can outline the algorithm that solves the targeting problem as follows:

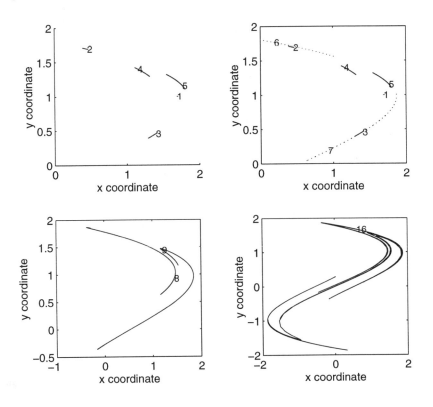

Figure 1: This figure shows the reachable sets calculated with initial point $(0.1, 0.1)$ and $P = (-0.01, 0.01)$. The numbers beside each set indicate the reachable set plotted.

Initial Perturbation Algorithm

- We are given a start point x_0 and a target set T.
  Set n to 1.

- **while** (S_n does not intersect T) **do**

    · Evaluate S_n.

    · Increment n.

- Choose x^* in the intersection of S_n and T.
  Set p^* to be the corresponding initial perturbation.

This algorithm will give us an initial perturbation p^* that will target the trajectory into the set T under no further perturbations to the system, provided there is no noise in the system and provided the target set intersects the attractor. The problems created by noise in the system will be discussed in the next section. When it comes to targeting points lying off the attractor, it is necessary to make further perturbations. As the trajectory is pushed onto the attractor with each iteration of the function, these perturbations should be made towards the end of the sequence so as to have the most effect. We discuss this in more detail in [1]. Clearly, however, if we continue to restrict the possible values of the control parameter to be in some small set P, the distance from the attractor that we can target is also restricted. To target points lying some distance from the attractor, a different method of control would be required. Another potential limitation to the algorithm is in targeting sets in systems of higher dimension. In this paper we give examples using a system whose phase-space is two-dimensional. Although the algorithms are applicable to higher-dimensional systems, it is clear that the number of points needed to approximate the reachable sets will increase as the dimension of the phase-space increases.

3 Targeting with Noise

While the Initial Perturbation Algorithm outlined in the previous section is effective for targeting in a noise-free system, once we introduce noise into the system, the algorithm is no longer effective. After a number of steps, the small errors caused by the noise will produce large changes in the resulting trajectory, due to the chaotic nature of the system. In other words, the system will wander from the targeting P-chain and, in most cases, fail to hit the target set. Clearly it is necessary to make further perturbations to the system to ensure that the target set is reached. One approach to this problem is to re-apply the targeting algorithm at every step of the process. This amounts to discarding the previous targeting P-chain and calculating a new one at each step. Although this is effective, it becomes time consuming, as we need to apply the Initial Perturbation Algorithm repeatedly.

An alternative solution to the problem of compensating for noise is to design an algorithm that will "track" or follow the targeting P-chain once it is found. In other words, we find a single P-chain that reaches the target set, and at each stage we make some perturbation to compensate for the noise that pulls the system away from this P-chain. Since the noise produces only slight deviations from the P-chain at any stage, we may make a correction for this error before the chaotic nature of the system causes the error to escalate.

We can design an algorithm that corrects for the noise quite simply, by using a form of targeting algorithm. Assume that the targeting P-chain has been calculated, and the points in the chain are stored. At each step, we want to correct for the error between the current step in the P-chain and the actual state in the system. We do this by targeting the next point in the P-chain. This gives us a perturbation that will direct the trajectory back on to the P-chain. Obviously, there will be further noise after this perturbation is implemented, but the effect of the noise at the previous step will have been removed. In this way, the errors due to noise are prevented from building up and causing the trajectory to move away from the targeting P-chain. We must be careful, however, to ensure that the perturbations that are made to the control parameter are sufficient to overcome this noise. Once the noise level reaches the order of the allowable perturbations, we may be unable to compensate for it by adjusting the control parameter. We give an outline of the correcting algorithm below.

Correcting Algorithm

- We are given a P-chain $y = \{y_t\}_{t=i}^M$ with $M \geq i+2$ such that the current state of the system, x_n, is in the neighborhood of y_i.

- Apply the Initial Perturbation Algorithm with $x_s = x_n$, a specified number of steps $N = 1$ and a target set, T, some small neighborhood of y_{i+N}.
  The algorithm produces the optimal initial perturbation p^*.

- Calculate $x_{n+1} = f(x_n, p^*)$, the new state of the system.

In the above algorithm, we have given the case where the perturbation is chosen so that we target the next point in the chain at the next step. We could alternatively aim to target a point two places further on in the chain in two steps of the system, by replacing N by 2. The algorithm may be applied with any choice of N, but is most effective when N is small. We can now outline a targeting algorithm that may be used in systems containing noise.

Targeting Algorithm

- We are given a start point $x_s = x_0$ and a target set T.

- Apply the Initial Perturbation Algorithm to find a P-chain $y = \{y_t\}_{t=0}^M$ with initial perturbation p^* such that $y_0 = x_s$ and y_M is in the set T.
  Set $p_0 = p^*$, set $i = 1$ and set $n = 1$.

- **while** $i < M$ **do**

   · Calculate $x_n = f(x_{n-1}, p_{n-1})$.
   · **if** $i < M - 1$
      · Apply the Correcting Algorithm to the P-chain $y = \{y_t\}_{t=i}^M$
      · This produces a perturbation p^*.
   · **else**
      · Apply the Initial Perturbation Algorithm with start point x_n and target set T.

· This produces a perturbation of p^*.
· Set $p_n = p^*$.
· Increment n. Increment i.
· **end while**

- Put $x_n = f(x_{n-1}, p_{n-1})$. Then x_n is in T.

The Targeting Algorithm is constructed in two stages: firstly a targeting P-chain is found and secondly this P-chain is tracked using the Correcting Algorithm. We recall that an alternative approach to the problem of compensating for noise was to re-apply a targeting algorithm at every step. If we compare this approach with our algorithm we see that, while we are also applying a targeting algorithm at each step, we are only targeting one step ahead, rather than a number of steps. Over a small number of steps, the reachable sets are nearly linear and thus may be approximated by a small number of points. Over a larger number of steps, however, these reachable sets become complicated, as shown in Figure 1. We need considerably more points in this case to get a reasonable approximation for the sets in order to calculate optimal perturbations. Thus an algorithm that need only calculate reachable sets a few steps ahead is more efficient than one that needs to look a number of steps ahead. For this reason, the Targeting Algorithm is an efficient method of targeting in the presence of noise.

Let us consider some examples using the system defined by equation (1). Figure 2 shows two examples where a point has been targeted with and without compensation being made for the noise in the system. In each of these examples, the point marked by a circle has been targeted from the point $(1,1)$. The system includes a noise term uniformly distributed with magnitude less than or equal to 0.005 and the maximum allowed perturbation in this case is 0.01 (that is, $P = (-0.01, 0.01)$). Firstly, the point has been targeted by the Initial Perturbation Algorithm without compensating for the noise, and the point reached indicated by a plus. Secondly, the point has been targeted by the Targeting Algorithm, which uses the Correcting Algorithm at each step, and the final point indicated by an asterisk. An approximation of the attractor of the unperturbed system is plotted in the background. Clearly, without the Correcting Algorithm, we are unable to target the point.

4 Stabilizing Periodic Orbits

A frequent objective when controlling a chaotic system is to stabilize a periodic orbit in the system, as discussed in [4]. Generally, a periodic orbit is chosen because it displays behavior that is required of the system. This behavior could be that of avoiding some area of phase space or it could be required that the system visit the neighborhood of a particular point periodically. When a periodic orbit is stabilized, we fix the behavior of the system indefinitely, and can thus ensure that the required conditions will always be met. Mostly we are not able to make any periodic orbit globally stable, and must then wait until the system enters the neighborhood of one of the points on the orbit. Once the system is close to one of these points, a stabilization algorithm is used to maintain the system on the orbit. We may reduce the time required for the system to enter this region by applying the Targeting Algorithm to target such a point.

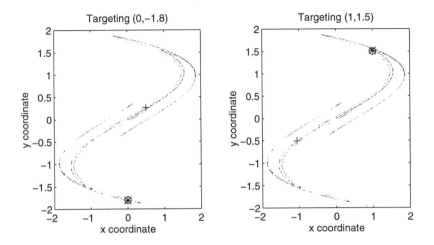

Figure 2: Two examples of points targeted from start point (1,1) by the Targeting Algorithm. The target point is marked with a circle, the point reached by the Targeting Algorithm is marked by an asterisk and the point reached if no correction is made for the noise is marked with a plus. An approximation of the attractor is plotted in the background.

The stabilization procedures that are used to maintain the system on a periodic orbit generally involve linearizing the system in the neighborhood of one of the points on the orbit and applying a form of feedback control (see [8]). These algorithms tend to become less effective for higher periodic orbits, especially in systems containing noise. The requirement for stabilization of a periodic orbit is that once the system is close to one point on the orbit, it should be close to the next point on the orbit at the next iteration. This is an obvious application for the Correcting Algorithm. Once we are close to a point on the periodic orbit, continually targeting the next point on the orbit will keep the trajectory on the periodic orbit, and thus stabilize it. Although we have described this stabilizing algorithm in terms of periodic orbits, we find that it is relevant to any P-chain whose points have been determined. The algorithm is outlined below:

P-Chain Stabilization Algorithm

- We are given a P-chain $y = \{y_t\}_{t=1}^{M}$. The current state of the system, x_n is in the neighborhood of y_1.

- **for** $i = 1$ to $M - 2$ **do**

    · Apply the Correcting Algorithm to target y_{i+1}.

    · The algorithm produces the optimal initial perturbation p^*.

    · Calculate $x_{n+1} = f(x_n, p^*)$, the new state of the system.

    · Set $n = n + 1$.

Since the Correcting Algorithm has been designed for systems containing noise, it follows that the P-Chain Stabilization Algorithm is also robust in noisy systems. We also notice that the algorithm is equally effective for stabilizing high period orbits as it is for stabilizing fixed points, as the algorithm is only considering the next point in the P-chain at any step. Thus the algorithm will work irrespective of the length of the periodic orbit or P-chain.

Although this algorithm is most useful for stabilizing high order periodic orbits, it may be used to stabilize orbits of any period, including fixed points. To test the algorithm on noisy systems, we will stabilize two period two points and a fixed point of the system given by (1). We include a noise term which is uniformly distributed with absolute value at most 0.005 and we restrict the perturbation set to $P = (-0.015, 0.015)$. In Figure 3, we plot the x coordinate of the system at each iteration as we target and stabilize points in the system. The circles in the plot indicate points at which the targeting algorithm is implemented, and the asterisks show where the P-Chain Stabilization Algorithm is turned off.

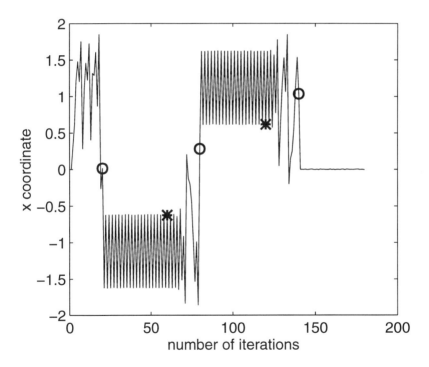

Figure 3: Graph showing the x coordinate of the state vector as the Targeting Algorithm and the P-Chain Stabilization Algorithm are applied to the system given by equation (1) containing a noise term. The points at which the targeting algorithm is applied are indicated by circles and the points at which the stabilization algorithm is discontinued are indicated by asterisks.

While the P-Chain Stabilization Algorithm was designed for stabilizing periodic orbits, it may be used to stabilize any form of P-chain. We find that by stabilizing P-chains with nontrivial perturbations, we can exert further control over the dynamics of the system.

5 Creating and Stabilizing P-Chains

We have already discussed the advantages of stabilizing periodic orbits to control the behavior of a system. We have also pointed out that the algorithms for stabilizing a periodic orbit could also be used for any form of P-chain. Thus, if we were to create a periodic P-chain, we may target and stabilize it in the same way we would target and stabilize a periodic orbit. By expanding our interest to the set of all periodic P-chains (of which the set of all periodic orbits is a subset) we have a greater variety of possible trajectories.

Creating periodic P-chains may be done very easily. Let us choose any initial point x_s on the attractor and target back to this point using the Initial Perturbation Algorithm. The resulting trajectory will not (in general) be a periodic orbit of the unperturbed system, as it will include some perturbations that allow it to repeat itself. It will, however, be a periodic P-chain. By applying the Initial Perturbation Algorithm, we may find these periodic trajectories very quickly, without having to search long sections of data for recurring behavior.

Let us now look at an example using system (1). Figure 4 shows two plots: the first gives 17 points in a periodic P-chain, and the second gives the x-coordinate of the system as this P-chain is targeted and stabilized. To generate the P-chain, the system was started at the point (-1, -0.5) and a small neighborhood of this point was targeted using the Initial Perturbation Algorithm. The points generated by this procedure are shown, with the initial point marked with a circle. The system was then started at the initial point, (-1,-0.5) and allowed to wander without control for 100 iterations. The x coordinate of the system is shown in the second plot. Clearly the behavior of the system at this point is not periodic when uncontrolled, and thus we have "created" the periodic behavior with the use of control. At the plus, the Targeting Algorithm is used to target a point on the P-chain, and it is then stabilized for over 100 iterations. For this example, the noise is uniformly distributed with maximum value 0.005 and we have $P = (-0.015, 0.015)$. Although the P-chain we have stabilized is not an orbit of the unperturbed system, there may be an (unperturbed) periodic orbit located near it. The significance of these P-chains, however, is that it is easier to create a chain with the behavior we require than it is to find an unperturbed trajectory that provides this behavior.

Of course, we are not restricted to stabilizing only one periodic P-chain — we may find and store points for a number of P-chains and then stabilize any one of them at any time. By constructing a number of periodic P-chains, each with different properties, we may select the type of behavior we want from the system and then stabilize the P-chain that produces this behavior. We may also change the behavior of the system by changing the periodic P-chain we are stabilizing. The simplest way of doing this is to stop stabilizing the first P-chain, use the Targeting Algorithm to move us to a point on the second P-chain, and then stabilize this second chain. However, should we need to move between the same periodic P-chains repeatedly, the targeting step of this process becomes inefficient. We will repeatedly locate a P-chain to connect

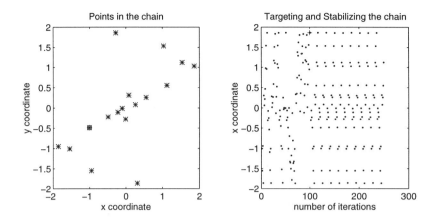

Figure 4: The first graph shows the 17 points in a periodic P-chain with the initial point marked with a circle. The second graph shows this chain being targeted and stabilized. The system is allowed to wander unperturbed initially, and the point at which the Targeting Algorithm is applied is indicated with a plus.

the same points. Since the P-Chain Stabilization Algorithm may be used for any P-chain, and not merely periodic chains, we may use it to stabilize a predetermined P-chain connecting these orbits. In other words, we determine P-chains that connect the periodic P-chains before we begin to control the system. The points in these connecting P-chains are stored in the same way that the periodic P-chains are stored. The process of moving between periodic P-chains is now achieved by changing the points that we stabilize. We stop stabilizing the first periodic P-chain and start stabilizing the connecting P-chain. When we reach the second periodic P-chain, we stop stabilizing the connecting chain, and begin stabilizing this periodic chain. As we have already remarked, this algorithm calculates the required perturbations very quickly, and may be applied in systems containing noise. A possible application of this is in the use of chaos in communications, as discussed in [2], where frequent switching between orbits is required.

We can demonstrate the above procedure with an example using system (1). Firstly we locate three periodic P-chains; each providing different behavior (see Fig. 5). The first has points distributed throughout phase space, the second chain consists only of points in the positive quadrant of phase space, and the third chain consists of points in the negative quadrant of phase space. We use the Targeting Algorithm to find P-chains connecting these periodic P-chains and store all points in these chains. We then stabilize the chains in turn. In this example, the system has a noise term of magnitude less than or equal to 0.005, and perturbations have magnitude at most 0.015. A plot of the x-coordinate at each iteration is given in Fig. 5.

Before we start the control, we let the system wander without control for 200 iterations. We then use the Targeting Algorithm to target a point on one of the P-chains. The point at which this is implemented is indicated with a plus. Once we reach the first periodic P-chain, the behavior of the system is completely controlled by

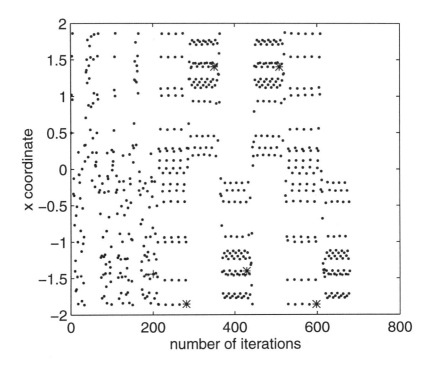

Figure 5: The x-coordinate at each iteration as three periodic P-chains are stabilized in turn. The Targeting Algorithm is started at the point marked with a plus and the asterisks indicate where the P-Chain Stabilization Algorithm switches from one periodic P-chain to the next.

the P-Chain Stabilization Algorithm in the manner outlined above. We stabilize the first periodic P-chain for 5 periods and then stabilize the P-chain connecting it to the second periodic P-chain. Once we reach the second chain, we begin stabilizing it, and so on. The points at which the system uses a connecting P-chain to switch periodic P-chains are indicated with asterisks. By storing both periodic and connecting P-chains in advance, we have reduced the amount of calculation that need be done during execution and thus increased the efficiency of the algorithms. Once we are in the neighborhood of any one of a number of previously determined chains, we require only the one algorithm to control the behavior of the system, even with a moderate amount of noise.

6 Conclusions

We have demonstrated that the flexibility present in a chaotic system may be exploited to control the behavior of that system. By making small perturbations to some control parameter, we are able to direct the trajectory of the system in a variety of different ways. The algorithms that are used to apply this control are all variants of a basic targeting algorithm, and are thus simple to implement. The algorithms do, however require knowledge of the equations governing the system. Since the perturbations that may be made to the control parameter are restricted, the possible behavior that may be produced from the system is largely confined to the attractor of the unperturbed system. Despite this, the large variety of orbits embedded in the attractor of the system allows us to stabilize many different types of behavior.

We have shown that the algorithms developed for targeting and stabilizing orbits are effective in systems containing dynamic noise. The quantity of noise that these algorithms will tolerate is largely dependent on the amount of perturbation that may be made to the control parameter. Although the examples considered in this paper have included very small noise terms, the quantity of noise added to the system could be increased if the neighborhood P was also increased. Once the quantity of noise present reaches the order of the maximum allowed perturbation, the algorithms become less effective. However, under these restrictions, these algorithms allow us to produce and stabilize many different trajectories of arbitrary length in a chaotic system.

References

[1] K. Glass, M. Renton, K. Judd, and A. Mees. Creating periodic orbits in chaotic systems. *Phys. Lett. A*, 203:107 – 114, 1995.

[2] S. Hayes, C. Grebogi, and E. Ott. Communicating with chaos. *Phys. Rev. Lett*, 70:3031 – 3034, 1993.

[3] E.J. Kostelich, C. Grebogi, E. Ott, and J.A. Yorke. Higher-dimensional targeting. *Phys. Rev. E*, 47:305 – 310, 1993.

[4] E. Ott, C. Grebogi, and J.A. Yorke. Controlling chaos. *Phys. Rev. Lett.*, 64:1196 – 1199, 1990.

[5] M. Renton. Controlling chaos with smart butterflies. Honours Dissertation, 1994.

[6] T. Shinbrot, C. Grebogi, E. Ott, and J.A. Yorke. Using small perturbations to control chaos. *Nature*, 363:411 – 417, 1993.

[7] T. Shinbrot, E. Ott, C. Grebogi, and J.A. Yorke. Using chaos to direct trajectories to targets. *Phys. Rev. Lett.*, 65:3215 – 3218, 1990.

[8] T. Shinbrot, E. Ott, C. Grebogi, and J.A. Yorke. Using chaos to direct orbits to targets in systems describable by a one-dimensional map. *Phys. Rev. A*, 45:4165 – 4168, 1992.

[9] T. Shinbrot, E. Ott, C. Grebogi, and J.A. Yorke. Using the sensitive dependence of chaos (the "butterfly effect") to direct trajectories in an experimental chaotic system. *Phys. Rev. Lett.*, 68:2863 – 2866, 1992.

Commentary by Y. Cohen

An interesting paper that deals with the "chaos" approach to control of chaotic systems (as opposed to the "optimal control" approach). I enjoyed reading the ms very much, and have only a few minor comments.

In their second paragraph of the introduction, the authors mention the long term un-predictability of chaotic systems. From an applied standpoint, in many systems we may be interested in the short term predictability. What light does the theory shed on short term predictability of dynamical systems?

P is defined as some neighborhood of $\hat{p}$. It may be interesting to define P in terms of the cost of control. A related question: What will be the control if we choose more than one step for p, before setting it to $\hat{p}$ (the nominal control)? This of course will depend on the objective criterion (e.g., time minimize, cost minimize).

One of the issues that needs to be addressed in "targeting control" is this: If you choose your target as a sub-region on the attractor, and you choose to ignore the problem of how long it will take you to reach your target, then it seems to me that controllability is not an issue (if you're willing to wait long enough).

The idea of setting the nominal control in noisy chaotic systems to counter the effect of noise is good; it is closely related to the traditional problem of feed-back control.

One extension to the paper that is intriguing to me: The authors deal with uniform noise. Because of non-linearities, what will be the effects of other probability distributions (of the noise) on the control of the system? This problem is important, for if (for example) the control takes a long time (as it may in this approach), and the noise is Normally distributed, then there is a likelihood that the system will blow up, unless the domain of attraction is the whole state space.

Commentary by E. Kostelich

This paper describes an approach to targeting in chaotic systems that is significantly different in detail from the approach by Kostelich et al. The key ingredient here is the construction of reachable sets, that is, all the points in the phase space that can be generated by all allowable perturbations of a system parameter. The geometrical structure of these sets can become quite complex as the map is iterated. The algorithms described in the paper use some simplified methods for approximating the reachable sets that work well for many maps of interest, such as the Henon map.

In some cases, this approach probably requires less computational time than that of Kostelich et al. In higher dimensions, however, more points may be needed to approximate the reachable sets, and more sophisticated data structures may be required to represent the surfaces or volumes that comprise the reachable sets. It is a challenging computational problem to identify all the periodic P-chains up to some period in a three- or higher-dimensional phase space.

Dynamical Systems, Optimization, and Chaos

John B. Moore
Department of Systems Engineering
and Cooperative Research Centre for Robust and Adaptive Systems
Research School of Information Sciences and Engineering
Canberra ACT 0200, Australia

Abstract

Much of engineering is concerned with the topic of optimization, and at the heart of much of our optimization is dynamical systems. Dynamical systems can be thought of as either non-linear continuous-time differential equations or difference equations. Chaos occurs in dynamical systems, and frequently in engineering we seek to avoid chaos. At times chaos becomes the central fascination.

This paper first introduces a situation in signal processing for neural systems in which chaos is the perhaps unexpected phenomena and the object of study. The focus then shifts to the topic of optimization of systems via dynamical systems, where traditionally chaos is avoided as much as possible. The essential dynamical system should converge in a very smooth manner to an optimal solution to some problem of interest. Our technical approach is summarized to optimization via dynamical systems is illustrated by an application in the area of robotics.

The key questions motivating this research are: Does the human brain exploit chaos for generating intelligence? Can our computing machines and control systems enhance their intelligence by a clever introduction of chaos?

1 Introduction

A Nobel prize-winning experiment in neurophysiology extracts very faint signals from synapses. A patch of the cell membrane with a gate molecule is studied. The one molecule acting as a gate opens and closes to allow various chlorine or potassium ions to flow through the cell membrane and activate the cell. The current flow is of the order of Femto amps. At these signal levels it is not surprising that noise due to thermal agitation of the molecule can dominate the measurement process. It is important to study these very small channel currents in order to not only understand signal processing in the brain in normal behaviour, but to study the effect of drugs for anesthesia, epilepsy and other conditions. A key question of interest is: Are the underlying processes at the cell membrane and synapse level governed by chaotic equations?

This paper points to research results which suggest that the underlying processes and synapse levels are in fact chaotic, see [9], and background material [3, 4, 6, 8, 10]. It is very difficult to be absolutely sure of such a conclusion because the signals are so much buried in noise. However, through experiments and signal processing, one can assess the self-similarity of the underlying signals at different resolution scales, and indeed estimate the fractal dimension of the underlying signals. Recall, that the fractal dimension is really a measure of the ruggedness of the underlying signals.

In the case of cell channel currents, the underlying signals appear to be currents which switch between a number of levels according to some transition probability law. The transition probabilities depend in a exponential manner on the time to the last transition. The longer the time since the last transition, the less likely there will be another transition. In cell channel currents, transitions occur in pico-seconds. It is clear that the inertia of the protein molecules forming the cell channel would be very small indeed.

It appears that in the process of evolution, there has been some advantage in exploiting chaos for the underlying processes within the human brain. Usually, chaos is avoided in performing a system design or optimization. The challenge before engineers is to somehow exploit the fascinating properties of chaos to enhance their system designs, and to further their ability to optimize and control their systems.

At this stage in our understanding of optimization, we do in fact exploit dynamical system behaviour, in particular discrete-time (recursive) systems for system optimization. The dynamical vector or matrix equations may be quite elegant and with the ability to flow on a constraint manifold towards an optimal solution. In the first instance, one is content that such dynamical systems converge to an optimal solution in a well-behaved manner. Subsequently, the motivation is to enhance the convergence capabilities of such algorithms by introducing non-smooth behaviour. There may be a deliberate introduction of ill conditioning into the equation or random perturbations to ensure the final desired outcome.

In Section 2 of the paper, we review the technical approach for investigating chaos in so called hidden Markov models for discrete-state systems and its application to study cell channel currents. In Section 3 we review the technical approach for system optimization via dynamical systems with an illustration from the application area of robotics. In the final Section 4, we discuss areas for current and future research.

2 Discrete-State Systems and Chaos

Let us consider dynamical systems which switch between discrete states, denoted $S_1, S_2, \ldots, S_N$. In discrete-time, the state of the system can be indicated by an indicator-vector X_k, where $k = 0, 1, \ldots$ and usually denotes a discrete-time sequence. The state X_k belongs to a discrete-set $\{e_1, e_2, \ldots e_N\}$ where e_i is the unit vector with unity in the ith element and zero otherwise. See Figure 1.

Two important properties of such indicator states X are as follows. Nonlinear functions of X are linear in X as

$$f(X) = f\left(\sum_{i=1}^{N} e_i X^{(i)}\right) = \sum_{i=1}^{N} f(e_i) X^{(i)} = FX. \tag{1}$$

where $F = [f(e_1), f(e_2) \ldots f(e_N)]$ and where $X^{(i)}$ denotes the ith element of X. This is readily checked since $X^{(i)} = 0$ for all i save some value j when $X^{(j)} = 1$ (i.e. $X = e_j$) and thus $f(X) = f(e_j)$.

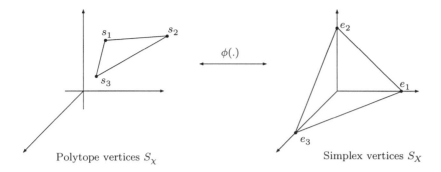

Figure 1: Depiction of state sets.

Also
$$E\left[X^{(i)}\right] = \sum_{j=1}^{N} e'_j e_i P(X = e_j) = P(X = e_i) \qquad (2)$$

Here $E[\cdot]$ is the expectation operator and $P(\cdot)$ denotes the probability. This result is immediate since $e'_j e_i = 0$ for $c \neq j$ and $e'_i e_i = 1$.

Consider that the system switches between states according to a probability law

$$X_{k+1} = AX_k + M_{k+1}, \qquad (3)$$

where A is a matrix of transition probabilities. That is, $E[X_{k+1}] = AE[X_k]$. It is immediately clear that M_{k+1} is a martingale increment process with the property

$$E[M_{k+1}|X_0, X_1, \ldots X_k] = 0 \qquad (4)$$

The transition matrix A of interest to us here will also depend upon the time to the last transition, denoted τ_k. Of course, $\tau_{k+1} = \tau_k + 1$ in the event that there has been no transition, and $\tau_k = 0$ in the event that there has been a state transition. The vector consisting of X_k and τ_k is seen to be first-order Markov, in that it depends only on the previous vector, X_{k-1}, τ_{k-1}, and not on any earlier such states. Thus the augmented state model is

$$\begin{bmatrix} X_{k+1} \\ \tau_{k+1} \end{bmatrix} = \begin{bmatrix} A(\tau_k) & 0 \\ 0 & \rho \end{bmatrix} + \begin{bmatrix} M_{k+1} \\ \rho \end{bmatrix}$$
$$\rho = \begin{cases} 0 & \text{if } X_{k+1} \neq X_k \\ 1 & \text{otherwise} \end{cases} \qquad (5)$$

The measurement process y_k belongs to continuous range, namely $\mathbb{R}^1$. Here we take

$$y_k = CX_k + w_k, \quad C = [c_1, c_2, \ldots c_n] \qquad (6)$$

where w_k is a discrete-time, identically and independently distributed noise process in a continuous range $\mathbb{R}^1$; here take w_k as zero mean, white and Gaussian with density $N[0, \sigma_w^2]$. In fact, we can think of the system switching between the N

states $S_i = Ce_i = c_i$ for $i = 1, 2, \ldots N$ with the measurements of the state of the system contaminated by the additive noise process w_k.

The above equations taken together denote what is termed a hidden Markov model. The word hidden refers to the fact that the states are hidden in noise. The term Markov indicates that there is an underlying state vector which summarizes all that we need to know about the past of the system in order to proceed in predicting its future. A simple situation is depicted in Figure 2.

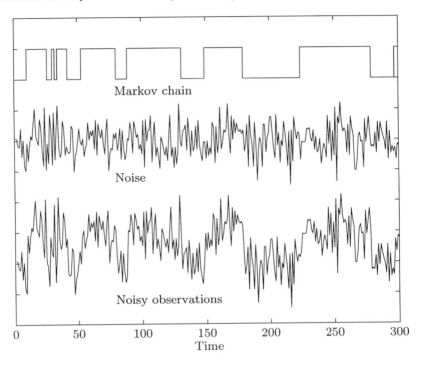

Figure 2: Binary Markov chain in noise.

A key signal processing task is to estimate the states of a hidden Markov model given the measurement data $y_0, y_1, \ldots y_k$. Ideally one would prefer a recursive signal processing scheme, which updates an estimate of the states at time $k - T$ given data up to time k, where T represents a delay in the processing so as to achieve improved estimates from future. In the first instance, signal processing algorithms assume knowledge of the transition probabilities A, the discrete-set states (parameters), namely $S_i = c_i$, and the statistics of the measurement noise process $w_k \sim N\left[0, \sigma_w^2\right]$. More sophisticated signal processing can simultaneously estimate both the parameters and the states of the model. In our situation where the transition probabilities can conceivably depend upon the time to the last transition, then one has to estimate this dependancy from the noisy data.

Of particular interest here is when elements of $A = (a_{ii})$ depend in an exponential manner on τ_k, as for example when for all i

$$a_{ii}(\tau) = a_{ii}(0) + (1 - a_{ii}(0))\left[1 - e^{-D\tau}\right] \quad (7)$$

Here D is taken as the fractal dimension of the signal y_k. Clearly, when $D = 0$, then $a_{ii}(\tau) = a_{ii}(0)$ is independent of τ, and as D increases $a_{ii}(\tau) \to 1$ in an exponential manner. That is, as D increases there is a greater tendency for X_k to stay in the same state e_i as τ_k increases, and conversely there is more likelihood of a state transition if τ_k is small.

With knowledge of the model $\{A(\tau_k), C, \sigma_w^2\}$ then the state estimates

$$\hat{X}_{k|k} = E\left[x_k|X_0, X_1, \ldots, X_k\right] = P\left(x_k|X_0, X_1, \ldots, D_k\right)$$

which evolve as illustrated in Figure 3, are given from

$$\begin{aligned} \alpha_{k+1} &= B\left(y_{k+1}, C\right) A(\tau_k)\alpha_k, \quad \alpha_0 \\ \hat{X}_{k|k} &= \left(\underline{1}'\alpha_k\right)^{-1}\alpha_k \end{aligned} \quad (8)$$

where $\underline{1}' = [1, 1, \ldots 1]$. Here α_k is an unnormalized version of the conditional density $p(X_k|X_0, X_1, \ldots X_k)$ so that α_0 is the *a priori* density $p(X_0)$.

Maximum likelihood estimation of D involves multi-passes, both forward and backwards, through the data [9, 10].

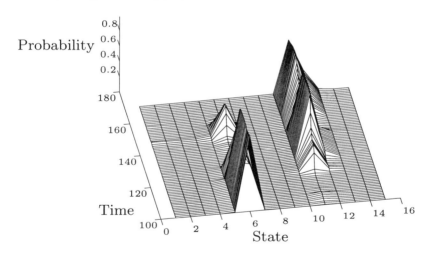

Figure 3: Evolution of state estimates.

On-line suboptimal methods based on recursive prediction error techniques can also be devised, see [6, 5]. The details are beyond the scope of this overview presentation.

In comparing how well various models and parametrizations fit the data, the key measure is simply the conditional probability p (model | data). In the absence of a priori information, the related probability p (data | model) is equivalent. In the neurophysiological model study, fractal models with fractal dimension in the range $D = 1.2$ to 1.6 were found to be most likely on the data tested, see [9].

3 Optimization via Dynamical Systems

In engineering applications, and in particular control applications, there is usually some underlying dynamical system description of a plant which has to be controlled by some control variable which, along with the states of the dynamical system, must satisfy certain constraints. For example, the control signals may be constrained so as not to exceed a certain magnitude. Since hardware is common to many industrial plants, the only competitive advantage of one plant over another is its control software. At the heart of this software are dynamical systems (recursive algorithms) which perform on-line optimization. These algorithms are driven by measurements from sensors placed on the process and their outputs drive the various actuators of the process. Contemporary research areas such as robotics, have brought to the fore novel control tasks. For example, in robotic dextrous hand-grasping, there are many fingers which must be co-ordinated so that in grasping and manipulating an object there is a balance of forces, excessive force is not used, and yet slipping is prevented.

Focusing on robotic hand-grasping, the existing optimization algorithms tend to use standard linear programming or non-linear programming techniques. Also, there are many ad-hoc devices supplied in the algorithms to achieve practical results. The challenge is to devise an on-line optimization which achieves well-conditioned optimal results, and rapid on-line calculations. For this task, we have proposed in [2, 1] that the friction constraints be viewed as the positive definiteness requirement of a certain matrix, while the force balancing constraints can be viewed as linear constraints on the elements of the positive definite matrix. The picture we have in mind then is of a cone, being the class of positive definite matrices, sliced by a hyper-plane, being the force balancing constraints, see Figure 4. The task is then to optimize the balancing of forces on this intersection of the cone and hyper-plane. Starting from an initial feasible solution at the intersection of the cone and hyper-plane, algorithms in the form of dynamical systems have been devised to converge to an optimal solution. It is important that the optimization be formulated so that there is a unique global minimum, and that the optimization is in essence a convex optimization task.

One of our first proposals for the dextrous hand-grasping problem requires a solution of a discrete-time Riccati equation modified to ensure projection of its solution into the constraint manifold. The index optimized is very similar to that which has been well studied for balancing controllability and observability in linear systems theory. It involves a term which penalises the forces at the fingertips, and a barrier function which prevents the constraints from being violated. There is a balance between these two requirements achieved in the optimization.

The optimization approach employed as expounded in [7] is to implement gradient flows of penalty functions on the smooth constraint manifolds of interest. Four key steps are formulations of the manifold, selecting cost function, choosing a Riemannian metric (or descent angle), and in discrete-time designing a step size. The "right" combination can result in elegant flow equations with linear (exponential) convergence properties to a global minimum. The "wrong" combination can result in "messy" equations which flow to local minima not the global minima.

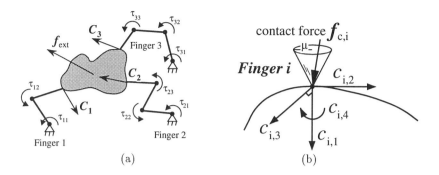

Figure 4: Object grasped by dextrous hand.

Decomposing the downhill search into two-dimensional geodesic searches, where step-size selection to achieve a minimum in the descent direction can be calculated analytically, results in highly efficient algorithms which are quadratically convergent. This is an area of current fruitful research.

To be more precise, consider the grasping situation depicted in Figure 4, with $c_{i,j}$ denoting the ith figure contact wrench (force) resolved in the j direction, then friction constraint requirements for N fingers are

$$\sqrt{c_{i,2}^2 + c_{i,3}^2} < \mu_i c_{i,1} \quad \text{for} \quad i = 1, 2, \ldots N \tag{9}$$

Here μ_i is the coulomb fraction coefficient and the $j = 1$ direction is normal to the finger contact. These inequalities are equivalent to the positive definitiveness of $P = \text{blockdiag}\{P_1, P_2, \ldots P_N\}$ where

$$P_i = \begin{bmatrix} \mu_i c_{i,1} & 0 & c_{i,2} \\ 0 & \mu_i c_{i,1} & c_{i,3} \\ c_{i,2} & c_{i,3} & \mu_i c_{i,1} \end{bmatrix} \quad \text{for} \quad i = 1, 2 \ldots N \tag{10}$$

The eigenvalues of P_i are

$$\lambda_i = \mu_i c_{i,1}, \quad \lambda_{2,3} = \mu_i c_{i,1} \pm \sqrt{c_{i,1}^2 + c_{i,3}^2} \tag{11}$$

The equilibrium or force balance constraint is in the form $f_{ext} = Wc$, where f_{ext} is the extended force vector, c is the vector of forces $c_{i,j}$ and W is the grip transformation matrix describing the geometric relation between contact wrench space and object co-ordinate frame [11]. This together with the structural constraints on P, i.e. $P_{i,21} = P_{i,12} = 0$ and $P_{i,11} = P_{i,22} = P_{i,23}$, and the blockdiagonal constraint can be represented as

$$A\text{vec}(P) = q \tag{12}$$

The index selected for optimization is

$$\phi(P) = tr\left(P + \rho P^{-1}\right) \tag{13}$$

for some scalar weighing $\rho > 0$.

The first term penalizes $c_{i,j}$ and the second is a barrier penalty function for the contraint (9). A very suitable Rienammenian metric for two vectors ξ_1, ξ_2 in the tangent space of the positive definite constraint manifold $P > 0$ is

$$\langle\langle \xi_1, \xi_2 \rangle\rangle = tr\left(P^{-1}\xi_1 P^{-1}\xi_2\right). \tag{14}$$

This leads to the gradient flow in the absence of the linear constraint (12) as

$$\dot{P} = \rho I - P^2 \tag{15}$$

and projecting into the hyperplane (12) we have the flow

$$\text{vec}(\dot{P}) = \left[I - A'(AA')^{-1}\right]\text{vec}(\rho I - P^2) \tag{16}$$

from which discrete-time flows can be derived, see [1, 2]. Figure 5 depicts the situation.

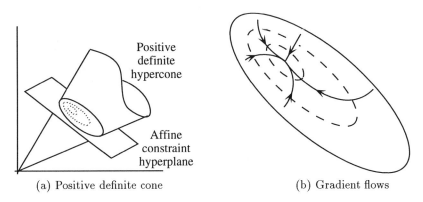

(a) Positive definite cone   (b) Gradient flows

Figure 5: Positive definite hypercone with affine constraints and gradient flows on the constraint hyperplane.

These gradient flows are appealing because of their mathematical elegance, but this is also their limitation. The flows are smooth, yielding exponential convergence. Quadratic convergence and faster convergence can be achieved using recursions such as

$$\text{vec}(P_{k+1}) = \left[I - A'(AA')^{-1}\right]\text{vec}\left(P_k - \alpha_k(\rho P_k^{-1} - P_k)\right)$$

for $0 < \alpha_k < 1$ which are more "violent". The α_k is selected to minimize the cost term (13) at each iteration.

In cases where the manifolds are not compact and convergence is not guaranteed, do flow equations exhibit chaos? Also, in more sophisticated optimization situations, perhaps with local minima not the global minima, must we use chaos itself to efficiently side-step a local minima?

4 Conclusions

This paper has summarized some recent research in signal processing and control in which chaos is the central fascination on the one hand and the allure to achieve improved results on the other hand. It seems clear from our studies that the human brain in its signal processing makes use of chaos for improved efficiency and performance. The challenge is for systems engineers working in the area of control applications to exploit the potential of chaos for enhanced control and on-line optimization. We have a long way to go.

References

[1] M. Buss, H. Hashimoto and J.B. Moore, "Dextrous Hand Grasping Force Optimization", *IEEE Transactions on Robotics and Automation*, to appear, see also *Proceedings of the IEEE Conference on Robotics and Automation*, pp 1034–1039, Nagoya, Japan, May 1995.

[2] M. Buss, L. Faybusovich, and J.B. Moore, "Dikin-Type Semidefinite Programming Algorithms for Dextrous Grasping Force Optimization", *International Journal of Robust and Nonlinear Control*, submitted.

[3] S.H. Chung, V. Krishnamurthy and J.B. Moore, "Adaptive Processing Techniques based on Hidden Markov Models for Characterizing Very Small Channel Currents Buried in Noise and Deterministic Interferences", *Philosophical Transactions of Royal Society*, Vol. 334, 1991, pp. 357-384.

[4] S.H. Chung, J.B. Moore, and L. Xia, P. Gage, L.S. Premkumar, "Characterization of Single Channel Currents using Digital Signal Processing Techniques based on Hidden Markov Models", *Phil. Trans. of Royal Society*, London B, Vol. 329, September 1990, pp. 265-285, see also *Proc. of Australian Physiological Society*, Newcastle, September 1989.

[5] I.B. Collings, V. Krishnamurthy and J.B. Moore, "On-line Identification of Hidden Markov Models via Recursive Prediction Error Techniques", *IEEE Transactions on Signal Processing*, vol 42, no 12, pp 3535-3539, see also *Proc. of IFAC World Congress*, Sydney, July 1993, vol V, 423-426.

[6] R.E. Elliott, L. Aggoun and J.B. Moore, *Hidden Markov Models: Estimation and Control*, Springer-Verlag, 1995, (361 pages).

[7] U. Helmke and J.B. Moore, *Optimization and Dynamical Systems*, Springer-Verlag, 1993, (390 pages).

[8] V. Krishnamurthy, J.B. Moore and S.H. Chung, "Hidden Markov Model Signal Processing in Presence of Unknown Deterministic Interferences", *IEEE Trans. on Automatic Control*, Vol 38, Jan 1993, pp.146-152, see also *Conf. on Dec. and Control*, Brighton UK, December 1991, pp. 662-667.

[9] V. Krishnamurthy, J.B. Moore and S.H. Chung, "On Hidden Fractal Model Signal Processing", Vol. 24, Issue No. 2, August 1991, pp. 177-191, see also *Proc. of Int. Symp. on Information Theory and its Applications*, Hawaii, December 1990, pp. 243-246.

[10] V. Krishnamurthy, J.B. Moore and S.H. Chung, "Signal Processing of semi-Markov Models with Exponentially Decaying States", *Conference on Dec. and Control*, Brighton UK, December 1991, Vol 3, pp. 2744-2749.

[11] M. Mason and J.Salisbury, *Robot Hands and the Mechanics of Manipulation*, Cambridge, Massachusetts: MIT Press, 1995.

Commentary by K. L. Teo

This paper first considers discrete dynamical systems which switch between states. The states are hidden in noise and there is an underlying state vector which summarizes all that are to know about the past of the system in order to proceed in predicting its future. This situation is termed Hidden Markov model. The technical approach is reviewed for investigating chaos in a hidden Markov model for discrete-state systems and its application to study cell channel currents.

The focus of the paper then shifts to the topic of optimization of system via dynamical systems. Usually, chaos is avoided in performing a system design or optimization. For an optimization problem in the area of robotic is used for illustration to a technical approach summarized in the paper.

The paper is both informative and interesting.

Commentary by T. Vincent

The author raises an interesting question in the abstract, "Can our computing machines and control systems enhance their intelligence by clever introduction of chaos?" One computing machine of interest to the author is the brain and he points to some results which suggest that the underlying processes and synapse levels are in fact chaotic. Another area of interest discussed by the author is optimization. While presenting an optimization approach, which uses gradient flows of penalty functions on smooth constraint manifolds, the question is raised if it is possible to use chaos to side-step local minima? Using chaos in this way to solve global optimization problems is a very attractive idea. No doubt we can look forward to progess in this area.

Six papers in this volume do address the authors question in regard to control systems. My own contribution uses the ergotic nature to chaos to eventually bring the system to a point in state space where some control action will be effective. The targeting methods discussed in the papers by Glass, Kostelich, and Ott represents a more 'clever' way to use chaos to accomplish the same objective. Mareels observes that chaotic behavior could be beneficial in achieving control objectives in adaptive control systems. Grantham shows that suppressing a single frequency can convert chaotic motion into a desired periodic one. I think the answer to Moore's question is a definite yes, but I also agree with his assessment, "We have a long way to go."

Combined Controls for Noisy Chaotic Systems

Mirko Paskota, Kok Lay Teo and Alistair Mees
Centre for Applied Dynamics and Optimization
Department of Mathematics
The University of Western Australia
Perth, 6907, Australia

Abstract

Consider a class of chaotic dynamical systems in a noisy environment. We propose a design method for the construction of a combined controller. There are two components involved in this combined controller: a directing controller and a local feedback correction. The directing controller is obtained by using a computational algorithm for solving open-loop optimal control problems. Its aim is to direct orbits of the dynamical system towards a desired target. The local feedback correction is to act on the dynamical system throughout the targeting process as a supplementary controller to counter the noisy effects. Numerical simulations are presented to illustrate the feasibility and efficiency of the proposed design method.

1 Introduction

There are two fundamental properties associated with any chaotic system. First, its orbits are extremely sensitive with respect to initial conditions. The second property is that its orbits are dense in the basin of attraction. In [13], it was demonstrated that the sensitivity and the denseness of orbits of a chaotic dynamical system can be utilized in the design of a stabilizing controller such that significant changes can be obtained by using only small, judiciously chosen perturbations. Other recent related articles are [5], [6], [8], [9], [11], [12], [14], [15], [16], [17], [18], [20], [21], [22], [25] and [27]. See also [4] — an up-to-date bibliography. From these articles, we see that some classical control engineering methods, such as linear state feedback control, can be used quite effectively to control chaos. By linearizing a nonlinear system around a desired target, we obtain a linearized system. Then, a linear state feedback control can be designed based on the linearized system. Due to the ergodicity property, the obtained local controller can be used to control orbits of a chaotic system to the desired target. However, the state may wander chaotically for a rather long time before entering the neighbourhood in which the local controller is effective. To overcome this problem, we propose to use a combined controller which consists of a directing controller with a feedback correction. The directing controller can be obtained by using the optimal control software package, DMISER3.1 (see [10] for details). This software was developed based on the discrete-time optimal control theory reported in [23] and [24]. The directing controller is to direct an orbit of the discrete-time chaotic dynamical system towards a pre-specified neighbourhood of the target, while the local feedback correction is to counter noise.

We consider only a class of nonlinear discrete-time dynamical systems where the analytical expression of the map is assumed available. Equivalently, we may assume that there is enough data for the reconstruction of the map together with its gradient.

The paper is organized as follows: In Section 2, the use of small perturbations to direct orbits of a nonlinear discrete-time dynamical system is posed as a discrete-time optimal control problem. In Section 3, a computational technique for solving a general class of discrete-time optimal control problems is briefly reviewed. The corresponding software package, DMISER3.1, is also discussed. In Section 4, a design method for the construction of the required local feedback correction is given. This local feedback correction acts as a supplementary controller to counter the effect due to the noisy environment. Numerical simulations are given to illustrate the efficiency of the proposed method.

2 Targeting Problem

Consider a class of chaotic systems described by the following system of nonlinear discrete-time difference equations:

$$\mathbf{x}(k+1) = \mathbf{f}(\mathbf{x}(k)), \quad k = 0, 1, 2, \ldots \quad (1)$$

where

$$\mathbf{x}(k) = [x_1(k), \ldots, x_n(k)]^T \in \Re^n$$

is the state, the superscript T denotes the transpose, and the function

$$\mathbf{f} = [f_1, \ldots, f_n]^T : \Re^n \to \Re^n$$

is continuously differentiable.

Let $\mathbf{x}^{(\mathrm{T})}$ be a desired 'target' state which is on or near the attractor of the dynamical system. The ergodic property of the chaotic dynamical system guarantees that any orbit in the basin of attraction eventually comes arbitrarily close to the specified target $\mathbf{x}^{(\mathrm{T})}$. However, it may take long time for this to happen in the absence of control. Thus, it is important to come out with a method which could use small perturbations to shorten the time needed for the orbit to reach a pre-specified neighbourhood of the target.

Let N be a fixed positive integer. Consider the following class of controlled nonlinear discrete-time dynamical systems:

$$\mathbf{x}(k+1) = \mathbf{f}(\mathbf{x}(k)) + \mathbf{u}(k), \quad k = 0, 1, 2, \ldots, N-1 \quad (2)$$

The control sequence

$$\{\mathbf{u}(0), \mathbf{u}(1), \ldots, \mathbf{u}(N-1)\}$$

is assumed to satisfy the following boundedness constraints:

$$\|\mathbf{u}(k)\| \leq \alpha \quad (3)$$

where $\|\cdot\|$ denotes the usual ℓ_2-norm, and α is a given constant. Any such control sequence is called an admissible control. Let $\mathcal{U}$ be the class of all such admissible controls.

The following is a list of three conflicting optimization criteria:

(i) the distance between $\mathbf{x}(N)$ and $\mathbf{x}^{(T)}$;

(ii) the number N of time steps; and

(iii) the bound α on the control.

Let $\mathcal{N}_\varepsilon$ be a neighbourhood around the target in which a local controller is effective. Then, we may pose an optimal control problem, where an admissible control $\mathbf{u} \in \mathcal{U}$ is to be chosen such that the number N of iterations needed for the controlled orbit to reach $\mathcal{N}_\varepsilon$ is minimized over $\mathcal{U}$.

Suppose it is only allowed to take, say κ, iterations. Then, we may choose to minimize the distance between $\mathbf{x}(\kappa)$ and the target with respect to $\mathbf{u} \in \mathcal{U}$.

The third optimal control problem is to minimize the bound α on the controls such that the distance from the target is less than or equal to some pre-specified small number $\varepsilon > 0$, where the number of time steps is restricted to be less than or equal to κ.

Among these three optimal control problems, we choose to formulate our problem formally as:

(P1) Let $N > 0$ be a given integer, and α a given constant. Subject to the dynamical system (1) with the initial condition

$$\mathbf{x}(0) = \mathbf{x}^{(S)}$$

find an admissible control $u \in \mathcal{U}$ such that the objective function

$$\left\|\mathbf{x}(N) - \mathbf{x}^{(T)}\right\|^2 = \sum_{i=1}^{n} \left(x_i(N) - x_i^{(T)}\right)^2$$

is minimized over $\mathcal{U}$.

This is a classical discrete-time optimal control problem. For brevity, we assume as in [14] and [15] that our control only acts on the first component of the vector-valued function $\mathbf{f}$.

With this restriction, our problem becomes:

(P2) Let $N > 0$ be a given integer, and α a given constant. Subject to the dynamical system

$$\begin{aligned} x_1(k+1) &= f_1(\mathbf{x}(k)) + u(k) & (4)\\ x_i(k+1) &= f_i(\mathbf{x}(k)), \quad i = 2, \ldots, n & (5) \end{aligned}$$

$k = 0, \ldots, N-1$, with initial condition

$$\mathbf{x}(0) = \mathbf{x}^{(S)}$$

find an admissible control $u \in \mathcal{U}$ such that the objective function

$$\left\|x(N) - \mathbf{x}^{(T)}\right\|^2 \qquad (6)$$

is minimized over $\mathcal{U}$.

3 Optimal Control Computation

The discrete-time optimal control problem (P2) can readily be solved by using the software package DMISER3.1. See [10] for details. This software was developed based on the algorithm reported in [23] and [24]. Let us give a brief review on this topic.

Consider the dynamical system

$$\mathbf{x}(k+1) = \mathbf{F}(k, \mathbf{x}(k), \mathbf{u}(k)), \quad k = 0, 1, \ldots, N-1 \qquad (7)$$

with the initial condition

$$\mathbf{x}(0) = \mathbf{x}^0 \qquad (8)$$

where

$$\mathbf{x}(k) = [x_1(k), \ldots, x_n(k)]^T \in \Re^n$$

is the state,

$$\mathbf{u}(k) = [u_1(k), \ldots, u_r(k)]^T \in \Re^r$$

is the control, and the function

$$\mathbf{F} = [F_1, \ldots, F_n]^T : \Re \times \Re^n \times \Re^r \to \Re^n$$

is continuous differentiable with respect to each of the components of $\mathbf{x}$ and $\mathbf{u}$.

Let $\mathbf{u}$ denote a control sequence $\{\mathbf{u}(0), \mathbf{u}(1), \ldots, \mathbf{u}(N-1)\}$ such that the following boundedness constraints are satisfied:

$$u_i^L \leq u_i(k) \leq u_i^U, \quad i = 1, \ldots, r; \quad k = 0, \ldots, N-1$$

where u_i^L and u_i^U, $i = 1, \ldots, r$, are given constants. Such a control sequence is called an admissible control. Let $\mathcal{U}$ be the class of all such admissible controls.

The state and control variables are assumed to satisfy, respectively, the following equality and inequality constraints in canonical form:

$$G_\iota(\mathbf{u}) = \Phi_\iota(\mathbf{x}(N)) + \sum_{k=1}^{N-1} g_\iota(k, \mathbf{x}(k), \mathbf{u}(k)) = 0, \quad \iota = 1, \ldots, M_e \qquad (9)$$

and

$$G_\iota(\mathbf{u}) = \Phi_\iota(\mathbf{x}(N)) + \sum_{k=1}^{N-1} g_\iota(k, \mathbf{x}(k), \mathbf{u}(k)) \geq 0, \quad \iota = M_e + 1, \ldots, M \qquad (10)$$

where Φ_ι and g_ι, $\iota = 1, \ldots, M$, are real-valued functions which are continuously differentiable with respect to each of the components of $\mathbf{x}$ and $\mathbf{u}$.

An admissible control, which satisfies constraints (9) and (10), is called a feasible control. Let $\mathcal{F}$ be the class of all such feasible controls.

A general class of discrete-time optimal control problems in canonical form may now be specified as follows:

Subject to the dynamical system (7)–(8), find a feasible control $\mathbf{u} \in \mathcal{F}$ such that the objective function

$$G_0(\mathbf{u}) = \Phi_0(\mathbf{x}(N)) + \sum_{k=0}^{N-1} g_0(k, \mathbf{x}(k), \mathbf{u}(k))$$

is minimized over $\mathcal{F}$, where Φ_0 and g_0 are real-valued functions which are continuously differentiable with respect to each of the components of $\mathbf{x}$ and $\mathbf{u}$.

Let this optimal control problem be referred to as problem (OP). In essence, we can view problem (OP) as a nonlinear mathematical programming problem in the control vectors. To solve it as a nonlinear optimization problem, we will need, at each iteration, the values of the objective function, all the constraint functions and their respective gradients. The objective function and all the constraint functions are treated in the same way as far as the computations of their values and respective gradients are concerned. To begin with, we first calculate the solution of the difference equations (7) with the initial condition (8) corresponding to a given admissible control $\mathbf{u} \in \mathcal{U}$. With this information, the values of G_ι, $\iota = 0, 1, \ldots, M$, corresponding to the $\mathbf{u} \in \mathcal{U}$ can easily be calculated by the following algorithm:

Algorithm For a given $\mathbf{u} \in \mathcal{U}$, compute the corresponding value of $G_\iota(\mathbf{u})$ from (9) (respectively, (10)) if $\iota = 0$ (respectively, $\iota = 1, \ldots M$).

For each $\iota = 0, 1, \ldots, M$, the gradient of the corresponding G_ι may be computed by using the following algorithm:

Algorithm Consider a given $\mathbf{u} \in \mathcal{U}$.

Step 1. Solve the co-state difference equations

$$(\lambda^\iota(k))^T = \frac{\partial H_\iota(k, \mathbf{x}(k), \mathbf{u}(k), \lambda^\iota(k+1))}{\partial \mathbf{x}(k)} \tag{11}$$

with terminal condition

$$\lambda^\iota(N) = \left(\frac{\partial \Phi_\iota(\mathbf{x}(N))}{\partial \mathbf{x}(N)}\right)^T \tag{12}$$

backward in time from $k = N$ to $k = 0$, where $\mathbf{x}(\cdot|\mathbf{u})$ is the solution the system (7) corresponding to $\mathbf{u} \in \mathcal{U}$; and H_ι, the corresponding Hamiltonian function for the objective function if $i = 0$ and the i-th constraint function if $\iota = 1, \ldots, M$, is defined by

$$H_\iota(k, \mathbf{x}(k), \mathbf{u}(k), \lambda^\iota(k+1))$$
$$= g_\iota(k, \mathbf{x}(k), \mathbf{u}(k)) + (\lambda^\iota(k+1))^T \mathbf{F}(k, \mathbf{x}(k), \mathbf{u}(k))$$

Let $\lambda^\iota(\cdot|\mathbf{u})$ be the corresponding solution of the co-state system (11)–(12).

Step 2. The gradient of G_i is computed from

$$\frac{\partial G_\iota(\mathbf{u})}{\partial \mathbf{u}} = \sum_{k=0}^{N-1} \frac{\partial H_\iota(k, \mathbf{x}(k), \mathbf{u}(k), \lambda^\iota(k+1))}{\partial \mathbf{u}(k)}$$

On the basis of these two algorithms, there are many efficient and reliable optimization techniques which we can use for solving the problem (OP). In DMISER3.1, the sequential quadratic programming technique, (for details, see [19]) is used for this purpose.

In this paper, we consider problem (P2) which is a special case of problem (OP). There are no constraints. More precisely,

$$M = 0, \quad g_0 = 0, \quad \Phi_0(\mathbf{x}(N)) = \left\|\mathbf{x}(N) - \mathbf{x}^{(\mathrm{T})}\right\|^2,$$

and

$$\mathbf{F}(k, \mathbf{x}(k), \mathbf{u}(k)) = \mathbf{f}(\mathbf{x}(k)) + \mathbf{u}(k) .$$

Here,
$$\mathbf{f} = [f_1, \ldots, f_n]^T \epsilon \Re^n$$
and
$$\mathbf{u} = [u, 0, \ldots, 0]^T \epsilon \Re^n .$$

The problem (P2) is solved by using DMISER3.1. The value of the objective function and its gradient with respect to the control vector at each iteration are done automatically. The user only needs to supply a set of explicit FORTRAN expressions for the functions Φ_0 and f_i and their derivatives with respect to $\mathbf{x}$ and $\mathbf{u}$. Choose an initial control estimate:

$$\mathbf{u} = \{\mathbf{u}^0(0), \mathbf{u}^0(1), \ldots, \mathbf{u}^0(N-1)\}$$

and specify the bound α on $\mathbf{u}$. Then, DMISER3.1 will generate a control:

$$\mathbf{u}^* = \{\mathbf{u}^*(0), \mathbf{u}^*(1), \ldots, \mathbf{u}^*(N-1)\} .$$

This may be a local minimum. In theory, we will be required to run DMISER3.1 from several different initial control estimates so as to obtain a 'global' optimal control. However, we have hardly ever encountered any problems with local minima in practice. A possible explanation for this outcome is that the problem admits only few local minima due to the strict bound on the control vectors.

Since our aim is to keep the number N of iterations small, we propose to solve the problem (P2) with $N = 1, 2, \ldots$ We stop at the smallest N for which

$$\left\| \mathbf{x}(N) - \mathbf{x}^{(T)} \right\|^2 < \varepsilon$$

where $\varepsilon > 0$ is a given constant.

To apply our method in practice, we specify a set of initial states. Then, we calculate off-line an open-loop controller corresponding to each of these initial states such that each of these controllers is to direct an orbit towards the desired neighbourhood of the target. We store these controllers. During a real control run, we would check the state of the dynamical system (4)–(5) and select an appropriate controller to apply.

4 Feedback Correction

It is well-known that the open-loop controllers lack robustness. The reason is quite obvious: the open-loop controller is calculated off-line, and is then used throughout the whole planning horizon regardless of the existance of disturbances.

We propose the following procedure: Subject to the dynamical system (4)–(5) with a given pair of initial and target states $\{\mathbf{x}^{(S)}, \mathbf{x}^{(T)}\}$, we choose parameters $\varepsilon > 0$ and $\alpha > 0$ (see Sections 2 and 3). Then, we solve the corresponding optimal control problem (P2) to obtain the optimal directing controller and the corresponding value of N. Let the obtained controller be denoted by

$$\mathbf{u}^* = \{\mathbf{u}^*(0), \mathbf{u}^*(1), \ldots, \mathbf{u}^*(N-1)\} .$$

$\mathbf{u}^*$ is called the (open-loop) directing controller. Assuming that there is no noise acting on the dynamical system (4)–(5), we calculate the corresponding trajectory. Let it be denoted by
$$\mathbf{x}^r = \{\mathbf{x}^r(0), \mathbf{x}^r(1), \ldots, \mathbf{x}^r(N)\} .$$
$\mathbf{x}^r$ is referred to as the reference trajectory. Now, at each iteration, we will add a feedback correction to supplement the directing controller. This feedback correction is proportional to the difference between the actual trajectory and the reference trajectory. In other words, we obtain a combined controller which is of the form

$$\widetilde{\mathbf{u}}(k) = \mathbf{u}^*(k) + K_k(\mathbf{x}(k) - \mathbf{x}^r(k)) \tag{13}$$

where $K_k \in \Re^{n \times n}$ is a feedback gain matrix.

With such controls, the dynamical system (4)–(5) reduces to:

$$\mathbf{x}(k+1) = \mathbf{f}(\mathbf{x}(k)) + \mathbf{u}^*(k) + K_k(\mathbf{x}(k) - \mathbf{x}^r(k)) \quad k = 0, 1, 2, \ldots, N-1 \tag{14}$$

Without relaxing much of the given bound on the controls, our aim is to find a feedback gain matrix K_k such that the actual trajectory would track the reference trajectory as closely as possible.

In view of the proposed combined control law given by (13), we see that when there is no random disturbance

$$\mathbf{x}(k) = \mathbf{x}^r(k)$$

and the last term on the right hand side of the difference equation (14) is zero. In other words, the combined controller will reduce to the open-loop directing controller in the noiseless environment.

In what follows, we shall develop a constructive method for the design of a desired local feedback correction

$$K_k(\mathbf{x}(k) - \mathbf{x}^r(k))$$

Clearly, this task is equivalent to finding a desired feedback gain matrix K_k.

Consider two successive points on the reference trajectory and their neighbourhoods (Figure 1). To prevent a trajectory from escaping from the neighbourhood of the reference trajectory, we should make sure that the proposed feedback correction will act as a supplementary controller to 'pull' the trajectory towards $\mathbf{x}^r(k+1)$ from all directions at each step. In other words, the feedback correction must be chosen such that the following condition is satisfied.

$$\mathbf{x}(k+1) - \mathbf{x}^r(k+1) = A_k(\mathbf{y} - \mathbf{x}^r(k+1)) \tag{15}$$

where the matrix A_k represents a contraction operator around $\mathbf{x}^r(k+1)$. A sufficient condition is that the eigenvalues of the matrix A_k lie inside the unit disc, i.e.,

$$\max |\text{eig}\{A_k\}| < 1 . \tag{16}$$

We now assume that a proper matrix A_k has been chosen such that the condition (16) is satisfied. Then, the left-hand side of (15) can be written as:

$$\begin{aligned}
&\mathbf{x}(k+1) - \mathbf{x}^r(k+1) \\
&= \mathbf{f}(\mathbf{x}(k)) + \mathbf{u}^*(k) + K_k(\mathbf{x}(k) - \mathbf{x}^r(k)) - (\mathbf{f}(\mathbf{x}^r(k)) + \mathbf{u}^*(k)) \\
&= \mathbf{f}(\mathbf{x}(k)) - \mathbf{f}(\mathbf{x}^r(k)) + K_k(\mathbf{x}(k) - \mathbf{x}^r(k))
\end{aligned}$$

while the right-hand side is:

$$A_k \left(\mathbf{y} - \mathbf{x}^r(k+1)\right) = A_k \left(\mathbf{f}(\mathbf{x}(k)) + \mathbf{u}^*(k) - (\mathbf{f}(\mathbf{x}^r(k)) + \mathbf{u}^*(k))\right)$$
$$= A_k(\mathbf{f}(\mathbf{x}(k)) - \mathbf{f}(\mathbf{x}^r(k)))$$

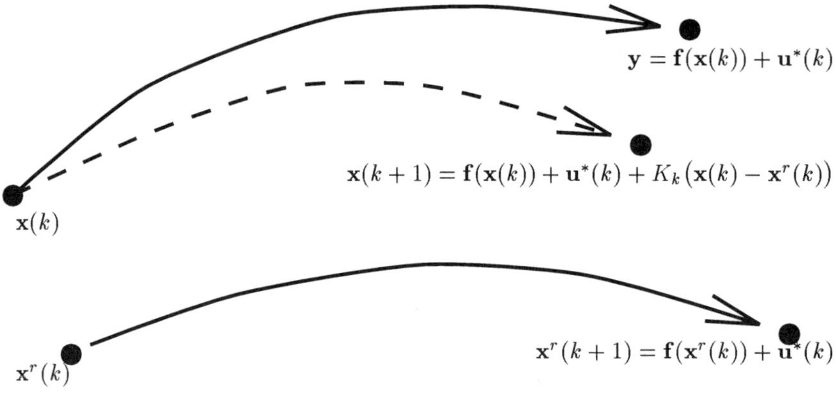

Figure 1: Two successive points on the reference trajectory.

Equating these two equations, we obtain

$$\mathbf{f}(\mathbf{x}(k)) - \mathbf{f}(\mathbf{x}^r(k)) + K_k(\mathbf{x}(k) - \mathbf{x}^r(k)) = A_k\left((\mathbf{x}(k)) - \mathbf{f}(\mathbf{x}^r(k))\right)$$

Simple re-arrangement yields:

$$K_k\left(\mathbf{x}(k) - \mathbf{x}^r(k)\right) = (A_k - I)\left(\mathbf{f}(\mathbf{x}(k)) - \mathbf{f}(\mathbf{x}^r(k))\right)$$

where $I \in \Re^{n \times n}$ is the identity matrix.

Now, due to the assumed closeness of $\mathbf{x}(k)$ to $\mathbf{x}^r(k)$, the corresponding linear approximation of the feedback correction can be written as:

$$\begin{aligned} K_k\left(\mathbf{x}(k) - \mathbf{x}^r(k)\right) &= (A_k - I)\left(\mathbf{f}(\mathbf{x}(k)) - \mathbf{f}(\mathbf{x}^r(k))\right) \\ &\cong (A_k - I)\left([D\mathbf{f}]_{\mathbf{x}^r(k)}(\mathbf{x}(k) - \mathbf{x}^r(k))\right) \end{aligned}$$

where $[D\mathbf{f}]_{\mathbf{x}^r(k)}$ denotes the Jacobian of $\mathbf{f}$ evaluated at $\mathbf{x}^r(k)$.

To simplify the notation, we quote a well known result in the following lemma.

Lemma 1. Let $B, C \in \Re^{n \times n}$. If $Bx = Cx$ for all $x \in \Re^n$, then $B = C$.

The proof is trivial.

Since the vector $(\mathbf{x}(k) - \mathbf{x}^r(k))$ is affected by the random noise, we must make sure that the feedback correction would work for any vector around $\mathbf{x}^r(k)$. Thus, in view of Lemma 1, we can simply write:

$$K_k = (A_k - I)[D\mathbf{f}]_{\mathbf{x}^r(k)} . \tag{17}$$

With such a feedback gain matrix, the dynamical system (14) reduces to:

$$\mathbf{x}(k+1) = \mathbf{f}(\mathbf{x}(k)) + \mathbf{u}^*(k) + (A_k - I)[D\mathbf{f}]_{\mathbf{x}^r(k)}(\mathbf{x}(k) - \mathbf{x}^r(k))$$

Let us now discuss how do we choose the appropriate matrix A_k at each step. Ideally, the best feedback correction is to push $\mathbf{y}$ straight back to $\mathbf{x}^r(k+1)$, i.e., to set $A_k = 0$ (all eigenvalues $= 0$, and hence obviously inside the unit disc). However, we may not be able to control all the states of the given dynamical system in a real situation. In light of this, and in accordance with the formulation of the problem (P2), where the directing controller acts only on the first component of the vector function $\mathbf{f}$, we need to modify the feedback correction as follows:

$$\widetilde{K}_k = B(A_k - I)[D\mathbf{f}]_{\mathbf{x}^r(k)} \tag{18}$$

where $B \in \Re^{n \times n}$ is the modification matrix given by

$$B = \begin{bmatrix} 1 & 0 & \cdots & 0 \\ 0 & 0 & \cdots & 0 \\ \vdots & \vdots & \ddots & \vdots \\ 0 & 0 & \cdots & 0 \end{bmatrix}$$

Furthermore, it is equally important that the negative Jacobians $(-[D\mathbf{f}]_{\mathbf{x}^r(k)})$ may fail to satisfy the total boundedness constraints at some points, where the total boundedness constraints are referred to as the total bound that we impose on the total control action: directing controller plus feedback correction.

Consider again the feedback gain matrix $\widetilde{K}_k$ given by (18). We need to bound the combined control action:

$$\widetilde{\mathbf{u}} = \mathbf{u}^*(k) + \widetilde{K}_k(\mathbf{x}(k) - \mathbf{x}^r(k))$$

where $\widetilde{K}_k$ is defined by (18). To continue, we note that $\mathbf{u}^*$ was subject to the boundedness constraints (3), where α is the bound. It remains to find a bound β for the feedback correction $\widetilde{K}_k$. Since

$$\|B\| = 1$$

it follows that

$$\left\|\widetilde{K}_k\right\| \leq \|K_k\|$$

where K_k is defined by (17).

Condition (16) is a sufficient condition for the contraction around $\mathbf{x}^r(t)$:

$$\max|\text{eig}\{A_k\}| < 1. \tag{19}$$

To simplify the analysis, we consider only matrices of the form:

$$A_k = \begin{bmatrix} \lambda_k & 0 & \cdots & 0 \\ 0 & \lambda_k & \cdots & 0 \\ \vdots & \vdots & \ddots & \vdots \\ 0 & 0 & \cdots & \lambda_k \end{bmatrix}$$

Let $\mathcal{M}$ be the class of all such matrices. Clearly, it is easy to ensure that elements of $\mathcal{M}$ satisfy the condition (19). We simply choose the parameters λ_k such that

$$|\lambda_k| < 1$$

Furthermore, the eigenvalues of any matrix in $\mathcal{M}$ are real. Thus, undesirable oscillation associated with complex eigenvalues are avoided. Last but not least, we can easily make $\|\widetilde{K}_k\|$ arbitrarily small. The reason for this assertion is really quite simple. In view of (17), we see that $\widetilde{K}_k$ is expressed in terms of $(A_k - I)$. Due to the diagonal structure of the matrix A_k in $\mathcal{M}$, the assertion follows readily.

These results are summarized in the following as a theorem.

Theorem 1. Let $\mathbf{f} : \Re^n \to \Re^n$ be a continuously differentiable function, let $\overline{\mathbf{x}} \in \Re^n$ be an arbitrary (fixed) vector, and let β be a fixed positive constant. Then, there exists a real number $\lambda \in [0,1)$ such that

$$\|(A - I)[D\mathbf{f}]_{\overline{\mathbf{x}}}\| \leq \beta$$

where

$$A = \begin{bmatrix} \lambda & 0 & \cdots & 0 \\ 0 & \lambda & \cdots & 0 \\ \vdots & \vdots & \ddots & \vdots \\ 0 & 0 & \cdots & \lambda \end{bmatrix}$$

and $I \in \Re^{n \times n}$ is the identity matrix.

Proof. Since

$$A = \lambda I$$

we have

$$\|(A - I)[D\mathbf{f}]_{\overline{\mathbf{x}}}\| = \|(\lambda I - I)[D\mathbf{f}]_{\overline{\mathbf{x}}}\| = \|(\lambda - 1)I[D\mathbf{f}]_{\overline{\mathbf{x}}}\|$$
$$= |\lambda - 1| \|[D\mathbf{f}]_{\overline{\mathbf{x}}}\| = (1 - \lambda) \|[D\mathbf{f}]_{\overline{\mathbf{x}}}\|$$

If $\overline{\mathbf{x}}$ is such that

$$\|[D\mathbf{f}]_{\overline{\mathbf{x}}}\| = 0$$

then, the inequality is trivially satisfied for any $\lambda \in [0, 1)$.

We now consider the case in which

$$\|[D\mathbf{f}]_{\overline{\mathbf{x}}}\| \neq 0 .$$

Since $\mathbf{f}$ is continuously differentiable, $\|[D\mathbf{f}]_{\overline{\mathbf{x}}}\|$ is a finite positive number. We wish to ensure that

$$(1 - \lambda) \|[D\mathbf{f}]_{\overline{\mathbf{x}}}\| \leq \beta$$

$\Longleftrightarrow$

$$(1 - \lambda) \leq \frac{\beta}{\|[D\mathbf{f}]_{\overline{\mathbf{x}}}\|}$$

$\Longleftrightarrow$

$$\lambda \geq 1 - \frac{\beta}{\|[D\mathbf{f}]_{\overline{\mathbf{x}}}\|}$$

Since $(\beta/[D\mathbf{f}]_{\overline{\mathbf{x}}})$ is a positive real number, $1 - (\beta/[D\mathbf{f}]_{\overline{\mathbf{x}}}) < 1$. Thus, we can always choose $\lambda \in [0,1)$ to satisfy this inequality. For example, let

$$\lambda = 0$$

if

$$1 - (\beta/[D\mathbf{f}]_{\overline{\mathbf{x}}}) < 0 .$$

Otherwise, we choose

$$\lambda = \frac{1}{2}\left((1 - (\beta/[D\mathbf{f}]_{\overline{\mathbf{x}}})) + 1\right) = 1 - (\beta/2[D\mathbf{f}]_{\overline{\mathbf{x}}}) .$$

The proof is complete.

5 Numerical Simulations

For illustration, let us consider the discrete-time dynamical system (1) described by the Hénon map:

$$\mathbf{f}: \Re^2 \to \Re^2, \quad \mathbf{f}(x_1, x_2) = [1 + x_2 - a(x_1)^2, bx_1]^T \qquad (20)$$

where the parameters a and b take their usual values, $a = 1.4$ and $b = 0.3$ (see [7] for details). Note that the numerical simulations have indicated that the results are not sensitive to the choice of $\mathbf{x}^{(S)}$ and $\mathbf{x}^{(T)}$.

Remark 1. Numerical studies were also carried out for other dynamical systems (e.g., the Ikeda map). We notice that the results obtained are very similar to those presented in this seciton. In fact, they are even slightly better. For the definition of the Ikeda map, see [2], [4], or [15]. Furthermore, good results are also obtained when the control is implemented differently in the dynamical system such as

$$\mathbf{f}(x_1, x_2, u) = [1 + (1+u)x_2 - 1.4(x_1)^2, 0.3x_1]^T$$

or

$$\mathbf{f}(x_1, x_2, u) = [1 + x_2 - (1.4+u)(x_1)^2, 0.3x_1]^T \ .$$

It is reasonable to expect that the number of time-steps required to reach the desired neighbourhood of the target is inversely proportional to the bound of the control. Numerical studies do confirm this expectation. If the control bound is larger, then the number of time-steps required is smaller. Since we are only interested in using small bounds, we take

$$N = 5, 6, \ldots, 10$$

and

$$\alpha = 0.01 \text{ to } 0.025 \ .$$

The fixed point of the Hénon map is taken as the desired target, i.e.,

$$\mathbf{x}^{(T)} = [0.63135, 0.18941]^T \qquad (21)$$

and we take

$$\mathbf{x}^{(S)} = [0.6, 0.2]^T \ . \qquad (22)$$

The size of the neighbourhood is chosen to be $\varepsilon = 0.02$.

After solving the optimal control problem, the distance between $\mathbf{x}(N)$ and $\mathbf{x}^{(T)}$ is the square root of the optimal value of the objective function associated with the problem (P2). Numerical results corresponding to various values of N and α are summarized in Table 1.

N	$\|\mathbf{x}(N) - \mathbf{x}^{(T)}\|$ $\|u(k)\| \leq 0.01$	$\|\mathbf{x}(N) - \mathbf{x}^{(T)}\|$ $\|u(k)\| \leq 0.025$
5	0.507	0.234
6	1.759	0.505
7	0.397	0.393
8	9.5×10^{-2}	8.9×10^{-2}
9	1.3×10^{-2}	9.9×10^{-3}
10	4.9×10^{-4}	4.1×10^{-9}

Table 1: Distances from the target with different control bounds

The numerical results reported in [14] suggest that the size of the neighbourhood in which a local stabilizing feedback controller is effective is about $\varepsilon = 0.02$. Thus, combining those results with the results from Table 1, we shall use 9 time-steps for our directing controller.

Numerical results, which show the effect of directing an orbit of the Hénon map towards $\mathbf{x}^{(T)}$, are depicted in Figure 2.

With all this in mind, we state our procedure as follows: Assume that we have the complete knowledge of the function $\mathbf{f}$ of the dynamical system (2) and its gradients with respect to state and control variables. Choose a value for the parameter $\varepsilon > 0$, the size of the neighbourhood in which a local feedback controller is effective. Choose a value for the parameter $\alpha > 0$, the bound on the control. For each pair $\{\mathbf{x}^{(S)}, \mathbf{x}^{(T)}\}$ of the initial and target states, set $N = 1$ and then run the program DMISER3.1 successively in the increasing order of N until $\left\|\mathbf{x}(N) - \mathbf{x}^{(T)}\right\| < \varepsilon$. Record the corresponding controller and the corresponding value of N.

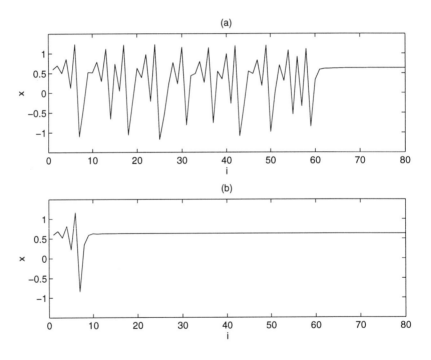

Figure 2: The time series of the first component of the Hénon map, with initial point $(0.6, 0.2)$ and $\varepsilon = 0.02$. (a) Without directing, the desired target region is reached in 61 time-steps, and a local stabilizing controller is then activated. (b) The directed orbit with the bound on perturbations $\alpha = 0.01$ reaches the desired target neighbourhood in only 9 time-steps.

For our problem which involves the dynamical system (2) with $\mathbf{f}$ defined by (20), the objective function defined by (6), and the bound on the control chosen to be $\alpha = 0.01$, the optimal controller that 'directs' the orbit in 9 time steps is:

$$u^* = \{-0.01, 0.002001874, -0.01, -0.01, -0.01, -0.01, -0.01, 0.01, 0.01\} \quad (23)$$

Examining the results presented in Table 1, we see that, for a fixed N, the distance between $\mathbf{x}(N)$ and $\mathbf{x}^{(T)}$ decreases as the control bound α increases. Furthermore, with the fixed value of ε, the number of time steps needed to reach the ε-neighbourhood of $\mathbf{x}^{(T)}$ decreases as α increases.

The best perturbations are those that bring the orbit onto the stable manifold of the target, and then keep it on or near the manifold. This is illustrated in Figure 3.

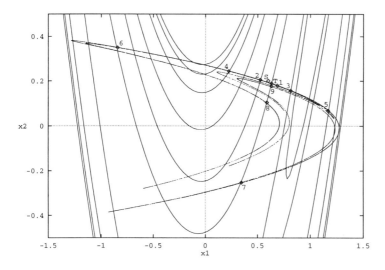

Figure 3: The attractor of the Hénon map, shown together with the stable manifold of the fixed point T. (All the solid lines together are parts of the same stable manifold of the fixed point T. The stable manifold repeatedly escapes towards infinity and then swerves back again.) The orbit directed from the source $S = [0.6, 0.2]^T$ towards the target T reaches the 0.02-neighbourhood of the target in 9 time-steps. The controlled trajectory is given by the points 1–9.

Geometrically speaking, we are stabilizing the stable manifold of the target. The name 'stable manifold' is unfortunately somewhat misleading. In fact, it refers to an unstable object, a set of all trajectories that move towards the target, but are themselves unstable. In other words, arbitrarily small perturbations would cause them to move off the manifold. In the absence of the control bound, it is possible to direct an orbit to the desired neighbourhood of the target in just a few time steps. With the presence of the control bound, it will limit the extent of the orbit directing, and hence it will take more time steps to reach the desired neighbourhood of the target. For the example considered in this section, we need to take 9 time-steps.

Interestingly, although the optimal control program DMISER3.1 does not make use of the special properties associated with chaotic dynamics and stable manifolds, it has generated an intuitively reasonable controlled trajectory depicted in Figure 3.

Suppose the target is a periodic orbit with period greater 1. Then, the situation is somewhat more complicated, because there are stable manifolds of several points to be considered. However, the principle remains the same.

In a chaotic system, its orbit is inherently sensitive to the changes of the initial conditions. Due to this property, it is possible to significantly change the orbit of the system in just a few time steps with only small bounded control. Non-chaotic systems do not exhibit such a sensitivity property. Furthermore, the ergodic property of a chaotic system ensures that a small neighbourhood of any state near or on the attractor is reachable from any other state near or on the attractor. Thus, chaotic properties turn out to be of advantageous rather than disadvantageous.

When only small perturbations are applied, the difference between the directed and undirected orbits is almost undistinguishable at the very beginning. However, the sensitivity property of the chaotic system soon causes the system to display two significantly different orbits (see Figure 2).

The significance of the directing control becomes more apparent when our main concern is to stabilize the higher periodic orbits of a chaotic system. If we find appropriate local controllers around all of the points of a higher periodic orbit, then it is irrelevant towards which of these points an orbit is going to be directed. Consequently, we consider each of those points as a possible target and calculate all the necessary directing controllers, from all the possible initial states. In a real control run, the system starts from a given initial state. We just choose that target which can be reached in the shortest time.

Another point which we wish to make is that with the Hénon map, the neighbourhood of the target is not really small. In some other cases, like higher periodic orbits or some other maps, the neighbourhood may be much smaller. As the size of the target neighbourhood decreases, the number of time-steps of the undirected system needed to reach that neighbourhood increases very rapidly. However, if small directing control is employed, the same decrease in the size of the neighbourhood gives rise to only a very slight increase in the number of time-steps needed to reach that neighbourhood. For details, see Table 2, where the initial and target points are given by (21) and (22).

	without directing			directing with $\alpha = 0.01$		
ε	0.02	10^{-3}	10^{-5}	0.02	10^{-3}	10^{-5}
N	61	60858	575047	9	10	11

Table 2: Number of time steps with and without directing.

6 Noise

Let us now consider the controlled system in a noisy environment. It is well known that open-loop controllers are noise-sensitive. Therefore, we calculate a controller which is optimal in the absence of noise, and apply it to find the reference trajectory of the noiseless system. After that, we add the feedback correction to the open-loop directing controller, and use this combined controller to direct the orbit of the controlled system in the noisy environment. Let us again consider the targeting problem given by (20)–(22). For comparison reasons, we consider the same pair $\{\mathbf{x}^{(S)}, \mathbf{x}^{(T)}\}$ of the initial and terminal states as before. The optimal directing controller, which is denoted by u^*, is given by (23). Using this optimal directing controller, we determine the following reference trajectory.

$$\begin{aligned}\mathbf{x}^r = \{&[0.60000, 0.20000]^T, [0.68600, 0.18000]^T,\\ &[0.52317, 0.20580]^T, [0.81261, 0.15695]^T,\\ &[0.22247, 0.24378]^T, [1.16449, 0.06674]^T,\\ &[-0.8417, 0.34935]^T, [0.34746, -0.2525]^T\\ &[0.58847, 0.10424]^T, [0.62943, 0.17654]^T\}\end{aligned}$$

Denoting

$$\mathbf{x}(k) = [x_1(k), x_2(k)]^T, \quad \mathbf{x}^r(k) = [x_1^r(k), x_2^r(k)]^T$$

and

$$\mathbf{u}^*(k) = [u^*(k), 0]^T$$

for the Hénon map (20), we have the following modified combined control law:

$$\begin{aligned}\mathbf{x}(k+1) = &\ \mathbf{f}(\mathbf{x}(k)) + \mathbf{u}^*(k) +\\ &\ B(A_k - I)[D\mathbf{f}]_{\mathbf{x}^r(k)}(\mathbf{x}(k) - \mathbf{x}^r(k)), \quad k = 0, \ldots, N-1\end{aligned}$$

where $N = 9$, $B = \begin{bmatrix} 1 & 0 \\ 0 & 0 \end{bmatrix}$, $A_k = \begin{bmatrix} \lambda_k & 0 \\ 0 & \lambda_k \end{bmatrix}$, and $[D\mathbf{f}]_{\mathbf{x}^r(k)} = \begin{bmatrix} -2.8x_1^r(k) & 1 \\ 0.3 & 0 \end{bmatrix}$.

After simplification, we obtain

$$\begin{aligned}x_1(k+1) = &\ 1 + x_2(k) - 1.4(x_1(k))^2 + u^*(k) +\\ &\ 2.8(1 - \lambda_k)x_1^r(k)(x_1(k) - x_1^r(k)) - (1 - \lambda_k)(x_2(k) - x_2^r(k))\\ x_2(k+1) = &\ 0.3x_1(k).\end{aligned}$$

To implement this combined control law, it remains to determine the real constants λ_k, $k = 0, 1, \ldots, 8$.

We shall do so such that

$$\left\| B(A_k - I)[D\mathbf{f}]_{\mathbf{x}^r(k)} \right\| \leq \beta$$

where $\beta \sim \mathcal{O}(1)$ is an appropriate positive constant. This, together with the term $(\mathbf{x}(k) - \mathbf{x}^r(k))$, ensures that the order of magnitude of the feedback correction is not higher than that of u^*. Therefore, the overall bound on the combined control does not change dramatically.

Notice that $A_k = \lambda_k I$. Thus, by appropriate manipulation, we obtain

$$\begin{aligned}\left\| B(A_k - I)[D\mathbf{f}]_{\mathbf{x}^r(k)} \right\| &= \left\| B(\lambda_k I - I)[D\mathbf{f}]_{\mathbf{x}^r(k)} \right\|\\ &= \left\| B(\lambda_k - 1)I[D\mathbf{f}]_{\mathbf{x}^r(k)} \right\|\\ &= \left\| (\lambda_k - 1)B[D\mathbf{f}]_{\mathbf{x}^r(k)} \right\|\end{aligned}$$

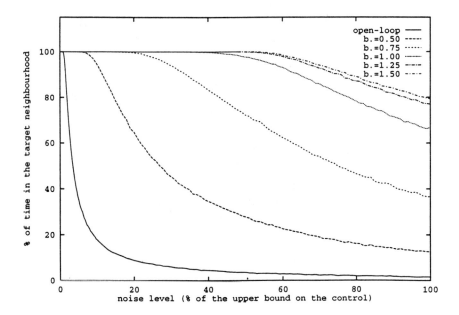

Figure 4: The comparison of the percentage of time the directed orbit of the Hénon map ends in the target neighbourhood, with and without the feedback correction. The initial and target points are given by (21) and (22), and the directing controller is given by (23). The target neighbourhood is given by $\varepsilon = 0.02$. The noise level of p% signifies uniformly distributed random noise with magnitude = p% of the bound on the control. For each value of β (the bound on the feedback correction gain matrix, denoted by 'b.' in the figure), the 'percentage of time' is found over 30000 runs for each noise level.

$$\begin{aligned} &= |\lambda_k - 1| \left\| \begin{bmatrix} 1 & 0 \\ 0 & 0 \end{bmatrix} \begin{bmatrix} -2.8 x_1^r(k) & 1 \\ 0.3 & 0 \end{bmatrix} \right\| \\ &= (1 - \lambda_k) \left\| \begin{bmatrix} -2.8 x_1^r(k) & 1 \\ 0 & 0 \end{bmatrix} \right\| \leq \beta \ . \end{aligned}$$

With the ℓ_2-induced matrix norm, we have

$$\left\| \begin{bmatrix} -2.8 x_1^r(k) & 1 \\ 0 & 0 \end{bmatrix} \right\|_2 = \sqrt{(2.8 (x_1^r(k))^2 + 1} \ .$$

Thus,

$$(1 - \lambda_k)\sqrt{(2.8(x_1^r(k))^2 + 1} \leq \beta$$

$\Leftrightarrow$

$$\lambda_k \geq 1 - \frac{\beta}{\sqrt{(2.8(x^r(k))^2 + 1}} \ . \tag{24}$$

Since it is desirable to make the feedback correction to be the strongest possible contraction mapping around the reference trajectory, we choose $\lambda_k \in [0, 1)$ to be as close to 0 as possible, subject to (24). Therefore, at each point along the reference trajectory, we set

$$\lambda_k = \max\left\{0, 1 - \frac{\beta}{\sqrt{(2.8(x^r(k))^2 + 1}}\right\}$$

The performances of the combined controllers corresponding to several different values of β are illustrated in Figure 4.

7 Conclusions

In this paper, a method for targeting orbits of chaotic dynamical systems was proposed. This method involves a combination of an open-loop directing controller with a local feedback correction. It was shown that open-loop directing controller, which is derived from optimal control theory, significantly reduces the number of time-steps needed to reach the target. Furthermore, it was shown that noisy environment can be handled successfully by introducing local contracting feedback corrections. The existence of bounded feedback corrections was established, and the results of numerical simulations were given to illustrate the success rate in the noisy environment with respect to control bounds.

This work complements some of the results on local control of chaos and targeting presented at the workshop.

8 Acknowledgments

This work was partially supported by OPRS, research grants from the Australian Research Council, and the Cooperative Research Centre for Broadband Telecommunications and Networking at the Curtin University of Technology, WA, Australia.

References

[1] B.D.O. Anderson and J.B. Moore. *Optimal Control: Linear Quadratic Methods.* Prentice Hall, Englewood Cliffs, NJ, 1990.

[2] M. Casdagli. Nonlinear Prediction of Chaotic Time Series. *Physica D*, 35:335–356, 1989.

[3] G. Chen. Optimal Control of Chaotic Systems. *International Journal of Bifurcation and Chaos*, 4:461–463, 1994.

[4] G. Chen. Control and Synchronization of Chaotic Systems (bibliography). Department of Electrical Engineering, University of Houston, TX, 1995. Available from uhoop.egr.uh.edu/pub/TeX/chaos.tex (login name and password: both 'anonymous').

[5] G. Chen and X. Dong. On Feedback Control of Chaotic Nonlinear Dynamic Systems. *International Journal of Bifurcation and Chaos*, 2:407–411, 1992.

[6] G. Chen and X. Dong. From Chaos to Order — Perspectives and Methodologies in Controling Chaotic Nonlinear Dynamical Systems. *International Journal of Bifurcation and Chaos*, 3:1363–1409, 1993.

[7] J. Guckenheimer and P.J. Holmes. *Nonlinear Oscillations, Dynamical Systems, and Bifurcations of Vector Fields.* Springer-Verlag, New York, 1983.

[8] E.A. Jackson. The Entrainment and Migration Controls of Multiple-Attractor Systems. *Physics Letters A,* 151:478–484, 1990.

[9] E.A. Jackson. On the Control of Complex Dynamical Systems. *Physica D,* 50:341–366, 1991.

[10] L.S. Jennings, M.E. Fisher, K.L. Teo and C.J. Goh. *MISER3.1 — Optimal Control Software: Theory and User Manual.* EMCOSS, Western Australia, 1990.

[11] E.J. Kostelich, C. Grebogi, E. Ott and J.A. Yorke. Higher dimensional targeting. *Physical Review E,* 47:305–310, 1993.

[12] G. Nitsche and U. Dressler. Controlling Chaotic Dynamical Systems Using Time Delay Coordinates. *Physica D,* 58:153–164, 1992.

[13] E. Ott, C. Grebogi and J.A. Yorke. Controlling Chaos. *Physical Review Letters,* 64:1196–1199, 1990.

[14] M. Paskota, A.I. Mees and K.L. Teo. On Control of Chaos: Higher Periodic Orbits. *Dynamics and Control,* 5:365–387, 1995.

[15] M. Paskota, A.I. Mees and K.L. Teo. Directing Orbits of Chaotic Dynamical Systems. *International Journal of Bifurcation and Chaos,* 5:573–583, 1995.

[16] M. Paskota, A.I. Mees and K.L. Teo. Directing Orbits of Chaotic Systems in the Presence of Noise: Feedback Correction. *Dynamics and Control,* to appear, 1995.

[17] K. Pyragas. Continuous Control of Chaos by Self-Controlling Feedback. *Physics Letters A,* 170:421–428, 1992.

[18] F.J. Romeiras, C. Grebogi, E. Ott and W.P. Dayawansa. Controlling Chaotic Dynamical Systems. *Physica D,* 58:165–192, 1992.

[19] K. Schittkowski. NLPQL: A Fortran Subroutine Solving Constrained Nonlinear Programming Problems. *Operations Research Annals,* 5:485–500, 1985.

[20] T. Shinbrot, E. Ott, C. Grebogi and J.A. Yorke. Using Chaos to Direct Trajectories to Targets. *Physical Review Letters,* 65:3215–3218, 1990.

[21] T. Shinbrot, E. Ott, C. Grebogi and J.A. Yorke. Using Chaos to Direct Orbits to Targets in Systems Describable by a One-Dimensional Map. *Physical Review A,* 45:4165–4168, 1992.

[22] I.M. Starobinets and A.S. Pikovsky. Multistep Method for Controlling Chaos. *Physics Letters A,* 181:149–152, 1993.

[23] K.L. Teo, Y. Liu and C.J. Goh. Nonlinearly Constrained Discrete Time Optimal Control Problems. *Applied Mathematics and Computation,* 38:227–248, 1990.

[24] K.L. Teo, C.J. Goh and , K.H. Wong. *A Unified Computational Approach to Optimal Control Problems.* Longman Scientific and Technical, Harlow, UK, 1991.

[25] T.L. Vincent and J. Yu. Control of a Chaotic System. *Dynamics and Control*, 1:35–52, 1991.

[26] S. Wiggins. *Introduction to Applied Nonlinear Dynamical Systems and Chaos.* Springer-Verlag, New York, 1990.

[27] T.H. Yeap and N.U. Ahmed. Feedback Control of Chaotic Systems. *Dynamics and Control*, 4:97–114, 1994.

Commentary by Y. Cohen

In this paper, the authors present an application of a well developed and tested numerical optimization software: DMISER3.1. The neat idea here is to use the ergodic property of discrete chaotic dynamical systems in designing an optimal controller that brings the system to a desired target.

The application uses two types of controls: one to bring the system to an orbit close enough to the target, and one to bring the system to the target once it is on that orbit.

The paper's contribution is mainly to the applied aspects of the conference, and as such, complements other papers, where control and chaos are treated theoretically. The lucid presentation would allow any practitioner to develop optimal control of such systems easily, and quickly, and in that, the paper presents significant contribution.

Several extensions come to mind, and would follow nicely from this work:

1. Although conceptually straightforward, the application of vector controls to such problems will present an interesting extension to the work here. It is interesting to examine, for example, in which of the model equations would the control(s) will be most effective according to the any of the criteria discussed in the paper.

2. How well would numerical solutions of optimal control problems such as those presented here will work if we remove the requirement to supply derivative of the system.

3. Can we develop optimal controls for systems that cannot be described by analytical expressions, but rather by simulation rules?

4. How does the approach here extend to continuous (as opposed to discrete) chaotic systems?

Commentary by J. Moore

This research exploits the nature of chaos and the strengths of both open-loop optimal control and feedback control to achieve desired objectives for chaotic systems. As expected there can be an arbitrary small amount of control energy required to achieve target objectives. The feedback allows some degree of robustness to noise. The ideas come together nicely and very convincingly in the examples studied. This work raises next the question of the degree of robustness to unmodelled dynamics or parametric uncertainty of the proposed schemes.

Complex Dynamics in Adaptive Systems

Iven M.Y. Mareels
Department of Engineering
Australian National University
ACT 0200, Australia

1 Introduction

The term adaptive systems was coined in the late fifties. It captures the intuitive concept of a system that is able to maintain a desired performance objective despite (significant) changes in the environment: the system adapts to its environment. This contribution aims to introduce in a partly tutorial manner some of the theory associated with adaptive systems, with a particular emphasis on the dynamics of adaptation. It does so in the context of control applied to discrete time finite dimensional linear systems.

In terms of control theory, the late fifties is characterised by a good understanding of designing (feedback) control laws for plants (systems to be controlled) whose dynamics are linear. Such plants can be modelled in the frequency domain via a transfer function. Frequency domain control design is essentially concerned with shaping the transfer function of the feedback interconnection of plant and controller. Unfortunately most physical/engineering systems are nonlinear. Only when the control task is concerned with maintaining an equilibrium does it seem sensible to approximate the plant by a linear system and to design the control law for the approximation rather than the actual plant. It follows then from the standard Hartman-Grobman theorem or Lyapunov's first stability theorem that the closed loop consisting of the actual plant and the control system behaves well for small external signals. However in the case the plant's operating range is not limited to a single equilibrium a nonlinear control and/or analysis methodology needs to be adopted. One such approach, known as adaptive control, germinated in the mid fifties [1] and was developed extensively over the next few decades see eg. [19, 2, 3, 4].

Another scenario where adaptation may be important may be described by the situation where many (more or less different) replicas of a similar plant need to be controlled. Rather than designing for each replica a specific controller, we could try to design a controller for the ensemble, which could adapt itself to the specific instance of the plant the controller is applied to. Such situations arise eg in the automotive industry where a single engine control unit can be used over the entire range of engine models of a manufacturer. Also in telephony a single adaptive transmission line termination unit may be used to terminate a whole range of different telephone lines [5]. Another well known application is the pressure control loop in kidney dialysis apparata, where the plant dynamics change with the patient as well as with the aging of the filter components [6].

The adaptive methodology that was developed, may be identified by describing the constituent components of the overall controlled system. A corresponding signal flow diagram is represented in Figure 1.

1. **The plant**, or object to be controlled. It is assumed that both the plant's input and the plant's output signals are available. One may actually define the plant as the dynamical system linking the input causally to the output. In this treatment we restrict ourselves to plants that may described by a linear, finite dimensional, discrete time system.

2. **The model class**. A parametrised ensemble of (linear) models used to represent the dynamical behaviour of the plant. The actual plant may or may not be a member of the model class. The former situation we describe as *ideal*. The latter is known as an *undermodelled* situation, which is rather the norm in practice. A particular member of the class is identified by a (finite dimensional) parameter vector.

3. **The controller class**. It is assumed that given the control objective that has to achieved, we are able to design a controller for each member of the model class, such that the feedback loop consisting of the model system and the corresponding controller achieves the control objective. The controller class may thus be viewed to be parametrised by the same parameter as the model class.

4. **The adaptation algorithm or parameter tuner**. Given the observed past input and output data, we have an algorithm (or mapping) that associates with the observed data a preferred model parameter and hence the preferred controller to be implemented.

The particular adaptation methodology we refer to above is known as *certainty equivalence*, because it associates with the identified preferred model immediately the controller designed for this model. Such an algorithm does not take into account that the model may be very poor. A situation which is bound to happen if very few data are available. Variants exists, but these will not be discussed here, see however [3].

Notice that the above methodology leads naturally to a nonlinear closed loop system, even when the actual plant, the members of the model class and their corresponding controllers are all linear input- output systems. Indeed the control signal is obtained by combining plant outputs and gains which are derived from previous outputs and inputs. This nonlinear interaction is indicative of rich dynamics.

It is clear that the above recipe for an adaptive system can hardly be said to constitute a definition of an adaptive system, but it is nevertheless useful in labelling a system as adaptive or not. A formal (mathematical) definition does not exist. Moreover, in our opinion, the boundary between the competing control paradigms robust control, nonlinear control and adaptive control does simply not exist from a formal point of view. Clearly adaptive control is a form of nonlinear control by construction and provided we can show that the above methodology actually works, it is by design a robust controller as it is able to control a whole class of systems. Robustness is viewed as a form of structural stability. We say that a (control) property is robust with respect to a class of systems if the particular (control) property is valid for each member of the class.

In this paper we are particularly interested in the nonlinear dynamical behaviour that may be exhibited by control systems of the form described above. Using a rather

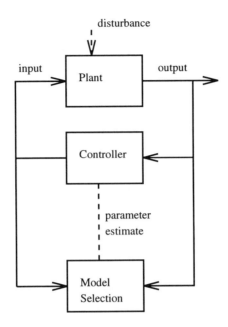

Figure 1: Adaptive System Structure.

simple example we demonstrate the richness of behaviours that ensues if the above adaptive methodology is implemented. Most of the results have been discussed in one form or another in a number of papers concerned with complex dynamics in adaptive systems. We refer the interested reader in particular to [5, 7, 8, 9, 10, 11, 12, 13, 14, 15, 16, 17]. A wealth of interesting and complex dynamics has been identified in adaptive system operating in the undermodelled regime. Many questions remain, not in the least a unifying framework to interpret the results.

It must be emphasised that we are not interested in the complex dynamics per se, but rather we want to unravel the essence of the driving force behind adaptation. This is the main motivation behind most of the cited studies. This grand goal is still elusive, as will transpire from the analysis results. Nevertheless, we may argue that the nonlinear and complex dynamics inherent in the adaptive systems' behaviour are at the core of the robustness properties enjoyed by these systems.

The paper is organised as follows. First we introduce, using the outlined standard adaptive methodology, an example on which the whole study is based. The example is chosen such that it may be variously interpreted as representing a number of different control or signal processing objectives. In this sense our results obtain a sense of universal legitimacy for adaptive systems. In Section 3 we present the basic results when the plant to be controlled is in the model class. We identify clearly sufficiency conditions for good adaptive system performance. In this example these conditions are also shown to be near necessary. Next we consider a number of plausible undermodelling scenarios. An incomplete bifurcation analysis is presented, which is nevertheless indicative of the behaviours we may expect to encounter in typical adaptive systems. In Section 5 we conclude, interpreting the observed phenomena from an adaptive system design perspective.

2 The example

In this section we introduce, following the outline presented in the introduction, an adaptive control system. The particular example is presented from a control system perspective. It may however equally well be interpreted as a simplified version of an echo cancellation algorithm used in telecommunication systems to suppress cross talk, see [5] and [18]. The system is a good representation for a simple chemical reactor as discussed in [12].

Let the plant to be controlled be represented by:

$$y(t+1) = ay(t) + by(t-1) + cu(t) + d \qquad (1)$$

Here t is an integer, the time index; $y(t)$ is the plant output at time t and $u(t)$ is the control input at time t. Both u and y are observed variables. The real variables a, b, c and d are (unknown) constant (time independent) plant parameters.

The control objective is to achieve regulation to a set point. More precisely given the past observations, $y(k)$ $k = 1, 2, \ldots t$ and $u(k)$ $k = 1, 2, \ldots t-1$, design the input $u(t)$ such that the plant output converges as time progresses to y^*. The adaptive methodology suggests the following scenario.

The control designer beliefs that the plant belongs to the class of models that may be represented as:

$$y(t+1) = \alpha y(t) + u(t) \qquad (2)$$

The plant belongs to the model class if $b = 0$, $c = 1$ and $d = 0$. The situation when these conditions are satisfied is referred to as the *ideal* case. When any of these conditions are violated we speak of an undermodelled problem. The parameter α uniquely identifies a member of the model class. Our estimate for the model parameter at time t is denoted as $\alpha(t)$.

According to the above model class, the designer selects a control law for the model with parameter α, see equation (2), as follows:

$$u(t) = -\alpha y(t) + y^* \qquad (3)$$

This control law is known as a *dead beat control law*. It assigns in the shortest time possible (one time step) the desired reference value to the model output.

In the adaptive system, the actual control input implemented is $u(t) = -\alpha(t)y(t) + y^*$ where $\alpha(t)$ is the present estimate for the *best* model parameter.

In order to select a model representative (obtain $\alpha(t)$) the control designer aims to minimise the prediction error between the model's predicted output given the past data and the actual observed output. The prediction error at time $t+1$ given the present model parameter estimate $\alpha(t)$ and the data $y(t)$, $u(t)$ is given by:

$$e(t+1) = y(t+1) - \alpha(t)y(t) - u(t)$$
$$= y(t+1) - y^* \qquad (4)$$
$$= (a - \alpha(t))y(t) + by(t-1) + (c-1)u(t) + d$$

The new parameter is selected in order to make the prediction error small(er). A frequently used update mechanism, known as *normalised least mean square* is represented by:
$$\alpha(t+1) = \alpha(t) + \frac{\mu e(t+1)y(t)}{\delta + y^2(t)} \qquad (5)$$

Here $\delta > 0$ ensures that no division by zero can occur and the parameter μ scales the step size. It is typically a small positive constant. (In the case that the normalising factor $\delta + y^2(t)$ is not implemented one refers to the algorithm as *least mean square*. It is often used in signal processing applications [5].)

The special selection $\mu = 1$ and $\delta = 0$ corresponds to the so called *projection algorithm* or *dead beat parameter tuner* as the new parameter estimate is chosen such as to null the prediction error:

$$\alpha(t+1) = \alpha(t) \qquad if \ y(t) = 0$$
$$= \frac{y(t+1) - u(t)}{y(t))} \quad if \ y(t) \neq 0 \qquad (6)$$

This can be equivalently written as
$$\alpha(t+1) = \alpha(t) \ if \ y(t-1) = 0 \ else$$
$$= a + (c-1)\frac{u(t)}{y(t)} + b\frac{y(t-1)}{y(t)} + d\frac{1}{y(t)} \qquad (7)$$

The complete adaptive system (using the normalised least mean square adaptation rule) can thus be summarised as follows:

The plant
$$y(t+1) = ay(t) + by(t-1) + cu(t) + d$$

The control law
$$u(t) = -\alpha(t)y(t) + y^*$$

The tuning error $\qquad (8)$
$$e(t) = y(t) - \alpha(t-1)y(t-1) - u(t-1)$$

The parameter tuner
$$\alpha(t+1) = \alpha(t) + \mu\frac{e(t)y(t-1)}{\delta + y^2(t-1)}$$

Introducing the variables $x(t) = a - \alpha(t)$, the parameter error and $d(t) = (1 - c)[-x(t)y(t) + ay(t) - y^*] + d + by(t-1)$, measuring the model-plant mismatch, we may rewrite the adaptive system as follows:

$$\begin{aligned} y(t+1) &= x(t)y(t) + y^* + d(t) \\ x(t+1) &= (1 - \mu\frac{y^2(t)}{\delta + y^2(t)})x(t) - \mu\frac{y(t)d(t)}{\delta + y^2(t)} \\ d(t) &= (1-c))[-x(t)y(t) + ay(t) - y^*] + d + by(t-1) \end{aligned} \quad (9)$$

In the sequel we refer to y as the plant state and x as the parameter (estimate) error.

Remark Notice that the prediction error (4) can equally be interpreted as the control error. This is important. Indeed, a necessary feature of a well chosen error variable to update the model parameter is that an identically zero prediction error implies that the control objective is achieved. If zero prediction error, which implies (here) no adaptation, did not imply zero control error, then the adaptive scheme is obviously flawed as this undesirably behaviour could not be altered. ○

3 Adaptive system behaviour: ideal case

Having introduced the adaptive system we now analyse its behaviour in the ideal case, where the plant (1) does belong to the model class (2) ($b = 0, d = 0, c = 1$). In this case the model-plant mismatch variable $d(t) \equiv 0$ and hence the adaptive system is given by the two-dimensional, discrete time, stationary nonlinear system:

$$\begin{aligned} y(t+1) &= x(t)y(t) + y^* \quad & y(0) = y_0 \\ x(t+1) &= (1 - \mu\frac{y^2(t)}{\delta + y^2(t)})x(t) \quad & x(0) = x_0 \end{aligned} \quad (10)$$

With this difference equation we may associate the map

$$F : (y,x) \to F(y,x) = \left(xy + y^*, (1 - \mu\frac{y^2}{\delta + y^2})x\right) \quad (11)$$

In the case the projection algorithm ($\mu = 1$ and $\delta = 0$) is used for the parameter tuner we have:

$$\begin{aligned} y(t+1) &= x(t)y(t) + y^* \quad & y(0) = y_0 \\ x(t+1) &= \begin{cases} 0 & \text{if } y(t) \neq 0 \\ x(t) & \text{if } y(t) = 0 \end{cases} \quad & x(0) = x_0 \end{aligned} \quad (12)$$

For the system with the dead beat parameter tuner 12 the control objective is attained in finite time. This follows from considering the possible trajectories.

Theorem 1 *Consider the adaptive system (12). If $y_0 y^* \neq 0$, then for all initial conditions x_0 we have that $x(k) \equiv 0$ and $y(k) \equiv y^*$ for all $k = 3, 4, \ldots$. If $y_0 y^* = 0$ then $x(t) \equiv x_0$ and $y(t) \equiv 0$ for all t.*

An analogous result can be derived for the adaptive system 10.

Theorem 2 *Consider the adaptive system (10). Let $\delta > 0$ and $0 < \mu < 2$. Then for all initial conditions y_0 and x_0 we have that all trajectories are bounded. Moreover $x(t)$ converges and $y(t)$ converges to y^*. Moreover if $y^* \neq 0$ then the parameter error $x(t)$ converges to 0.*

Proof Along the trajectories of the system (10) we have that:

$$x^2(k+1) - x^2(k) = -\mu(2 - \mu\frac{y^2(k)}{\delta + y^2(k)})\frac{x^2(k)y^2(k)}{\delta + y^2(k)}$$

This implies that $|x(k)| \leq |x_0|$ for all $k = 0, 1, \ldots$.
Furthermore it follows that $\mu(2-\mu)\sum_k \frac{x^2(k)y^2(k)}{\delta+y^2(k)} \leq x_0^2$.
Now consider

$$\begin{aligned}(y(k+1) - y^*)^2 &= (x(k)y(k))^2 \\ &= \frac{x^2(k)y^2(k)}{\delta + y^2(k)}[y^2(k) + \delta] \\ &\leq \frac{x^2(k)y^2(k)}{\delta + y^2(k)}[2(y(k) - y^*)^2 + 2y^{*2} + \delta]\end{aligned}$$

It follows then from the Bellman-Gronwall Lemma (using that $\sum_k \frac{x^2(k)y^2(k)}{\delta+y^2(k)}$ is bounded) that $y(k) - y^*$ is bounded and converges to zero.

If y^* is nonzero, we conclude that $x(k)$ converges to zero. (q.e.d)

Some remarks are in order:

Remark The above results can be generalised to the case that the plant to be controlled is a finite dimensional linear system belonging to the model class. See eg. [19]. The proofs follow along similar lines. ∘

Remark The condition $y^* \neq 0$ is known as a *persistency of excitation* condition. It ensures that the parameter error converges to zero.

Intuitively speaking, a signal is said to be persistently exciting if the outputs of any two systems in the model class when driven by the particular signal are necessarily distinct. For more information we refer the reader to [19]. ∘

Remark It is typical for control applications that the excitation is insufficient. Here, when $y^* = 0$ but $y_0 \neq 0$, we obtain that $x(t)$ converges to some point x_∞ in the open interval $(-1, 1)$. This may be interpreted in control terms as asymptotically the adaptation leads to a time invariant linear stable system: $y(t+1) = x_\infty y(t)$. This may alternatively be interpreted as stating that the initial condition y_0 is itself a (weak) source of information for the adaptive algorithm, sufficient to achieve the control objective, insufficient to identify the particular plant under control. ∘

Remark The true adaptive nature of the closed loop systems 10 or 12 may be appreciated by observing that the dynamical behaviour of the systems is independent of the actual plant (belonging to the model class) that has been controlled. The particular instance of plant controlled (identified by the parameter a) only affects the transient, via the initial condition of the parameter error $x_0 = a - \alpha(0)$.

Alternatively we could state that the control objective is robust with respect to the complete model class. ∘

Complex Dynamics in Adaptive Systems

Remark The conditions $\delta > 0$ and $2 > \mu > 0$ are the normal conditions for which we can show that the control objective is achieved, regardless of initial conditions. $\delta < 0$ does not make sense. $\mu < 0$ implies global instability as now $x(t)$ diverges which forces also $y(t)$ to diverge. ○

Remark Although $\mu = 0$ defeats the purpose of adaptation ($x(t) \equiv x_0$), it is instructive to see how the trajectories of the corresponding system actually behave. ○

Remark Clearly, small $1 \gg \mu > 0$ and $|x_0| > 1$ may cause $y(t)$ to become very large during the transient. A very conservative lower bound on the transient behaviour for $x_0 > 1$, $y^* \geq 0$ and $y_0 > 0$ is obtained from

$$x(t) \geq (1-\mu)^t x_0.$$

This yields

$$y(t) \geq (1-\mu)^{\frac{t(t+1)}{2}} x_0^t y_0.$$

These lower bounds follow directly from the system equations (10).

Clearly large x_0, which is equivalent to stating that the initial model was a very poor approximation for the plant, may lead to extremely large peak values in the system state $y(t)$.

In general it is very hard to provide quantitative results about the transients in adaptive systems. The above lower bounds are testimony to the "simplicity" of this adaptive system. ○

Although, normally not of interest in adaptive system design, we now consider the situation where $y^* \neq 0$ and $\mu > 2$. In this case global stability is lost, but some pointers to rich dynamics appear.

Theorem 3 *Consider the adaptive system (10). Let $\delta > 0$, $y^* \neq 0$ and $\mu > 2$. Define*

$$\Omega = \{(y_0, x_0) : \frac{y_0^2}{\delta + y_0^2} > \frac{2}{\mu} \quad |x_0| > 1 + \frac{|y^*|}{|y_0|}\}$$

Trajectories with initial conditions in Ω diverge.

Proof On Ω we have that $|y(k+1)| > |y(k)|$ and $|x(k+1)| > |x(k)|$ for all $(y(k), x(k)) \in \Omega$. Hence Ω is invariant, ie. $F(\Omega) \subset \Omega$ (see equation (11)). (q.e.d)

This result is sufficient motivation not to consider $\mu > 2$ in adaptive system design. Nevertheless, the equilibrium $(y^*, 0)$ remains locally asymptotically stable:

Theorem 4 *Consider the adaptive system (10). Let $\delta > 0$, $y^* \neq 0$ and $2 \leq \mu < 2(1 + \frac{\delta}{y^{*2}})$. The equilibrium $(y^*, 0)$ is locally asymptotically stable.*

Proof The result is immediate from the Jacobian of the mapping F at $(y^*, 0)$:

$$DF(y^*, 0) = \begin{pmatrix} 0 & y^* \\ 0 & 1 - \mu \frac{y^{*2}}{\delta + y^{*2}} \end{pmatrix}$$

(q.e.d)

For $\mu = 2(1 + \frac{\delta}{y^{*2}})$ the equilibrium $(y^*, 0)$ undergoes a flip bifurcation. A locally stable 2-periodic orbit comes into existence. A numerically obtained bifurcation diagram (Figure 2), obtained using Matlab, illustrates the period doubling sequence induced by increasing the step size parameter μ. The diagram represents the y component of the periodic orbit against the value for μ for the adaptive system (10) with $\delta = 1$ and $y^* = 1$. The period doubling sequence commences at $\mu = 4$. The asymptotic behaviour of the orbits is displayed for $\mu = 4 + 0.025 * n$, $n = -4, 1, \ldots$. Beyond $\mu > 5.4$ the domain of attraction of the attractor is so small that numerical simulation becomes virtually impossible.

From this it is clear that the complex dynamics present in adaptive systems (at least when operating in the ideal case) can be viewed as a consequence of adapting "too fast" i.e. μ too large.

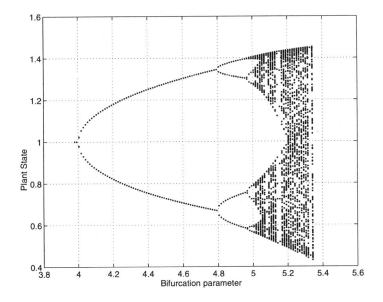

Figure 2: Period doubling in the ideal adaptive system.

4 Adaptive system behaviour: undermodelled case

We are now to exemplify that the nonlinear adaptive system (9) is a highly nonlinear system with a rich variety of dynamical behaviours. We consider several scenarios of undermodelled adaptive systems. Attention is paid to both the dead beat parameter tuner as well as the normalised least mean square variant.

First we concentrate on the situation where the undermodelling amounts to an error in the signal environment of the plant $b = 0$, $c = 1$ but $d \neq 0$. The constant d

can be viewed as a perturbation on the input side of the plant (1), neglected in the model (2). Although not very interesting from a control point of view, this particular situation is highly relevant when interpreting our example as a simplified model of the echo cancellation problem in telephony [5]. (The problem is not interesting from a control point of view because the linear pole placement control law proposed would normally contain integral action if such a constant disturbance were anticipated.)

Next we consider the adaptive system in the situation where the input gain is incorrectly modelled $b = 0$, $c \neq 1$ and $d = 0$. This situation is particularly relevant as often in adaptive control systems one purposely chooses not to adapt one parameter in the model (here the gain 1 in front of the input) in order to avoid a problem known as parameter drift. See [6] for a motivating discussion.

Finally we consider the adaptive system in the situation where the dynamics of the plant itself do not belong to the model set $b \neq 0$, $c = 1$ and $d = 0$.

In essence we present an incomplete bifurcation analysis for the overall adaptive system (9). A complete analysis has never been performed.

4.1 Under modelling due to signal error: $d \neq 0$, $b = 0, c = 1$.

In this situation the adaptive system (9) is represented by:

$$y(t+1) = x(t)y(t) + y^* + d \qquad y(0) = y_0$$
$$x(t+1) = (1 - \mu \frac{y^2(t)}{\delta + y^2(t)})x(t) - \mu \frac{y(t)d}{\delta + y^2(t)} \qquad x(0) = x_0 \tag{13}$$

With the above two dimensional, nonlinear and time invariant discrete time system we associate the two dimensional map:

$$F_d : (y, x) \rightarrow F_d(y, x) = \left(xy + y^* + d, x - \mu \frac{y(xy + d)}{\delta + y^2} \right) \tag{14}$$

In the case we use the dead beat identification algorithm $\mu = 1, \delta = 0$ we get the following discontinuous system instead:

$$y(t+1) = x(t)y(t) + y^* + d \qquad y(0) = y_0$$

$$x(t+1) = \begin{cases} x(t) & if\, y(t) = 0 \\ -\dfrac{d}{y(t)} & if\, y(t) \neq 0 \end{cases} \qquad x(0) = x_0 \tag{15}$$

The relevant parameter range is limited to $0 < \mu < 2$ and $\delta \geq 0$ as for $\mu > 2$ or $\mu < 0$ global stability is lost.

Theorem 5 *Consider the adaptive system (13). Let $\delta > 0$, $y^* \neq 0$ and $\mu > 2$. Define*

$$\Xi = \{(y_0, x_0) : (\mu - 2)y_0^2 - \mu|y_0 d| - 2\delta \geq 0 \qquad |x_0| \geq 1 + \frac{|y^* + d|}{|y_0|}\}$$

Trajectories with initial conditions in Ξ diverge.

Proof On Ξ we have that $|y(k+1)| > |y(k)|$ and $|x(k+1)| > |x(k)|$ for all $(y(k), x(k)) \in \Xi$. Hence Ξ is invariant $F_d(\Xi) \subset \Xi$ (see equation (14)). (q.e.d)

For $\mu \in (0, 2)$ it is conjectured that the adaptive system is well behaved in the sense that:

Conjecture 1 *Consider the adaptive system (13). Let $\delta > 0$ and $0 < \mu < 2$. For all $d > -1$, $y^* \neq 0$, and all initial conditions y_0, x_0, the trajectories are bounded. For all $d \leq -1$, $y^* \neq 0$ almost all solutions remain bounded.*

The conjecture is known to be false if the case of regulation $y^* = 0$ is allowed, or when general time varying disturbances d are considered. Boundedness of all solutions can be demonstrated if one modifies the adaptation law such that the parameter estimate and hence the parameter error x is forced to be bounded [20]. We believe that the enforcement of the boundedness of x is not necessary for $d > -1$. For $d \leq -1$ we conjecture the existence of a single smooth manifold along which solutions diverge.

4.1.1 Dead beat parameter tuner

Let us first consider the adaptive system with dead beat parameter tuner (15).

Under the condition $-1 < d/y^* < 0.5$ the system response (apparently) converges to the equilibrium solution $(y(t), x(t)) \equiv (y^*, -d/y^*)$ regardless of the initial conditions. Figure 3 displays a typical trajectory for $d = -0.8, y^* = 1$, Figure 4 presents an orbit for $d = 0.4, y^* = 1$.

Local analysis, ignoring the division by zero event, indicates that this equilibrium is asymptotically stable when the external signals are in the range $-1 < d/y^* < 0.5$. The equilibrium undergoes a flip (or period doubling) bifurcation for $d/y^* = 0.5$ and a degenerate Hopf bifurcation at $d/y^* = -1$. (For $d/y^* = -1$ apparently every trajectory degenerates into a periodic orbit, determined solely by the initial conditions.)

In the Appendix A the convergence to the equilibrium and a domain of attraction are established using a Lyapunov argument when the parameter are such that $-1 < d/y^* \leq 0$.

A few remarks are in order:

Remark Despite the fact that the plant is no longer in the model set $d \neq 0$, and despite the fact that no static control law from the control set is able to achieve the control objective (output regulation) in the presence of a disturbance $d \neq 0$, the adaptation does achieve regulation for all disturbances in the range $-1 < d/y^* < 0.5$. We may say that the regulation property is robust with respect to the plant class described by the difference equation $y(k+1) = ay(k) + u(k) + d$ with $a \in R$ and $-1 < d/y^* < 0.5$. This is strong indicator for the robustness properties enjoyed by adaptive systems. ○

Remark The fact that at the equilibrium the control objective is realised, is a direct consequence of the property that zero prediction error e implies that the control objective is achieved see equation (4). It remains nevertheless surprising that the adaptive scheme in this particular situation indeed possesses an equilibrium! ○

Remark Although neither the Hopf bifurcation nor the flip bifurcation is catastrophic from a stability point of view, these bifurcations do form the boundary of the useful range of the adaptive system. Notice that from an engineering perspective an ability to tolerate at least 50% error in the signal environment may be considered excellent. ○

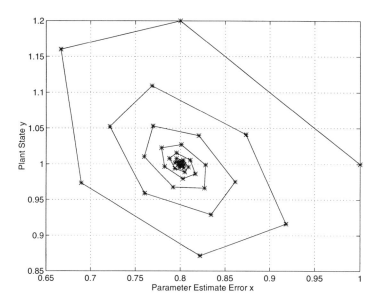

Figure 3: Typical trajectory with signal disturbance $d = -0.8$.

Remark The result is also a strong pointer to the importance of persistently exciting signals. Indeed if $y^* = 0$ the adaptive system does not tolerate any disturbance!   ○

4.1.2 Normalised least mean square parameter tuner

Let us now consider the continuous map (14).

The map F_d has a unique fixed point $(y^*, -d/y^*)$. The Jacobian of F_d at the fixed point is given by:

$$\begin{pmatrix} \dfrac{-d}{y^*} & y^* \\ \dfrac{\mu d}{y^{*2} + \delta} & 1 - \dfrac{\mu y^{*2}}{\delta + y^{*2}} \end{pmatrix} \qquad (16)$$

The equilibrium is locally asymptotically stable for $\mu \geq 0$ and $-1 < d/y^* < 1 - \dfrac{\mu y^{*2}}{2(\delta + y^{*2})}$.

At the boundary $-1 = d/y^*$ with $4 > \dfrac{\mu y^{*2}}{\delta + y^{*2}}$ the equilibrium undergoes a Hopf bifurcation, possibly degenerate. The Hopf bifurcation may be both super critical as well as subcritical.

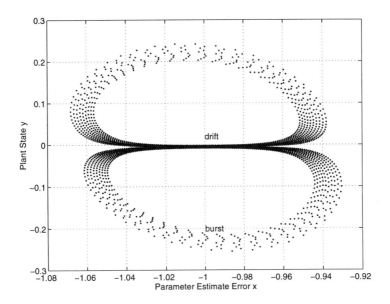

Figure 4: Typical trajectory with signal disturbance $d = 0.4$.

The conditions $-1 < d/y^* = 1 - \frac{\mu y^{*2}}{2(\delta + y^{*2})}$ correspond to the equilibrium experiencing a period doubling bifurcation.

The flip bifurcation was already present in the ideal adaptive system. The difference is that we now observe that the onset of instability due to high step size μ is influenced by the signals. Moreover whereas in the ideal system the flip bifurcation signalled the end of the usefulness of the adaptation (global stability loss), this is no longer the case here. Indeed the flip bifurcation indicates a graceful decay in adaptive system performance. The plant state's cesaro mean is indeed very close to y^*, whilst the root mean square error $y(k) - y^*$ steadily increases with increasing d.

Considering the disturbance parameter d as a bifurcation parameter (fixing $\mu = 1$, $y^* = 1$ and $\delta = 1$), the initial flip bifurcation is the start of a period doubling sequence of bifurcations, which for increasing parameter d spells more complicated behaviour and steady deterioration of the adaptive system performance. A numerically obtained bifurcation diagram is presented in Figure 5. The figure displays the plant state y component of the asymptotically stable orbit for increasing values of d. The initial flip bifurcation occurs at $d = 0.75$. The diagram represents the asymptotic attractor for values of $0.7 \leq d \leq 10$ in steps of 0.025.

The Hopf bifurcation indicates a more severe loss of performance. Depending on the value of μ and δ both sub critical and super critical as well as degenerate and non degenerate Hopf bifurcations take place. For sufficiently small values of μ the Hopf bifurcation is super critical. Do notice that the presence of sub critical Hopf

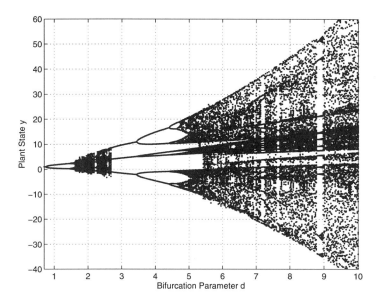

Figure 5: Bifurcation diagram (adaptive system with disturbance).

bifurcations indicate that the equilibrium can not be globally attractive.

For sufficiently small μ (e.g. $\mu = 0.1$, $y^* = 1$ and $\delta = 1$) decreasing the bifurcation parameter $d < -1$ (but close to -1) leads to complicated behaviour sometimes referred to as *bursting with drift*. (Small μ implies that the observed bifurcations are super critical.) The trajectories are characterised by a slow drift phase followed by a burst. In the drift phase, the plant state is regulated towards zero, which indicates a severe loss of performance, whilst the parameter estimate error x drifts till it destabilises the plant $x > 1$. This leads to the burst phase: the plant state explodes and enforces a re-identification of the parameter estimate, leading to a new drift phase. The complete orbit is aperiodic. A particular orbit for $\mu = 0.1$, $y^* = 1$, $\delta = 1$ and $d = -1.01$ is displayed in Figure 6. In Figure 7 the corresponding time trajectories are displayed against time. On this Figure 7 it is easy to recognise the *drift* and *burst* phases.

As d decreases further, the orbit approximates more and more the $y = 0$ axis, extending over the complete $-1 < x < 1$ range, whilst the bursts become more pronounced with larger escapes away from $y = 0$.

4.1.3 Summary $d \neq 0$

Important points to notice are:

- The adaptive control scheme tolerates a significant amount of undermodelling.

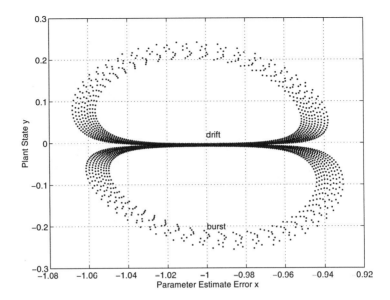

Figure 6: Burst and drift: aperiodic orbit after Hopf bifurcation.

- Slow adaptation allows for larger undermodelling errors.

- When the equilibrium undergoes a Hopf bifurcation the control performance is severely deteriorated.

- When the equilibrium undergoes a flip bifurcation the control performance degrades gracefully.

4.2 Under modelling due to gain mismatch $d = 0, b = 0\ c \neq 1$

In this situation the adaptive system is represented by the two dimensional nonlinear difference equation:

$$\begin{align}
y(t+1) &= (1-c)ay(t) + cx(t)y(t) + cy^* \\
x(t+1) &= x(t) - \mu \frac{y(t)(cx(t)y(t) + (1-c)ay(t) - (1-c)y^*)}{\delta + y^2(t)}
\end{align} \quad (17)$$

With the above difference equation we associate the map F_c:

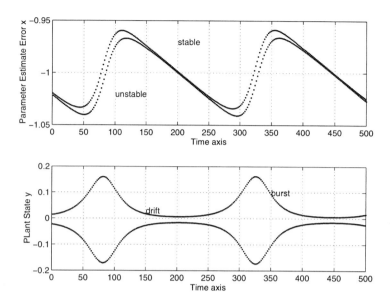

Figure 7: Burst and drift: aperiodic orbit after Hopf bifurcation.

$$F_c : (y,x) \to F_c(y,x) \qquad = ((1-c)ay + cxy + cy^*,$$
$$x - \mu \frac{y(cxy + (1-c)ay - (1-c)y^*)}{\delta + y^2})$$
(18)

In the case of dead beat parameter tuning we have the discontinuous difference equation:

$$y(t+1) = (1-c)ay(t) + cx(t)y(t) + cy^* \qquad y(0) = y_0$$

$$x(t+1) = \begin{cases} x(t) & if \, y(t) = 0 \\ (1-c)(x(t) - a + \dfrac{y^*}{y(t)}) & if \, y(t) \neq 0 \end{cases} \qquad x(0) = x_0$$
(19)

4.2.1 Dead beat parameter tuner

For the dead beat parameter tuning (19) case we have the following facts.

- For $c < 0$ or $c > 2$ global stability is lost, moreover almost all trajectories diverge .

- If $y^* = 0$ and $y_0 = 0$ then $(y(k), x(k)) \equiv (y_0, x_0)$ for all $k = 0, 1 \ldots$. The control objective is achieved. But clearly no adaptation takes places.

- If $y^* = 0$ but $y_0 \neq 0$ then we have for $0 < c < 2$ that all trajectories are well defined $(y(t) \neq 0$ for all $t)$ and converge exponentially fast to the fixed point $(0, \frac{-(1-c)a}{c})$. For any other value of c all trajectories diverge.

- If $y^* \neq 0$ there is a unique equilibrium $(y(t), x(t)) = (y^*, (1-c)(1-a)/c)$ which is locally asymptotically stable for $0 < c < 4/3$.

- At $c = 0$ a degenerate Hopf bifurcation takes place, with almost all trajectories diverging for $c = 0$.

- For $c = 4/3$ a flip bifurcation takes place, global stability is lost, but a locally stable 2-periodic orbit survives. The global behaviour is rather more complicated and it appears that aperiodic orbits co-exist with the 2 periodic orbit. (This is a consequence of the discontinuity in the dead beat parameter estimator.)

The above facts indicate that the adaptive system is capable of achieving the desired control objective (at least locally) despite a 30% error in the assumed plant input gain. Clearly underestimating the plant gain carries a worse penalty than overestimating.

Simulation evidence suggests that the equilibrium is globally stable whenever it is locally stable. A Lyapunov analysis as in the previous case enables one to partially confirm this observation.

4.2.2 Normalised least mean square update

The adaptive system (17) allows for an equilibrium, at which the control objective is necessarily satisfied: $(y(t), x(t)) \equiv (y^*, (1-c)(1-a)/c)$.

Although the equilibrium does depend on the actual plant parameter, the Jacobian of the map F_c (18) evaluated at the equilibrium is independent of it and given by:

$$\begin{pmatrix} 1-c & cy^* \\ \frac{-\mu(1-c)y^*}{\delta + y^{*2}} & 1 - \mu \frac{cy^{*2}}{\delta + y^{*2}} \end{pmatrix} \qquad (20)$$

This equilibrium is locally asymptotically stable if $4 - 2c - c\frac{\mu y^{*2}}{\delta + y^{*2}} > 0$ and $c > 0$.

At $c = 0$ a degenerate Hopf bifurcation occurs whilst the limit $c = 4/(2 + \frac{\mu y^{*2}}{\delta + y^{*2}})$ corresponds to a flip bifurcation.

4.2.3 Summary $c \neq 1$

- Slow adaptation $\mu << 1$ increases significantly the parameter range for which the control objective is achieved: the robustness margin improves with slow adaptation.

- The Hopf bifurcation again indicates a more severe loss of performance than the period doubling bifurcation.

- Estimating the domain of attraction for the fixed point is non trivial. (No non trivial estimate is known.) This is a serious problem in applications. Obviously a finite domain of attraction indicates that some prior knowledge about the plant to be controlled is required before the adaptive system can be applied.

4.3 Dynamic undermodelling $b \neq 0$, $c = 1$, $d = 0$

Here we consider a situation in which the control objective is achieved by the adaptive system with a non stationary control law.

For more details we refer the reader to [8] and [9].

We first consider the dead beat parameter tuner ($\mu = 1, \delta = 0$) with regulation to $y^* = 0$. The previous subsections serve to indicate that this situation may capture the essential features, albeit not the full story.

The we consider briefly the normalised least mean square tuner.

4.3.1 Dead beat tuner with $y^* = 0$

The system description (see equation (9)) with $\mu = 1, \delta = 0$ and $y^* = 0$ may be represented by:

$$y(k+1) = -b\frac{y(k-2)}{y(k-1)}y(k) + by(k-1) \quad if y(k-1) \neq 0 \tag{21}$$

This recursion (21) is obviously only valid if $y(k-1) \neq 0$, and as we are interested in driving y to zero, the above equation (21) can only capture the transient behaviour.

Assuming therefore that we are only considering the trajectories for which $y(k) \neq 0$ for all $k = 0, 1, \ldots$ we introduce the variable $r(k) = \frac{y(k)}{\sqrt{|b|}y(k-1)}$. We have then:

$$r(k) = sign(b)\left(\frac{1}{r(k-1)} - \frac{1}{r(k-2)}\right)$$

$$y(k) = \sqrt{(|b|)}r(k)y(k-1) \tag{22}$$

$$x(k) = -\sqrt{(|b|)}\frac{1}{r(k)}$$

Clearly the variable $r(k)$ captures the dynamic behaviour completely, (for as long as we are willing to ignore the division by zero event).

In this particular situation, no equilibrium exists. This is a consequence of considering regulation to zero $y^* = 0$. The situation $y^* \neq 0$ leads to a considerably more complicated analysis. Yet the essential features are captured by the present analysis.

We consider separately $b > 0$ and $b < 0$.

Case 1: $b < 0$

This situation is pertinent in the case the discrete time plant to be controlled is arrived at from a continuous time system via sampled data control.

For $b < 0$, the r recursion (22) is described by

$$r(k) = -\left(\frac{1}{r(k-1)} - \frac{1}{r(k-2)}\right) \tag{23}$$

It is easy to verify that this recursion (23) has a two periodic orbit characterised by $\{\ldots, \sqrt{2}, -\sqrt{2}, \ldots\}$.

Moreover a linear analysis indicates that this periodic orbit is locally asymptotically stable. In the Appendix B a Lyapunov argument confirms this and provides an estimate for its domain of attraction. The following result can be demonstrated [9]:

Theorem 6 *Consider the adaptive system (9). Let $\mu = 1, \delta = 0, y^* = 0$ and $b < 0$. Then for all initial conditions, except for a set of Lebesgue measure zero:*

- *The parameter estimate error is attracted towards a two-periodic solution:*

$$x(k) \to \pm(-1)^k \sqrt{\frac{|b|}{2}} \text{ as } k \to \infty$$

- *The plant state is regulated towards $y^* = 0$ exponentially fast if $-0.5 < b < 0$.*

- *The plant state diverges exponentially fast if $b < -0.5$.*

- *The plant state becomes eventually two-periodic if $b = -0.5$, the particular orbit being determined by the initial conditions.*

Some remarks are in order:

Remark No time invariant linear controller, regardless of its complexity, could achieve regulation for all plant systems in the class regulated by the adaptive system, i.e. arbitrary a and $-0.5 < b \leq 0$. ○

Remark If we had known the plant parameter a exactly, the proposed control law could have achieved regulation for all $|b| < 1$. The fact that the adaptive algorithm is unable to identify a (this is due to the undermodelling of course) leads to a reduced tolerance margin. ○

Remark The two periodic stabilisation observed here can be recognised in the transient behaviour of the corresponding normalised least mean square scheme. Indeed for large signals in the loop, the normalised least mean square algorithm approximates arbitrarily well the dead beat tuning scheme. This is an important indicator for how complex the transient analysis of an adaptive scheme may be. ○

Case 2 $b > 0$

For $b > 0$, the r recursion (22) is described by

$$r(k) = \left(\frac{1}{r(k-1)} - \frac{1}{r(k-2)}\right) \quad (24)$$

This recursion (24) leads to more complex dynamics than $b < 0$ described by equation (23). One can verify that:

- There are no periodic orbits of period $1, 2$ and 3.

- There is a periodic orbit of period 4, given by

$$\{\ldots, \sqrt{2+\sqrt{2}}, -\sqrt{2-\sqrt{2}}, -\sqrt{2+\sqrt{2}}, \sqrt{2+\sqrt{2}}, \ldots\}$$

Local analysis reveals that this periodic orbit is of saddle type. Numerically one can establish that the stable and unstable manifold intersect transversally which suggest the existence of horseshoe like chaos. This is indicative of at least transient chaos.

- Numerical analysis leads to the identification of a series of periodic orbits of periods, larger than 4 (we verified the existence of all periodic orbits up to period 100), all of which are of saddle type.

Assuming that along the chaotic attractor the cesaro mean of a function evaluated on the trajectories is well defined, and leads to the definition of a natural measure on the attractor, we arrive at the following conjecture:

Conjecture 2 *Consider the adaptive system (9). Let $\mu = 1, \delta = 0, y^* = 0$ and $b > 0$. Then for all initial conditions, except for a set of Lebesgue measure zero:*

- *The parameter estimate error is attracted towards an erratic solution, satisfying:*

$$\lim_{N \to \infty} \frac{1}{N} \sum_{t=1}^{N} \log(|x(t)|) = \frac{1}{2} \log(\frac{b}{2})$$

- *The plant state is regulated towards $y^* = 0$ almost surely exponentially fast if $0.5 > b > 0$.*

- *The plant state diverges almost surely exponentially fast if $b > 0.5$.*

A typical trajectory for $b = 0.4$ is represented in Figure 8. Observe the aperiodicity of the parameter estimate x.

Notice that the adaptive scheme regulates the plant state, as required, but the actual parameter estimate error, or the controller gain, behaves in an erratic fashion for $0.5 > b > 0$. It follows that for all $0.5 > b > -0.5$ the adaptive scheme succeeds in regulating the plant, despite the unmodelled dynamics.

It appears that there is a less than one dimensional attractor in the planar phase space of the adaptive system, and this attractor is a subset of the parameter estimate axis. From a practical point of view this has the important implication that the strange dynamics live inside the computer whereas in the plant everything seems to behave nicely, at least asymptotically.

4.3.2 Normalised least mean square parameter tuner $b \neq 0$

We complement the previous more global considerations with a local analysis in the case of a dynamic disturbance $d(t) = by(t-1)$ together with a reference signal $y^* \neq 0$.

In this case the adaptive system (9) is represented by:

$$\begin{aligned} y(t+1) &= x(t)y(t) + y^* + by(t-1) \\ x(t+1) &= (1 - \mu \frac{y^2(t)}{\delta + y^2(t)})x(t) - \mu \frac{y(t)by(t-1)}{\delta + y^2(t)} \end{aligned} \quad (25)$$

With this recursion we associate the map F_b:

$$F_b : (y, z, x) \to \left(z, xz + by + y^*, x - \mu \frac{z(xz + by)}{\delta + z^2} \right) \quad (26)$$

The system (25) allows for an equilibrium, at which the control objective is achieved: $(y, z, x) = (y^*, y^*, -b)$. The Jacobian at this equilibrium for the map F_b (26) is given by:

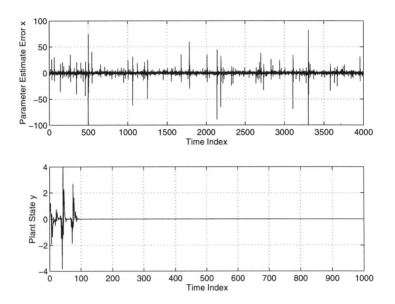

Figure 8: Chaotic regulation

$$\begin{pmatrix} 0 & 1 & 1 \\ b & -b & y^* \\ -\mu\dfrac{by^*}{\delta+y^{*2}} & \mu\dfrac{by^*}{\delta+y^{*2}} & 1-\mu\dfrac{y^{*2}}{\delta+y^{*2}} \end{pmatrix} \quad (27)$$

The characteristic polynomial of the matrix (27) is given by $p(\lambda) = \lambda^3 + \lambda(-1 + b + \nu) + \lambda(-2b) + b$, where $\nu = \mu y^{*2}/(\delta + y^{*2})$.

The equilibrium is hence locally stable provided

$$\nu > 0 \text{ and } b < 0.5 - 0.25\nu \text{ and } b > -1/(1+\nu)$$

The limit $\nu = 0$ corresponds to a global bifurcation, ($\nu < 0$ implies complete instability). The boundary $b = 0.5 - 0.25\nu$ indicates a flip bifurcation, whilst $b(1+\nu) = -1$ indicates a Hopf bifurcation.

We again observe that slow adaptation (small $\nu = \mu y^{*2}/(\delta + y^{*2})$, i.e. small step size and/or small reference signal) leads to the largest robustness margin (at least for local stability).

The Hopf bifurcation typically indicates a more severe loss of performance than does the flip bifurcation. This observation holds for both the domain of attraction, which is significantly smaller in the case of the Hopf bifurcation, as well as the asymptotic performance. Again the Hopf bifurcation can be both degenerate or regular, sub critical or super critical. For sufficiently small ν the Hopf bifurcation is super critical.

The flip bifurcation for $b = 0.5 - 0.25\nu$ is again the first in a series of period doubling bifurcations which leads to complicated dynamics and steady loss of control performance.

5 Conclusions

Adaptive systems, even in the context of controlling a linear plant, display a wealth of nonlinear dynamics. The heuristic approach to the design of an adaptive system presented here leads to adaptive systems characterised by a highly nonlinear interaction between model tuning and control, which in turn leads to remarkably robust control properties.

In particular we exemplified that the control objective can be achieved, arbitrarily well, despite significant undermodelling errors, even when the parameter estimate behaves in an erratic/chaotic or periodic fashion.

One of the difficulties of adaptive systems as presented here, is the lack of a suitable Lyapunov function, capable of dealing with undermodelling, to analyse and estimate the region in state space where adaptation is useful. Our insistence on understanding global behaviour stems from the fact that in designing adaptive systems both asymptotic and transient dynamics need to be understood. This observation has prompted many researchers to rederive an adaptive system methodology from a Lyapunov perspective [3].

The (local) bifurcation analysis, even for this rather simple example is far from being complete. Further work is ongoing in order to unravel some of the more generic observations that we pointed out: slow adaptation clearly increases robustness in the presence of signal disturbances, but fast adaptation, although not as able to deal with undermodelling, provides better transient behaviour. The importance of excitation, especially in conjunction with slow adaptation is well understood, but its implications for transient behaviour and its influence on more global dynamical characteristics needs further work. The important difference between the Hopf bifurcation and the period doubling bifurcation is believed to be generic in adaptive systems as introduced here.

A more complete understanding of the bifurcation mechanism may eventually lead to an adaptive algorithm that can detect (severe) undermodelling by detecting the nature of the relevant bifurcation and is consequently capable of adjusting the model class/ controller structure/tuner as to cope with the plant and control objective. No such algorithm has been designed to date.

A global dynamical analysis for the present example is another avenue for further work. Here the emphasis is on understanding the transients in the adaptive system's response. Clearly a difficult problem, but of significant importance from a practical point of view. Nice asymptotic dynamics is not enough for a successful application. A method of estimating the domain of initial conditions from which the adaptive system may achieve the control objective is vital in applications.

It is clear that we are compelled to conclude that adaptive systems are essentially nonlinear in their behaviour. The "learning" achieved by adaptive systems as described here may exhibit periodic, aperiodic and chaotic transients as well as steady state behaviour. Obviously, there remains a lot of fun research to be done before we understand how adaptive systems actually do learn and how we could design better adaptive systems.

Acknowledgement The author is grateful to the organisers of the Chaos and Control workshop for making the event happen. The financial support of DITAC, Australia and NSF, US is gratefully acknowledged.

It is the author's pleasure to acknowledge Prof. R. Bitmead for introducing him to the topic of dynamical system analysis for adaptive systems.

References

[1] H.P. Whitaker, An adaptive systems for control of the dynamics performance of aircraft and spacecraft, *Inst. Aeronautical Sciences*, Paper 59-100, 1959.

[2] S. Sastry, M. Bodson, *Adaptive Control*, Prentice-Hall, 1989.

[3] M. Kristic, I. Kanellakopoulos, P. Kokotovic, *Nonlinear and Adaptive Control Design*, Wiley, Inter-Science, 1995.

[4] B.D.O. Anderson, R.R. Bitmead, C.R. Johnson, P. Kokotovic, R. L. Kosut, I.M.Y. Mareels, L. Praly, B. Riedle, *Stability of adaptive systems, Passivity and averaging analysis*, MIT Press, 1986.

[5] W. Sethares, I.M.Y. Mareels, Dynamics of an adaptive hybrid, *IEEE Trans on Circuits and Systems*, January 1991, Vol 38, No 1, pp 1-12.

[6] K.Astrom, B. Wittenmark, Adaptive Control, *Addison-Wesley*, 1989.

[7] I.M.Y. Mareels, *Dynamics of adaptive control*, Australian National University PhD thesis, 1986

[8] I.M.Y. Mareels, R.R. Bitmead, Nonlinear dynamics in adaptive control: chaotic and periodic stabilisation, *Automatica*, Vol 22, No 6, pp 641-655, 1986.

[9] I.M.Y. Mareels, R.R. Bitmead, Nonlinear dynamics in adaptive control: chaotic and periodic stabilisation, II Analysis*Automatica*, Vol 24, No 4, pp 485-497, 1988.

[10] I.M.Y. Mareels, R.R. Bitmead, Bifurcation Effects in Robust Adaptive Control, invited contribution, *IEEE Transactions on Circuits and Systems*, Vol.CAS 35, July 1988, No. 7, pp 835-842.

[11] M.P. Golden, B. E. Ydstie, Bifurcation in model reference adaptive control systems, *Systems&Control Letters*, Vol 11, pp413-430, 1988.

[12] M.P. Golden, B.E. Ydstie, Small amplitude chaos and ergodicity in adaptive control, *Automatica*, Vol 28, No1, pp 11-25, 1992.

[13] M.G. Kush, B.E. Ydstie, Drift instability and chaos in forecasting and adaptive decision theory, *Physica D*, vol 72, No 4, pp 309-323.

[14] R. Adomaitis and I. Kevrekidis, On the global bifurcations characteristics of adaptive systems, Proc. *11th IFAC World Congress*, 1990, Tallinn, pp 299-304

[15] F.M.A. Salam, S. Bai, Disturbance generated bifurcation in a simple adaptive control system: simulation evidence, *Systems&Control Letters*, Vol 7, pp269-280, 1986.

[16] M.D. Espana, L.Praly, On the global dynamics of adaptive systems: a study of an elementary example, *SIAM Journal on Control and Optimisation*, Vol 31, No 5, pp 1143- 1166, 1990.

[17] J.B. Pomet, J.M. Coron, L. Praly, On periodic solutions of adaptive systems in the presence of periodic forcing terms, *Mathematics of Control, Signals and Systems*, Vol 3, No4, pp373-399, 1994.

[18] J. Homer, I.M.Y. Mareels, Echo canceller performance analysis in 4-wire loop systems with correlated AR subscriber signals, *IEEE Trans Information Theory.*, Jan 1995, pp 322-329

[19] G.C. Goodwin, K.S. Sin, *Adaptive filtering, prediction and control*, Prentice Hall, 1984.

[20] L. Praly, *Commande lineaire adaptative: solutions bornees et leurs proprietes*, PhD thesis, Ecole Nationale Superieure des Mines de Paris, 1988

Appendix A

The system (15) can be represented by a single difference equation in terms of the parameter error $x(k)$ after eliminating the plant state $y(k)$ as follows:

$$x(k+2) = \frac{x(k+1)}{x(k) + (\frac{y^*+d}{-d})x(k+1)}$$

Here we neglected consideration of the event $y(k) = 0$. Defining $\alpha = (y^*+d)/(-d)$ and $v(k) = x(k)/(1+\alpha)$, this recursion can be rewritten as follows:

$$v(k+2) = \frac{(1+\alpha)v(k+1)}{\alpha v(k+1) + v(k)} \tag{28}$$

The parameter range $d/y^* \in (-1,0)$ corresponds to $\alpha \in (0,\infty)$.

Furthermore with $v(0), v(1) \in (0,\infty)$, it is clear that $v(k) \in (0, \frac{1+\alpha}{\alpha})$ for all k. (From which it is clear that on this domain the division by zero event, neglected above, can never occur.)

Consider now the recursion (28), we show that for all $\alpha \in (0,\infty)$ all trajectories starting from initial conditions $v(0), v(1) \in (0,\infty)$ converge to the equilibrium solution $v(k) \equiv 1$.

Let $V(k) = max(1-v(k), \alpha(v(k)-1), 1-v(k+1), \alpha(v(k+1)-1))$ for $k = 0, 1, \ldots$. Along the solutions of the above recursion (28) $V(k+1) - V(k) \leq 0$. From which we conclude that $V(k)$ converges to zero, hence $v(k)$ converges to 1.

That $V(k+1) - V(k) \leq 0$ along the solutions of (28) may be established from considering:

- Assume $V(k) = \alpha(v(k+1) - 1)$
  In this case $v(k+1) \geq 1$, $v(k+1) \geq v(k)$ and $v(k+1) + v(k) \geq (1+\alpha)$. It follows that $v(k+2) \geq 1$ and $v(k+2) \leq v(k+1)$, hence $V(k+1) = V(k)$.

- Assume $V(k) = 1 - v(k)$

    In this case we have $v(k) \leq 1$, $v(k) \leq v(k+1)$ and $v(k) + v(k+1) \leq 1 + \alpha$. It follows that $v(k+2) \geq v(k)$ as well as $v(k+2) + v(k) \leq 1 + \alpha$, hence $V(k+1) \leq V(k)$.

- The other cases $V(k) = 1 - v(k+1)$ and $V(k) = v(k) - 1$ follow along similar lines.

Appendix B

Consider the recursion

$$w(k+1) = \frac{1}{2}(\frac{1}{w(k)} + \frac{1}{w(k-1)}) \qquad (29)$$

This recursion is related to the r recursion (23), by the time varying Lyapunov transformation: $x(k) = (-1)^k \frac{r(k)}{\sqrt{2}}$.

For the recursion (29) the fixed points correspond to $w(k) \equiv 1$ or $w(k) \equiv -1$. (This corresponds to the two periodic orbit in the $r(k)$ recursion (23).)

Consider $w(0) > 0$ and $w(1) > 0$. Obviously, $w(k) > 0$ for all $k = 0, 1, \ldots$.

Consider the evolution of the positive quantity

$$V(k) = \max(w(k), w(k-1), \frac{1}{w(k)}, \frac{1}{w(k-1)}) - 1$$

along the solutions of the recursion (29) for $w(0) > 0$ and $w(1) > 0$.

Because for all $\alpha, \beta > 0$ we have that

$$\max(\frac{1}{\alpha}, \frac{1}{\beta}, \alpha, \beta) \geq \max(\alpha, \frac{1}{\alpha}, \frac{\alpha+\beta}{2\alpha\beta}, \frac{2\alpha\beta}{\alpha+\beta}) \geq 1$$

and also

$$\min(\frac{1}{\alpha}, \frac{1}{\beta}, \alpha, \beta) \leq \min(\alpha, \frac{1}{\alpha}, \frac{\alpha+\beta}{2\alpha\beta}, \frac{2\alpha\beta}{\alpha+\beta}) \leq 1$$

we conclude that $V(k+1) - V(k) \leq 0$, with $V(k+1) \equiv V(k)$ iff $w(k) \equiv 1$.

From this it follows that all solutions of the recursion (29) starting with $w(0) > 0$, $w(1) > 0$ converge to the solution $w(k) \equiv 1$.

By implication, all solutions of the recursion (23) with $r(0)r(1) < 0$ converge to the two periodic solution $\ldots, 1, -1, 1, \ldots$.

Commentary by E. Ott

Very well-written and clear. Nicely pedagogical. I was able to benefit from the paper, although I had no prior knowledge of adaptive control. The paper is very appropriate for the volume subject.

Hitting Times to a Target for the Baker's Map

Arthur Mazer
Dept. of Mathematics and Statistics
Utah State University
Logan, UT 84322, USA

Abstract

A targetting theorem for ergodic systems without control is presented. It states that the average hitting time depends only on the size of the target, not its location. An algorithm is then presented to compute average hitting times for ergodic processes.

Finally hitting times for the Baker's Map with a specified target are calculated with and without control. The average hitting time without control is 57. With a small control the average hitting time is 3.63.

Introduction

Those who were fortunate enough to attend the conference would be baffled by this title because it appears to be totally unrelated to the talk that the author presented. This is in fact the case and is due to the far sighted philosophy of the organizers. This philosophy encouraged the participants to write and present their complete papers well after the conference was held so that a sufficient amount of time could elapse for reflection. The organizers encouraged the participants to write papers that would address themes that emerged during the conference. In this light the author felt that the best thing to do would be to scrap the presented material. Indeed motivation for this paper came from other talks presented at the conference as well as a comment made by Walter Grantham which was a reaction to the author's talk. This comment will be remarked upon below.

A recurring theme at the conference was the targetting of a trajectory toward a specified set. In many cases targetting is considered as a precursor to employing a feedback control which only works in the targetted set. This was discussed by many participants in connection with OGY control. Also in this context is a novel application of chaos presented by Tom Vincent. In his work, a control to produce a chaotic attractor that includes the targetted set is presented. There are two ways to target, passively or actively. The passive mode is in fact to do nothing and let the trajectory proceed upon its ergodic path until it enters the target set. The active mode is to apply a control which hastens the hitting time. In either case, one is interested in hitting times.

There are many examples in which the system performance depends upon the hitting time and one wishes to assure that the hitting time meets specified conditions. Tom Vincent showed a video of control of a bouncing ball in which his approach works because the hitting time is reasonable. Eric Kostelich presented an example concerning satellite control as well as a control algorithm for targetting and calculating hitting times. Kathryn Glass as well as Teo Kok also presented algorithms to accomplish these goals. All of these methods are computationally expensive. This is not a criticism but is a consequence of the fact that accomplishing this task for an actual system requires a lot of computational power.

The emphasis in the conference was in designing algorithms that could be applied to actual problems. The logic in this approach is unimpeachable. Afterall, we all want to design implementable controls. In this paper hitting times are investigated for an abstract system, "the Baker's Map". The system is so simple that all computations can be made analytically. This should alert the reader that such a system could not possibly exist in nature. Although the warning has been issued there is no cause for alarm. Indeed the simplicity which allows for analytic solutions might provide the insight needed to remove some thorns from this problem.

The organization of this paper is as follows. In section 1 concepts and notation that are used throughout the paper is introduced. Definitions for ergodicity and expected hitting time are given along with an algorithm for computing the expected hitting time. Finally a theorem relating the expected hitting time to the measure of the targetted set is stated. Section 2 gives a proofs of the theorem that is stated in section 1. Section 2 requires an elementary knowledge of ergodic theory and may be bypassed.

The bulk of the paper is in the analysis of the Baker's Map. This is given in section 3. Hitting times for sets of a specified size are analytically calculated. A small control is then introduced and hitting times are once more calculated. This allows one to assess the effect of the control on the hitting times.

The conclusion follows section 3. The conclusion includes the author's opinion of what features in the specialized problem are applicable to more general systems. A remark on Walter Grantham's comment also appears as well as some open questions.

1 Algorithms

We start by introducing the notation which is used throughout the paper. Let (X, μ, ϕ) represent a space, X, along with a mapping, ϕ and a measure μ. The space X is bounded. One can think of X as an attractor. ϕ maps points of X back onto points of X. There is a measure, μ, which gives the size of a sufficiently large class of subsets of X. If A is a subset of X, then $\mu(A)$ is its size. Here one must use one's imagination to interpret the word size since the measure may not conform to the Euclidean notion of size. However, the measure has nice properties that we associate with Euclidean size. For example two such properties are; if set A is contained in set B then $\mu(A) \leq \mu(B)$, if sets A and B are disjoint then $\mu(A \cup B) = \mu(A) + \mu(B)$.

There are further properties that (X, μ, ϕ) must satisfy.

i. $\mu(X) = 1$.

ii. Let A be a subset of X. Then $\mu(\phi(A)) = \mu(A)$.

Property i. is for convenience. Any measure satisfying property ii. can be normalized to satisfy property i. Property ii. determines the measure that is of interest to us. In general to satisfy this property we must consider measures that do not correspond with Euclidean size of a set. At first glance it looks restrictive,

Hitting Times to a Target for the Baker's Map 253

however any piecewise continuous map from a bounded set onto itself has a unique measure satisfying properties i. and ii.. In particular, if X is an attractor for a set of O.D.E.'s and ϕ is the map that is associated with integrating on the attractor for a specified unit of time, then a measure can be found to satisfy properties i. and ii.. The triple (X, μ, ϕ) is called a dynamical system.

Suppose we are presented with a dynamical system along with some starting point $x \in X$. The state of the system moves along an orbit given by the map ϕ. By this it is meant that after k units of time, where k is an integer, the state has evolved from the point x to the point $\phi^k(x)$. Suppose additionally that there is a target set, $A \in X$, and we wish to know an expected time that the state will land in A (hit A). We will investigate this issue for a class of systems known as ergodic systems. The definition of ergodicity is as follows.

Definition: (X, μ, ϕ) is *ergodic* if for any integrable function, f the following equality holds.

$$\int_X f(x)\,d\mu = \lim_{n \to \infty} (1/n+1) \sum_{k=0}^{n} f \circ \phi^k(x) \qquad (1)$$

The integral on the left is taken over the domain X with respect to the measure, μ. The sum on the right is over an orbit of a typical point, x, which is being moved to new locations by iterates of the map ϕ. Ergodicity states that the spatial average is equal to the time average. Or another way to put it is that the orbit of a point distributes itself uniformly over the domain so that the limit of the averages of a function calculated over the trajectory represents the spatial average. Chaotic maps on attractors are examples of ergodic dynamical systems.

Next explicit definitions of hitting time and expected hitting time are presented.

Definition: (X, μ, ϕ) be a dynamical system. Let $A \in X$. The *hitting time* of x to A is denoted by $h_A(x)$. $h_A(x) = min_{k \geq 0}$ such that $\phi^k(x) \in A$; k ranges over the nonnegative integers.

Definition: The *expected hitting time* to A is denoted by EH_A.

$$Ef_A = \int_X h_A(x)\,d\mu$$

Remark: This definition is equivalent to the following:

$$EH_A = \sum_{k=1}^{\infty} k\mu(C_k),$$

where

$$C_k = \{x \mid h_A(x) = k\}.$$

This definition for expected hitting time is an averaged hitting time where averaging occurs over the entire set X. The definition of ergodicity provides a simple way to design an algorithm to determine the expected hitting time. If A is a target subset for the system, take $f(x)$ in the definition of ergodicity to be the function $h_A(x)$. The expected hitting time is the integral. This can be approximated by truncating the sum on the right hand side at a sufficeintly large value of n.

$$EH_A = \frac{1}{n+1} \sum_{k=0}^{n} h_A(\phi^k(x))$$

This formula says that we must track the hitting time for each point in the orbit. This would require storing a lot of data. We can however reduce the amount of data that is necessary to keep. Let $a_1, a_2, a_3, \ldots, a_m$ represent all the iterates in which the orbit is in A, i.e. $\phi^{a_p}(x)$ is in A for p from 1 to m. This groups the trajectory into m groups. A group ends when the trajectory lands in A. The next element is the beginning of the next group and $\phi^0(x) = x$ is the beginning of the first group. Let g_j be the number of elements in each group. Then the above formula is identical to the following.

$$EH_A = \frac{1}{n+1} \sum_{j=1}^{m} s_j,$$

$$s_j = \sum_{p=1}^{g_j} p$$

Example 1: Let 7, 12, 25, 26, 32, 51, 61, 62 be the a_m's. This means that $\phi^7(x)$, $\phi^{12}(x)$, $\phi^{25}(x)$, etc. are in A. The g_j's are 7, 4, 12, 0, 5, 8, 9, 0. The s_j's are 28, 10, 78, 0, 15, 36, 45, 0. In this case n is 62. This gives the average hitting time as 3.37.

Remark: The method used in the example reduces the data that is necessary to save. Suppose we were to directly use the formula involving h_A. The h_A would have to be saved for all points in the orbit. Let $h_k = h_A(\phi^j(x))$; $h_0 = 7$, $h_1 = 6$, $h_2 = 5$, $\ldots$, $h_7 = 0$, $h_8 = 12$, $\ldots$. Indeed all h_k would have to be saved from $k = 0$ to $k = n$. The method from the example allows us to reduce the amount of data required and provides the same answer.

Remark: This algorithm has a straightforward extension for getting the moments of the hitting time as well.

Another point of intereset for the control theorist is the probability of the hitting time being greater than a critical number, t. By this it is meant what is the measure of the set of all points in X whose members have the property that their hitting time is larger than t. We call this set C and wish to find $\mu(C)$. Let $I_C(x)$ be the indicator function of C. Recall the indicator function is as follows.

$$I_C(x) = \begin{cases} 1 & x \in C \\ 0 & x \notin C \end{cases}$$

Then by definition, $\mu(C) = \int I_C(x) \, d\mu$. Substituting $I_C(x)$ for $f(x)$ into the definition of ergodic systems leads to an algorithm for approximating the size of C. This is given by the following formula.

$$\frac{1}{n+1} \sum m_j,$$

$$m_j = \max(0, g_j - t)$$

g_j is defined just as it is in Example 1..

Example 2: Let $t = 8$ and g_j be as in the preceeding example. The m_j are 0, 0, 4, 0, 0, 0, 1, 0. An approximation for $\mu(C)$ is 0.08.

The above algorithms can be carried out numerically provided that one is able to represent the map ϕ reasonably. (Phil Diamond presented some problems with numerical representation of maps that apply to the Baker's Map.) A question always arises as to how many iterates are necessary before a good approximation is reached. The most simple minded answer to that question is to keep increasing the iterates until the answer settles down.

A nice feature about ergodic systems is that EH_A depends only on the measure of A, not its location in X. This means that if EH_A is calculated for a single set A, then one has actually calculated the expected hitting time to all sets with the same measure as A. This result is stated formally in the following theorem and proven in section 2.

Theorem 1 *Let (X, μ, ϕ) be an ergodic dynamical system. Let A and B be measurable subsets of X with $\mu(A) = \mu(B)$. Then $EH_A = EH_B$.*

Theorem 2 *Let (X, μ, ϕ), A, and B be as above. Also let t be an integer and let $C_A = \{x \in X \mid h_A(x) > t\}$ and similarly for C_B. Then $\mu(C_A) = \mu(C_B)$.*

2 Proofs of theorems

In this section the condition a.e. will not be explicitly stated. Equality of functions and sets means equality a.e.. We also use the fact that a Borel Field is associated with the measure space (X, μ) and that ϕ is a measurable map.

Before proving the main result, some lemmas are needed.

Lemma 1 *Let (X, μ, ϕ) be an ergodic dynamical system and A be a measurable subset of X. Then $h_A(x)$ is a measurable function.*

Proof: This follows from the definition of $h_A(x)$ and the fact that /phi is /mu measurable.

Lemma 2 *Let (X, μ, ϕ) be given. Also let A and B be two sets in X with $mu(A) = mu(B)$. Then there exists an invertible, measurable map g with the following properties.*

i. $g(A) = B$

ii. $g(x) = x$ for all x not in $A \cup B$.

Proof: The proof is constructive. Let $g(x) = x$ for all $x \notin A \cup B$ so that property (ii) holds. Let $A_k = \{x \in A \mid \phi^k(x) \in B \setminus \cap_{i<k} A_i\}$. Let $B_k = \{x \in B \mid x \in \phi^k(A) \cap B \setminus \cap_{i<k} B_i\}$. Note that $\phi(A_k) = B_k$. Now for $x \in A_k$ define $g(x) = \phi^k(x)$ and let k assume values over all the nonnegative integers.

To show that property (i) holds it is necessary to show that $\cup_k B_k = B$. This holds since ϕ is ergodic. The map g is measure preserving and invertible since ϕ is. QED.

Lemma 3 *Let (X, μ, ϕ), A, B and g be as above. Define the map $f(x)$ as follows.*

$$f(x) = \begin{cases} g \circ \phi(x) & \phi(x) \in A \text{ and } x \notin B \\ \phi \circ g^{-1}(x) & x \in B \text{ and } \phi(x) \notin A \\ g^{-1} \circ \phi(x) & \phi(x) \in B \text{ and } x \notin A \\ \phi \circ g(x) & x \in A \text{ and } \phi(x) \notin B \\ g^{-1} \circ \phi \circ g(x) & \phi(x) \in B \text{ and } x \in A \\ \phi(x) & otherwise \end{cases}$$

then $f(x)$ is measure preserving and ergodic.

Note: All that f does is reroute a ϕ trajectory from A to B and vice versa as directed by the map g. Otherwise it leaves the trajectory unchanged. Proof: f is measure preserving since g and ϕ are. To show that it's ergodic we show that any invariant set is trivial. Let C be an invariant set. Let C_A be the intersection of C with A and similarly for C_B and let C_X be the remaining part of C. Then the set C_X in union with $g(C_A)$ and $g^{-1}(C_B)$ is an invariant set with respect to the map ϕ. But this set must be trivial since ϕ is ergodic. Hence C is also trivial. QED

Proof of theorem 1: Construct g and f as above. Using the definitions of ergodicity and hitting times the following equalities hold provided that EH_A is in $L^1(X)$.

$$EH_B = \lim_{n\to\infty} 1/n \sum_k h_B \circ f^k(x) \qquad (2)$$

$$= \lim_{n\to\infty} 1/n \sum_k h_A \circ \phi^k(x) \qquad (3)$$

$$= EH_A. \qquad (4)$$

If the sums above are unbounded then both EH_A and EH_B are unbounded. QED

Proof of theorem 2: Construct g and f as above. Let $C_k = \{x \in C_A \,|\, \phi^k \in A \setminus \cup_{i<k} C_i\}$ and $f^k(C_k)$ be the image of C_k under the mapping f^k. C_B is the union of all $f^k(C_k)$ taken over all values of $k > t$. The conclusion follows since f is measure preserving. QED

3 The Baker's Map

In this section hitting times are calculated for a specific system which has been coined "The Baker's Map". First the dynamics of the Baker's Map are described. Then a target set is chosen and the average hitting time is calculated. A control is then added to the dynamics and the average hitting time is recalculated.

The domain, X, for the Baker's Map is the unit square. One representation for a point in the domain is to give the coordinates (x, y) where both x and y range from 0 to 1. The measure, μ, in this case is the standard area. The map ϕ is given below.

$$\phi(x,y) = \begin{cases} (2x, 1/2y) & 0 \leq x < 1/2 \\ (1-2x, 1/2y + 1/2) & 1/2 \leq x < 1. \end{cases} \qquad (5)$$

The Baker's Map is a simple example of a chaotic map. It has Lyapunov exponents 2 and .5. For our purposes a more useful way to define the dynamics is as follows. Expand x and y into their base 2 representations.

$$x = .x_1 x_2 x_3 \ldots$$

$$y = .y_1 y_2 y_3 \ldots$$

The x_j's and y_j's are either 1 or 0 and the representation without an infinite string of 1's in the tail is used.

The base 2 representation for the x and y components of $\phi(x, y)$ are:

$$\phi_x(x,y) = .x_2 x_3 x_4 \ldots$$

$$\phi_y(x,y) = .x_1 y_1 y_2 \ldots.$$

Often one will see this set of equations written in the following manner. Express the pair (x, y) in their base 2 representation as follows:
$\ldots y_3 y_2 y_1 . x_1 x_2 x_3 \ldots$. Then

$$\phi(x, y) = \ldots y_2 y_1 x_1 . x_2 x_3 x_4 \ldots \quad (6)$$

The above equation shows that the Baker's map is equivalent to the shift operator.

Next a target set, A, is selected and the hitting times, $h_A(x, y)$. are calculated. Let $A = \{(x, y) \mid 0 \leq x < 1/32\}$. Notice that a point is in A when the first 5 digits of the base 2 expansion of the x component are all 0; $A = \{(x, y) \mid x_1 = x_2 = x_3 = x_4 = x_5 = 0\}$. This and equation 5 gives a means to calculate the hitting time, $h_A(x, y)$. Assume that the first string of 5 zeroes of the base 2 expansion of x commences with x_k, i.e. $x_k = x_{k+1} = x_{k+2} = x_{k+3} = x_{k+4} = 0$ and no string of 5 zeros occurs before this one. Then $h_A(x, y) = k - 1$.

We are now in a position to calculate the average hitting time. The starting point is the definition of average hitting time.

$$EH_A = \sum_{j=0}^{\infty} j\mu(C_j)$$

$$C_j = \{(x, y) \mid h_A(x, y) = j\}$$

The following formula for $\mu(C_j)$ which is presented without proof.

$$\mu(C_j) = c_j/d_j \quad (7)$$
$$c_{-4} = c_{-3} = c_{-2} = c_{-1} = 0, c_0 = 1 \quad (8)$$
$$c_j = c_{j-5} + c_{j-4} + c_{j-3} + c_{j-2} + c_{j-1}, j > 0 \quad (9)$$
$$d_j = 2^{j+5}. \quad (10)$$

Plugging the expression for $\mu(C_j)$ into the expression for EH_A and using MAPLE to sum over the first 500 terms gives an approximation of $EH_A = 57$. Note that this is a lower bound.

One would like to know if the hitting time can be significantly improved by adding a small control. This is the next objective. The map with control is denoted by ϕ_u, where u represents the control. The control is introduced into the map as follows.

$$\phi_u(x, y) = \phi(x, y) + (u, 0) \quad (11)$$
$$|u| \leq 1/64 \quad (12)$$
$$x + u = (1 + x + u) \bmod 1 \quad (13)$$

Once again above expression for EH_A is used to calculate the average hitting time. To find the sets one looks for the preimages of A with the control set at both $u = 1/32$ and $u = -1/32$. This gives the following formulas for C_j.

$$C_0 = A \quad (14)$$
$$C_j = \{(x, y) \in \phi_{1/64}^{-j}(A) \cup \phi_{-1/64}^{-j}(A) \setminus \cup_{i<j} C_i\} \quad (15)$$
$$\phi_u^{-1}(x, y) = \phi^{-1}(x, y) + (u, 0) \quad (16)$$

The preimages have been calculated and their union covers the entire domain with $j = 5$. This means that all points in the domain can be driven to A within 5 iterations of the map. Since in this case there are so few C_j's, all of their areas along with the expected hitting time are given below.

$$\mu(C_0) = 2/64 \tag{17}$$
$$\mu(C_1) = 4/64 \tag{18}$$
$$\mu(C_2) = 7/64 \tag{19}$$
$$\mu(C_3) = 15/64 \tag{20}$$
$$\mu(C_4) = 31/64 \tag{21}$$
$$\mu(C_5) = 5/64 \tag{22}$$

Average hitting time: 3.63.
The improved performance with control is obvious.

Conclusion

Two theorems relating the size of the domain to the hitting time have been presented. These demonstrate that for an ergodic system the average hitting time depends only upon the size of the set, not its location. Note the theorems apply to the and ergodic system. One wonders if the minimum average hitting time is also only a function of set size when a control which destroys ergodicity is applied.

The algorithms to determine the average hitting time and the probability that the hitting time will be beyond a critical value also apply to ergodic systems. This is of interest to one who is using the approach of Tom Vincent to target. Also it allows one to determine whether a control is necessary for targetting when the uncontrolled system is ergodic.

The example of the Baker's Map demonstrates how dramatically a small control can enhance system performance for chaotic systems. Furthermore, it shows that enhancement comes about when the control is aligned with the unstable direction of the map. In this example, the asymptotic unstable direction or what is known as the unstable manifold is aligned with the local unstable direction. (The local unstable direction is the direction which takes neighboring points furthest away from a given travelling point after one iterate of the map.) One question that the author attempted to raise debate upon at the conference was which hyperbolic system, the local or asymptotic, is more relevant in these control problems? Kathy Glass uses the local stucture while Eric Kostelich uses the asymptotic one. One more comment on the Baker's Map. It would be straight forward to perform this type of analysis on an arbitrary Axiom A map.

In the author's original talk mixing times of chaotic systems was discussed. Walt Grantham surmised that there is a relationship between the system improvement in average hitting times after a control is introduced to a chaotic system and the mixing time of that chaotic system. This is a reasonable notion that one might wish to investigate.

References

1. Arnold, V.I.; Avez, A.; Ergodic Problems in Classical Mechanics; W.A. Benjamin: New York, 1968.

2. Bowen, R.; Ruelle, D.; The Ergodic Theory of Axiom A Flows.; Invent. Math. 1975, 79, 181-202.

3. Mazer, A.; A Model for Investigating Mixing Times in Suspended Flows.; Physica D 1992, 57 (122), 226-237.

4. Nagata, S.; Yanagimoto, M.; Yokoyama, T.; A Study of the Mixing of High Viscosity Liquid.; Kagaku Kogaku 1975b, 21, 278-286 (in Japanese).

5. Ruelle D.; Resonances for Axiom A Flows.; L. of Differential Geometry 1987, 25, 99-116.

6. The author also referred to the works of papers presented at the conference throughout the paper.

Commentary by K. Judd

I am a little disappointed that Arthur Mazer has not described something closer to his conference presentation, since I thought it interesting and relevant to the conference.

This paper makes rigorous a fact about ergodic systems: that the hitting time to a target does not depend on its location, only on its effective size. If one knows something about ergodic systems, then this statement is intuitively what one would expect, almost a tautology, although the proof that Arthur gives, being beautifully neat, highlights subtleties that are not obvious. If one is not familiar with ergodic theory, then I think Arthur's exposition should provide an intuition about the nature of these systems.

The paper ties together nicely some of the early material on OGY control, in which there is only stabilization within a target region that one must wait to hit, and the new emerging methods that employ global control to quickly arrive at the target set, such as those of Katie Glass, Eric Kostelich and Kok-Lay Teo. Arthur raises in his conclusions what I consider to be a significant question: Does the average hitting time with global control still depend only on the effective size and not on the location of the target set? A result such as this would enable more confident application of these new methods.

Controllable Targets Near a Chaotic Attractor

Thomas L. Vincent
Aerospace and Mechanical Engineering
University of Arizona
Tucson, Arizona 85721, USA

Abstract

A nonlinear system with a chaotic attractor produces random like motion on the attractor. As a consequence, there must exist a neighborhood about any given point on or near the attractor such that the system will visit the neighborhood in finite time. This observation leads to a very simple "Chaotic control" algorithm for bringing nonlinear systems to a fixed point. Suppose that the system to be controlled is either naturally chaotic or that chaotic motion can be produced by means of open loop control. Suppose also, a neighborhood of the desired fixed point can be found, such that, using standard feedback control techniques, the system is guaranteed to be driven to the fixed point. If this neighborhood also has points in common with the chaotic attractor it is then called a "controllable target" for the fixed point. The chaotic control algorithm consists of first using, if necessary, open loop control to generate chaotic motion and then wait for the system to move into the controllable target. At such a time the open loop control is turned off and the appropriate closed loop control applied. The basic requirement with this approach is, of coarse, being able to determine a large controllable target. The following method is used here: The system is first linearized about the desired fixed point solution. If necessary, a feedback controller is then designed so that this reference solution has suitable stability properties. A Lyapunov function is then obtained based on this stable linear system. Through simulation, a level curve for the Lyapunov function is determined, such that, whenever the state of the nonlinear system is within this level curve, the feedback controller will drive the nonlinear system to the desired equilibrium solution. Such a level curve defines a controllable target. A novel approach for determining the Lyapunov level curve is presented which helps in finding large controllable targets for discrete systems. Two discrete control problems are examined. The first example is the well known Hénon map and the second example is the bouncing ball. Application of the method to both an actual bouncing ball system as well as its computer simulation are compared.

1 Introduction

Since 1990, there have been a number of papers dealing with the control of chaotic systems [Ott, et al., 1990, Vincent and Yu, 1991, Chen and Dong, 1993, Paskota, et al.1994, 1995]. These references tend to focus on the problem of designing a stabilizing controller for system which, without control, would be chaotic. Here we include systems which are not necessarily chaotic but as in [Vincent, et al., 1994] chaos can be created as a part of the total control design. We make use of the fact that for many nonlinear systems chaos is easy to create using open-loop control.

In the **chaotic control algorithm** given below, two essential ingredients are needed: a chaotic attractor and a controllable target. It is assumed that chaotic motion is either "naturally" present in the system or that it can be created using open loop control. A **controllable target** is any subset of the controllable set to a fixed point (corresponding to a specific feedback control law) which has a non empty intersection with the chaotic attractor. If we start the system at any point within the domain of attraction of the chaotic attractor, the resulting chaotic motion will, through time, ultimately arrive in the controllable target. The chaotic control algorithm simply has to keep track of when the system enters the controllable target. When it does, any open loop control used to create chaos is turned off and at the same time the closed-loop feedback control is turned on. The closed-loop controller may be designed using the linearized version of the nonlinear system about a specified equilibrium point.

The chaotic control algorithm requires that a controllability target be specified. One method for determining such a target for discrete systems is given in [Paskota, et. al., 1995] however, it is not applicable to the bouncing ball problem. Instead, a Lyapunov function approach is used here to obtain the controllable target. The same approach can be used for continuous systems as well.

2 Chaotic Control Algorithm

Consider the class of dynamical systems which can be described by nonlinear difference equations subject to control of the form

$$\mathbf{X}(j+1) = \mathbf{F}(\mathbf{X}(j), \mathbf{U}(j)) \tag{1}$$

where $\mathbf{F} = [F_1 \cdots F_{N_X}]$ is an N_X dimensional vector function of the state vector $\mathbf{X} = [X_1 \cdots X_{N_X}]$, and control vector $\mathbf{U} = [U_1 \cdots U_{N_U}]$. The notation (j) indicates current time and $(j+1)$ is one time unit latter. The functions F_i are assumed to be continuous and continuously differentiable in their arguments. The control will, in general, be bounded and we will assume that at every time step, the control $\mathbf{U}(j)$ must lie in a subset of the control space $\mathcal{U}$ defined by

$$|U_i| \leq U_{i_m} > 0$$

for $i = 1 \cdots N_U$.

Assume that for a specified control input $\widehat{\mathbf{U}}(j) \in \mathcal{U}$, the system has a chaotic attractor and that for a specified fixed control $\bar{\mathbf{U}} \in \mathcal{U}$,

$$\bar{\mathbf{X}} = \mathbf{F}(\bar{\mathbf{X}}, \bar{\mathbf{U}})$$

is a fixed point of interest which is near the chaotic attractor.

Henceforth, the notation (j) will be dropped when indicating current time. The system is first linearized about the target point to obtain the perturbation equations

$$\mathbf{x}(j+1) = \mathbf{A}\mathbf{x} + \mathbf{B}\mathbf{u} \tag{2}$$

where
$$\begin{aligned} \mathbf{x} &= \mathbf{X} - \bar{\mathbf{X}} \\ \mathbf{u} &= \mathbf{U} - \bar{\mathbf{U}} \\ \mathbf{A} &= \left.\frac{\partial \mathbf{F}}{\partial \mathbf{X}}\right|_{\bar{X},\bar{U}} \\ \mathbf{B} &= \left.\frac{\partial \mathbf{F}}{\partial \mathbf{U}}\right|_{\bar{X},\bar{U}}. \end{aligned}$$

In general (2) will not be a stable system. However, the discrete **LQR** method [Ogata, 1987] can be used to find a feedback control law of the form

$$\mathbf{u} = -\mathbf{K}\mathbf{x} \qquad (3)$$

where $\mathbf{K} = \left(\mathbf{R} + \mathbf{B}^T \mathbf{S} \mathbf{B}\right)^{-1} \mathbf{B}^T \mathbf{S}$, which will stabilize the linear system about the fixed point. The matrix **S** is determined by solving the discrete Riccati equation

$$\mathbf{S} = \mathbf{Q} + \mathbf{A}^T \mathbf{S} \mathbf{A} - \mathbf{A}^T \mathbf{S} \mathbf{B} \left(\mathbf{R} + \mathbf{B}^T \mathbf{S} \mathbf{B}\right)^{-1} \mathbf{B}^T \mathbf{S} \mathbf{A}$$

for a particular choice for the positive definite matrices **R** and **Q**. The controlled system is then given by

$$\mathbf{x}(j+1) = \widehat{\mathbf{A}} \mathbf{x} \qquad (4)$$

where

$$\widehat{\mathbf{A}} = \mathbf{A} - \mathbf{B}\mathbf{K}.$$

A Lyapunov function of the form

$$V = \mathbf{x}\mathbf{P}\mathbf{x}^T \qquad (5)$$

may now be determined for the linear stable controlled system (4) using the discrete Lyapunov equation

$$\mathbf{P} = \widehat{\mathbf{Q}} + \widehat{\mathbf{A}}^T \mathbf{P} \widehat{\mathbf{A}}$$

where $\widehat{\mathbf{Q}}$ is a positive definite matrix (Luenberger, 1979). For the linear system, starting from any point in state space, a solution will have the property that $V_{j+1} < V$ for every point of the linear map (4) except at the origin where $V = 0$.

The Lyapunov function (5) may now be used to determine a controllable target for the nonlinear system (1). Under the LQR control we would use the control law given by (3) however since the control must lie in $\mathcal{U}$, we use instead **saturating LQR** control defined by

$$\begin{aligned} &\mathbf{U} = \bar{\mathbf{U}} - \mathbf{K}\left(\mathbf{X} - \bar{\mathbf{X}}\right) \\ &\text{IF } |U_i| > U_{i_m} \\ &\text{THEN} \\ &\quad U_i = U_{i_m} \text{SIGN}(U_i) \\ &\text{END IF}. \end{aligned} \qquad (6)$$

Using this control law, we seek the largest level curve $V = V_{\max}$ such that

$$\Delta V = V(j+1) - V < 0$$

for all points on the level curve. In fact, since it is desired to get the largest possible controllable target, we need not look just one time step ahead, but may look several time steps ahead. In particular we will seek the largest level curve $V = V_{\max}$ such that

$$\Delta V(s) = V(j+s) - V < 0$$

where
$$V = (\mathbf{X} - \bar{\mathbf{X}})^T \mathbf{P} (\mathbf{X} - \bar{\mathbf{X}})$$
and
$$V(j+s) = (\mathbf{F}(j+s) - \bar{\mathbf{X}})^T \mathbf{P} (\mathbf{F}(j+s) - \bar{\mathbf{X}}) .$$

All points which lie within the level curve $V = V_{\max}$ are guaranteed to be driven to the fixed point under the control law (6). The advantage of using $s > 1$ is demonstrated in the examples which follows.

Let F_c be a control flag. This flag is first set equal to zero
$$F_c = 0 .$$
The chaotic control algorithm given by.

$$
\begin{aligned}
&\text{IF } V > V_{\max} \text{ AND } F_c = 0 \\
&\text{THEN} \\
&\quad \mathbf{U} = \widehat{\mathbf{U}} \\
&\text{ELSE} \\
&\quad \mathbf{U} = \bar{\mathbf{U}} - \mathbf{K}(\mathbf{X} - \bar{\mathbf{X}}) \\
&\quad \text{IF } |U_i| > U_{i_m} \\
&\quad \text{THEN} \\
&\quad\quad U_i = U_{i_m} \text{SIGN}(U_i) \\
&\quad \text{END IF} \\
&\quad F_c = 1 \\
&\text{END IF}
\end{aligned}
\qquad (7)
$$

is to be applied at each time step. Under this algorithm, if the system is started at a point where $V > V_{\max}$ then the system will run under the chaotic control $\mathbf{u} = \widehat{\mathbf{U}}$ until it lands on a point where $V \le V_{\max}$ at this time the control is switched to saturating LQR control and it will maintain this control for all future time. Note that the flag F_c is necessary, since the system my leave the level region $V \le V_{\max}$, after first entering, for up to $s - 1$ iterations.

3 Hénon Map

The Hénon map [Hénon, 1976] given by
$$\begin{aligned} X_1(j+1) &= -1.4 X_1^2 + X_2 + 1 \\ X_2(j+1) &= 0.3 X_1 \end{aligned} \qquad (8)$$

is a well known map with the fixed point
$$\bar{X} = \begin{bmatrix} 0.6314 & 0.1894 \end{bmatrix}^T \qquad (9)$$

and chaotic attractor illustrated in Figure 1. This figure was obtained by starting the system at $\bar{X} = \begin{bmatrix} -0.3 & 0 \end{bmatrix}^T$, labeled "1" and iterated 200 times. The first iterate is labeled "2" with the system moving very quickly to the attractor after that. The fixed point is denoted by a "+". Since the system is "naturally" chaotic, there will be no need to add an open-loop controller in order to produce chaos ($\widehat{u} = 0$).

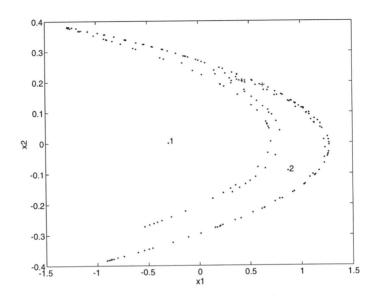

Figure 1: The Hénon map.

Paskota, et al. (1995) examined the control of this map to the fixed point by adding an unbounded control term to each equation in (8). Using a local stabilizing linear state feedback controller and applying the mean value theorem in vector spaces they were able to find a sufficient condition for the feedback gains so that an ε-neighborhood of the target point will be attractive. For this problem they obtained a circular controllable target with a radius $\varepsilon = 0.357$. This result provides a useful standard for comparison. Here the Hénon map will be controlled in a much more restrictive way. Control is added only to the first equation so that the map becomes

$$\begin{aligned} X_1(j+1) &= -1.4X_1^2 + X_2 + 1 + U \\ X_2(j+1) &= 0.3X_1 \end{aligned} \qquad (10)$$

furthermore the control is assumed to be bounded by

$$|U| \leq U_m \ .$$

Using the chaotic control algorithm of Section 2, it will now be shown that even under bounded control, a controllability target larger than a circle of radius $\varepsilon = 0.357$ is possible. The effect of varying the magnitude of U_m is also examined.

Linearizing (10) about the fixed point (9) yields

$$\mathbf{A} = \begin{bmatrix} -1.7678 & 1 \\ 0.3 & 0 \end{bmatrix}$$

$$B = \begin{bmatrix} 1 \\ 0 \end{bmatrix}.$$

Using the discrete LQR design with

$$Q = \begin{bmatrix} 1 & 0 \\ 0 & 1 \end{bmatrix}$$

$$R = 1$$

yields the feedback gains

$$K = \begin{bmatrix} -1.5008 & 0.7974 \end{bmatrix}$$

so that

$$\widehat{A} = \begin{bmatrix} -0.2670 & 0.2026 \\ 0.3000 & 0 \end{bmatrix}.$$

Solving the discrete Lyapunov equation using the same Q as above yields

$$P = \begin{bmatrix} 1.1902 & -0.0686 \\ -0.0686 & 1.0489 \end{bmatrix}.$$

Using $s = 10$ to calculate $V_{\max}$ we obtain

$$V_{\max} = 0.1 \quad \text{when} \quad u_m = 0.3$$
$$V_{\max} = 0.25 \quad \text{when} \quad u_m = 0.4.$$

The $V_{\max}$ level curves, along with the chaotic attractor, are illustrated in Figure 2. Note that the axis are drawn at different scales so that these elliptical curves are actually nearly circular. Also shown, between the two Lyapunov level curves is the circular domain of attraction obtained by Paskota, et al.(1995).

The $V_{\max}$ level curves are obtained by trial and error as follows: An initial point $x(0)$ is chosen such that from (5) $V = V_{\max}$, then a Lyapunov level curve is generated by solving the differential equations

$$\dot{x}_1 = \frac{\partial V}{\partial x_2}$$

$$\dot{x}_2 = -\frac{\partial V}{\partial x_1}$$

for a sufficiently long period of time so that one loop around the level curve is made. At every integration point of the level curve, $\Delta V(s)$ is calculated and plotted as a function of time. In this way it is easy to check if $\Delta V(s) < 0$ at every point of the level curve. The advantage of using $s > 1$ is quite apparent. For example, for the $u_m = 0.4$ case, if $s = 1$ is used, one obtains $V_{\max} = 0.1$ (instead of 0.25) and for the $u_m = 0.3$ case, if $n = 1$ is used, obtains $V_{\max} = 0.06$ (instead of 0.1). Further increasing u_m will produce larger controllable targets, (e.g. $u_m = 1$, yields $V_{\max} = 0.77$, with $s = 10$) however the controllable target can not be made arbitrarily large since the system starting at $x(0) = [-3\ 0]^T$ is unstable under the LQR control design even with unbounded control.

Figures 3 illustrates use of the control algorithm with the system starting at $x(0) = [-3\ 0]^T$ with $u_m = 0.3$ and the controllable target defined by $V_{\max} \leq 0.1$. The initial point is labeled "1" and we see that seven iterations are required before

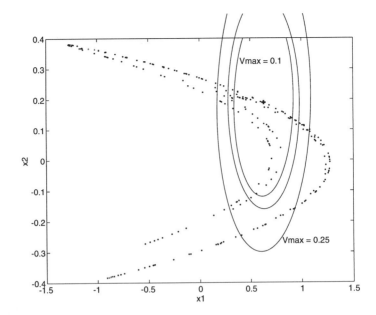

Figure 2: Controllable targets for the Hénon map fixed point.

the system, under the control $u = 0$, lands on the controllable target at the point labeled "8". At this point closed loop control is turned on, according to (7), moving the system to the target in just a few more iterations, without ever leaving the controllable target. Using the same starting point, but increasing u_m to 0.4 the system jumps into the controllable target ($V_{\max} \leq 0.25$) in just one iteration. It is evident that the Hénon map can be driven to the fixed point with limited control in just a few iterations.

4 Bouncing Ball

A ball bouncing on a vibrating plate is perhaps one of the simplest physical systems which can produce chaotic motion [Tufillaro, et al., 1992]. The objective here is to design a control algorithm for the plate so that the ball, started at rest on the plate, can be bounced up to and maintained bouncing at a prescribed maximum height. Complications include the fact that not all heights represent stable motion, more that one cycle of the plate may be necessary to achieve a given height, multiple height solutions are possible, and chaotic motion can result at certain height-frequency combinations. Keeping the amplitude of the vibrating plate fixed, we will use the frequency of the plate as a control input. Two state variables define the motion. One state variable is the phase angle of the plate at the time of bounce and the other, related to the maximum height of the ball, is the phase angle change between the current bounce

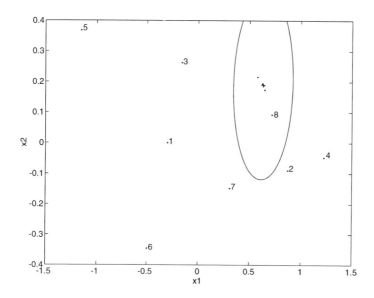

Figure 3: The Hénon map under the chaotic control algorithm with $u_m = 0.3$ and the controllable target defined by $V_{\max} \leq 0.1$.

and the next one.

A theoretical study of the bouncing ball is given by Guckenheimer and Holmes (1993). They consider a ball bouncing on an oscillating plate of infinite mass and showed that the general nature of the motion of the ball is chaotic. This is due to the fact that the map for the system is a "horseshoe". For certain frequencies of the plate and certain initial conditions of the ball it is possible to get periodic motion for any period between one and infinity. We will analysis the same system here even though the real system which we wish to control has somewhat different properties. The general character of the motion will be the same and we will be able to use insights gained from the simplified model to design a controller for the actual system.

4.1 A simple model for the bouncing ball

Let the motion of the plate is given by

$$y = A\sin(\omega t) \qquad (11)$$

where y is the displacement of the plate in the vertical direction, A is the amplitude of the plate, ω is the frequency of the plate, and t is time. If we let U be the velocity of the ball just before impact, V be the velocity of the ball just after impact, and

$$W = A\omega \cos(\omega t) \qquad (12)$$

be the velocity of the plate, then by conservation of momentum and the definition of the coefficient of restitution

$$V = \left(\frac{m-e}{1+m}\right)U + \left(\frac{1+e}{1+m}\right)W \tag{13}$$

where m is the mass ratio of the ball to the plate and e is the coefficient of restitution. In addition we have that from one bounce to the next

$$U_{j+1} = -V \tag{14}$$

$$t_{j+1} = t + \frac{2V}{g} \tag{15}$$

where g is the acceleration of gravity. Note that using (15) to evaluate the time of the next bounce assumes that A is negligible compared with the height of the bounce. Evaluating (13) at $j+1$ and substituting (14) yields

$$V_{j+1} = a_2 V + a_1 W_{J+1} \tag{16}$$

where $a_1 = \left(\frac{1+e}{1+m}\right)$ and $a_2 = \left(\frac{m-e}{1+m}\right)$. It follows from (12) and (15) that (16) may be written as

$$V_{j+1} = -a_2 V_j + a_1 A w \cos\left[w\left(t + \frac{2V}{g}\right)\right] \tag{17}$$

If we let

$$\begin{aligned} \phi &= wt \\ \psi &= \frac{2wV}{g} \end{aligned}$$

where ϕ represents the current phase angle of the plate and ψ the phase angle change which will take place between the current bounce of the ball and the next. It follows from (15) and (17) that

$$\begin{aligned} \phi_{j+1} &= \phi + \psi \\ \psi_{j+1} &= -a_2 \psi + \hat{a}_1 w^2 \cos(\phi + \psi) \end{aligned} \tag{18}$$

where $\hat{a}_1 = \frac{2Aa_1}{g}$. We will refer to (18) as the **ball map**.

4.2 The actual ball system

The experimental system differs from the model developed above in several ways. First there are additional forces acting on the ball besides gravity. In addition to air resistance, "uncertain" forces are introduced due to a restraining fish line used to keep the ball aligned with the plate. Second, the plate is, in fact, a voice coil actuator. Even though a feedback control loop has been placed around the actuator so that (11) can be used as a command input, the output of the actuator, while close to the input will not be exactly the same. Thus the voice coil actuator, along with its controller, would have to be included for an accurate simulation of the actual system.

The experimental bouncing ball system is defined by the following parameters

$$\begin{aligned} m &= 1/26 \\ e &= 0.8 \\ A &= .013 \text{ meters} \end{aligned}$$

so that
$$a_1 = 1.733333$$
$$a_2 = -0.733333$$
$$\hat{a}_1 = 0.004594 \text{ sec}^2.$$

4.3 Periodic solutions

There are many possible periodic equilibrium solutions to (18). For example the ball can bounce to a fixed height at every n cycles of the plate. It can also bounce to m different heights at $n \times m$ cycles of the plate before repeating the pattern. Only the case of $m = 1$ will be examined here.

We seek equilibrium solutions $\bar{\phi}$ and $\bar{\psi}$ such that

$$\begin{aligned} \bar{\phi}_{j+1} &= \bar{\phi} + \bar{\psi} \\ \bar{\psi}_{j+1} &= -a_2\bar{\psi} + \hat{a}_1 w^2 \cos\left(\bar{\phi} + \bar{\psi}\right) \end{aligned} \tag{19}$$

In order to obtain equilibrium solutions from (19) the right hand side of the first equation must be evaluated modulo 2π, in which case we obtain

$$\begin{aligned} \bar{\psi} &= 2n\pi \\ \cos\bar{\phi} &= \frac{2n\pi(1+a_2)}{\hat{a}_1 w^2} \end{aligned} \tag{20}$$

where n is the number of cycles of the plate between bounces. Linearizing (18) about the (20) yields

$$\begin{aligned} x_{1_{j+1}} &= x_1 + x_2 \\ x_{2_{j+1}} &= a_{12}x_1 + a_{22}x_2 \end{aligned} \tag{21}$$

where
$$\begin{aligned} x_1 &= \phi - \bar{\phi} \\ x_2 &= \psi - \bar{\psi} \\ a_{21} &= -\hat{a}_1 \omega^2 \sin\bar{\phi} \\ a_{22} &= -a_2 + a_{21} \end{aligned}$$

Examining the eigenvalues of (21) we obtain information regarding stable periodic motion which is summarized in the following two tables.

n	Frequency, rad/sec	Phase	Height, m
1	19.0977	0°	0.1327
2	27.0082	0°	0.2655
3	33.0782	0°	0.3982
4	38.1954	0°	0.5309
5	42.7038	0°	0.6637
6	46.7796	0°	0.7964

Table 1. Minimum value of ω for which a stable equilibrium solution exists with the corresponding phase angle ϕ and bounce height.

n	Frequency, rad/sec	Phase	Height, m
1	28.9505	64.2045°	0.0578
2	32.3966	45.9717°	0.1845
3	36.4575	34.5929°	0.3278
4	40.5278	27.3504°	0.4716
5	44.4249	22.4799°	0.6132
6	48.1123	19.0260°	0.7529

Table 2. Maximum value of ω for which a stable equilibrium exists and the corresponding phase angle ϕ and bounce height.

For each value of n there are both minimum and maximum frequencies for which there exists a stable solutions to (21). These frequencies are given in Table 1 and 2 along with the corresponding phase angle ϕ at which the bounce takes place and the corresponding height of the bounce. It follows from these tables that there are gaps in the bounce heights for the $m = 1$ case in which stable solutions can be found. For example, with $n = 1$, stable bounce heights, in meters, are restricted between $0.0578 \leq y \leq 0.1327$ with the next bounce height, corresponding to $n = 2$, starting at $y = 0.1845$. Other more complicated solutions are possible, with $m > 2$.

4.4 Controlling the Ball

Consider now controlling the ball to a steady state constant height solution with $n = m = 1$ starting with the ball at rest on the plate. This objective avoids having to control more complex motion associated with $n > 1$ or $m > 1$ so as to focus on the idea of using chaos as a part of the control algorithm and to see if such an approach will work for a real laboratory system. In fact if we choose a height between $0.0578 \leq y \leq 0.1327$ we need not even apply feedback control within the controllable target. By way of example let us choose a frequency which should provide at least a small region of stability on either side of the equilibrium phase angle. We see from the tables that a frequency with a phase angle midway between 0° and 64° should be a good choice. Working with round numbers, we choose $\omega = 22$ rad/sec. This choice results in a stable steady state height of 0.100 m at a phase angle of 41.1°.

Can we simply set the ball on the plate and then start the plate in motion at $\omega = 22$ in order to achieve the desired result? No, since the plate must have an acceleration greater than g in order for the ball to leave the plate. The plate will have an acceleration greater than the acceleration of gravity only when $\omega > 27.47$ rad/sec. If we started the plate in motion at a frequency somewhat higher than this, say $\omega = 28$ rad/sec then we might expect to achieve periodic motion at a height of $y = 0.062$ m, but without passing thorough our desired height. In order to achieve the design objective without using any control, other than changing the frequency of the plate, we must put the ball in motion with a plate frequency such that the resultant motion in state space (ϕ and ψ) is guaranteed to land in the controllable target to the fixed point given by $\bar{\phi} = 41.1°$, and $\bar{\psi} = 360°$. Once at the controllable target, the driving frequency is changed to $\omega = 22$ rad/sec and the desired steady state motion should be achieved.

4.4.1 The ball map at a frequency of $\omega = 45$ rad/sec

Outside of the controllable target we need a frequency which will produce chaotic motion. We see from the tables that, without further analysis, a good choice would be a frequency greater than 32.3966 rad/sec that does not lie within the frequencies which can produce periodic motion. For example the following frequencies will not support $m = 1$ periodic motion.

$$
\begin{aligned}
32.3966 &< \omega < 33.0782 \\
36.4575 &< \omega < 38.1954 \\
40.5278 &< \omega < 42.7038 \\
44.4249 &< \omega < 48.1123 \ .
\end{aligned}
$$

Choosing a frequency in this range will either produce chaotic motion or a chaotic transient to a periodic motion with $m > 1$. In what follows we use a frequency of $\omega = 45$ rad/sec to produce the chaotic motion.

Figure 4 illustrates the ball map (18) using the parameters for the actual experimental system at a frequency of $\omega = 45$ rad/sec. Also illustrated in this figure is the target point and controllable target to be discussed shortly. The map was obtained by starting the system at $\phi(0) = \psi(0) = 0$ (not plotted) and run for 1000 iterations. The resulting chaotic attractor lies between an upper bound defined by

$$\psi = 2\hat{a}_1\omega^2 + \hat{a}_1\omega^2 \cos\phi$$

and the lower bound defined by

$$\psi = \hat{a}_1\omega^2 \cos\phi$$

except where ψ goes negative, in which case, ψ is set equal to zero. The upper bound is proportional to the height the ball would bounce, if it could bounce periodically at the phase angle ϕ and the lower bound is proportional to the height the ball would achieve if were to impact the plate at the phase angle ϕ with zero velocity. The points on the map which form the parabola arc is an artifact resulting from the fact that the ball map is only approximation to an actual bouncing ball. Under the ball map (18) negative values for ψ are possible (corresponding to a negative bounce height). In order to correct for this, whenever the map returns a negative value for ψ it is replaced by

$$\psi = -\hat{a}_1\omega^2 \cos\phi$$

4.4.2 Estimating the domain of attraction

Choosing $n = 1$, $\omega = 22$ rad/sec, we have from (20) that $\bar{\phi} = 41.1°$ and $\bar{\psi} = 360°$. In this case (21) is a stable linear system of the form

$$\mathbf{x}_{j+1} = \mathbf{A}\mathbf{x} \qquad (22)$$

with

$$\mathbf{A} = \begin{bmatrix} 1 & 1 \\ -1.4617 & -0.7283 \end{bmatrix}.$$

The eigenvalues for this A matrix are $0.1358 \pm 0.8455i$, which lie inside the unit circle. Thus there must exist a 2×2 symmetric positive definite solution for $\mathbf{P}$ satisfying the discrete Lyapunov equation

$$\mathbf{P} = \mathbf{Q} + \mathbf{A}^T\mathbf{P}\mathbf{A}$$

where $\mathbf{Q}$ is any 2×2 symmetric positive definite matrix such that

$$V = \mathbf{x}^T\mathbf{P}\mathbf{x}$$

is a Lyapunov function for the system. Choosing $\mathbf{Q} = \mathbf{I}$, we obtain

$$P = \begin{bmatrix} 8.6394 & 4.6968 \\ 4.6968 & 5.9586 \end{bmatrix}.$$

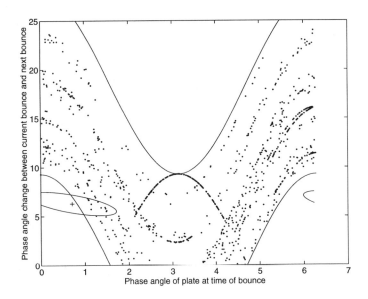

Figure 4: The ball map and controllable target.

The only problem which remains is to determine the controllability target which is defined by
$$V \leq V_{\max} . \qquad (23)$$
Using trial and error, with $n = 10$, a usable value of $V_{\max}$ was found to be $V_{\max} = 5$ which is the elliptical level curve illustrated in Figure 4. Note that, in this case, the target point lies off of the chaotic attractor. If $n = 1$ were used, not only would the resulting controllability target be small, it would not even intersect any of the points shown on the chaotic attractor.

4.4.3 Chaotic controller

Using V and $V_{\max}$ as defined above, the chaotic control algorithm is given simply by

$$\begin{aligned}
&F_c = 0 \\
&\text{IF } V > V_{\max} \text{ AND } F_c = 0 \\
&\text{THEN} \\
&\quad \omega = 45 \text{ rad/sec} \\
&\text{ELSE} \\
&\quad \omega = 22 \text{ rad/sec} \\
&\quad F_c = 1 \\
&\text{END IF}
\end{aligned} \qquad (24)$$

5 Bouncing Ball Results

The above chaotic control law was used with the ball map, the actual ball system, and a continuous simulation of the actual ball system. The continuous simulation is a better approximation to the laboratory system than the ball map used for analysis, since both the bouncing ball and the control system used to drive the plate are simulated using appropriate differential equations. Since a small A assumption is not made, the impact takes place at the proper height. However when the ball impacts the plate the velocity vector is adjusted according to (13).

5.1 The ball map under chaotic control

Figure 5 illustrates the ball map (18) under the chaotic control law (24) with the system starting at $\phi = \psi = 0$. After some two hundred iterations of the map with the control set at $\omega = 45$ rad/sec, the system finally lands in the controllability target. It then takes about 15 more iterations with the control at $\omega = 22$ rad/sec before the ball goes into steady state bouncing at the prescribed height and bounce frequency. Different starting conditions can reduce or increase the number of "chaotic" bounces before the controllability target is reached. However once in the controllability target, the system is always driven to the target.

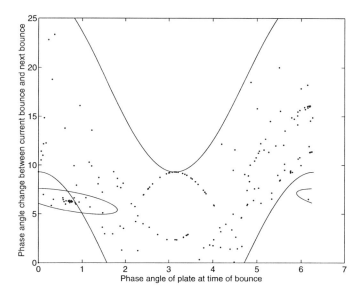

Figure 5: The ball map under the chaotic control algorithm.

5.2 The actual system under chaotic control

Two sensors are used in the laboratory setup. One is an linear optical encoder attached to the actuator which is used to measure the position of the top of the plate. Information from this actuator is used to both control the plate and to determine the phase angle of the plate ϕ when the ball strikes it. The second sensor is a microphone attached to the plate with a digital processor to provide a "spike" voltage whenever the ball strikes the plate. This allows for the determination of the time interval between bounces from which ψ can be estimated. Thus both state variables can be measured directly from the experiment and these values can be used without further processing to calculate V. The control signal sent to the actuator controller is determined directly from (24). All information gathering and processing used in the experiment was also used in the continuous simulation. In fact, almost the same code is used.

It was found that for both the laboratory system and the continuous simulation, the chaotic controller (24) resulted in very satisfactory performance. The system is always started with the ball at rest on the plate. Generally fewer bounces under the $\omega = 45$ rad/sec control were needed to enter the controllability target than might be expected from the ball map results given above. After the control is switched to the frequency of $\omega = 22$ rad/sec, the ball would always settle down to the periodic solution. In the continuous simulation this would be the end of the control procedure as equilibrium is forever maintained. In the laboratory system, the results were very similar except that now and then the fish line guiding the ball would kink resulting in a loss of the equilibrium solution. Since the chaotic control algorithm does not account for this possibility, it continues to maintain the $\omega = 22$ rad/sec frequency resulting in the ball eventually just riding on the plate (no bouncing). However, it is easy to tell when this happens from the microphone data. Consequently, the chaotic control algorithm for the experimental system was modified so that when ever it was detected that the ball was riding on the plate, the frequency is changed back to $\omega = 45$ rad/sec. This makes for a much more interesting demonstration, as the controller now seems much more "intelligent".

Data taken from the actual ball system under the chaotic controller is shown in Figure 6. This represents a run with the ball starting at rest on the plate. After about 50 bounces at $\omega = 45$ rad/sec the system enters the controllable target and after a few bounces at $\omega = 22$ rad/sec, the ball is bouncing nearly periodically. The data run was terminated before any kinking of the fish line resulted in the loss of this equilibrium solution. Note that the steady state phase angle obtained is shifted to the right from that obtained using the ball map. This shift to the right was also obtained using the continuous simulation. In fact, the continuous simulation results (not shown) are very similar to the data plotted in figure 6. The major difference is in the number of points which lie below the lower boundary. This is attributed to the occasional kinking of the fish line (which was not simulated). The scattering of the data about the equilibrium solution and the occasional data point above the upper boundary is attributed to the approximate $\pm 15°$ maximum error in estimating the phase angles.

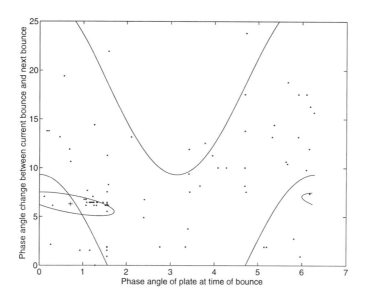

Figure 6: Data for the actual ball system under the chaotic control algorithm.

6 Discussion

One of the advantages with the approach presented here is that the bounds on the control are included as a part of the total control design. Even if u_m is quite small (in the spirit of several other papers in this volume) as long as it is known, a controllable target can be found. The controllability target should prove useful with other control algorithms as well. It should be useful particularly in conjunction with the targeting method presented by Kostelich.

"Control and chaos," our volume title, does not exclude "control of chaos", but the latter title would give a false impression. Of coarse we can control chaos, the heavy hand of control can make almost any system do most any thing we want. That is not the point. What we do want, is to be able to use the nonlinear nature of the system in conjunction with moderate control effort to achieve targets which would otherwise require a great deal of control.

The conference participants could be roughly classed as **system dynamics types** and **controls types**. It was interesting and informative to observe the various approaches to control and chaos. It appeared to me that the "system dynamics" participants brought control into chaotic systems theory with the small control view very much in mind. However they tended not to specify what is allowed in control, only that it should be small. They also seem to focus mainly on periodic solutions. In contrast, the "controls" participants seem to have brought chaotic systems to controls as a problem which should be solvable using the machinery of control theory. As such, the bounds on control are known, large or small and it becomes a problem of "how to do it" with the control available. They also tend to focus on fixed point solutions. Being a "controls" type myself, there are a couple of system dynamic ideas I am now interested in applying to the bouncing ball problem. Certainly it is rather unsatisfactory to simply wait for the ball to reach a controllable target. The targeting approach should be applicable to this problem and should make for a better control strategy. Also it would be of interest to see if the controllable target idea can't be used in conjunction with higher periodic orbits.

References

[1] G. Chen, G. and X. Dong, 1993. "From Chaos to Order - Perspectives and Methodologies in Controlling Chaotic Nonlinear Dynamical Systems," *International Journal of Bifurcation and Chaos* 3(6), pp. 1363-1409.

[2] Guckenheimer, J. and P. Holmes, 1993. *Nonlinear Oscillations, Dynamical Systems and Bifurcations of Vector Fields*, Springer-Verlag, New York.

[3] Hénon, M., 1976, "A Two Dimensional Map with a Strange Attractor," Commun. Math. Phys. 50,69.

[4] Luenberger, D.G., 1979. *Introduction to Dynamic Systems*, Wiley, New York,

[5] Ogata, K, 1987, *Discrete-Time Control Systems*, Prentice Hall, Englewood Cliffs, New Jersey.

[6] Ott, E., C. Grebogi, and J.A. York, 1990. "Controlling Chaos," *Physical Review Letters* 64, pp. 1196-1199.

[7] Paskota, M., A.I. Mees, and K.L. Teo, 1994 "Stabilizing Higher Periodic Orbits," *International Journal of Bifurcation and Chaos* 4(2), pp. 457-460.

[8] Paskota, M., A.I. Mees, and K.L. Teo, 1995. "On Local Control of Chaos, the Neighborhood Size," Manuscript, personal communication.

[9] Tufillaro, N.B., T. Abbot, and Reilly, A., 1992. "An Experimental Approach to Nonlinear Dynamics and Chaos," Ch 1, Addison Wesely.

[10] Vincent, T.L. and J. Yu, 1991. "Control of a Chaotic System," *Dynamics and Control* 1, pp. 35-52.

[11] Vincent, T.L., T.J. Schmitt, and T.L. Vincent, 1994. "A Chaotic Controller for the Double Pendulum," *Mechanics and Control*, edited by R.S. Guttalu, Plenum Press, New York, pp.257-273.

Commentary by M. Watanabe and K. Aihara

This paper takes an interesting approach to controlling with chaos. Usually when we think of chaotic control, the system is chaotic in nature. However, in this paper, a non-chaotic system is controlled to be chaotic, and chaotic control is applied positively. This approach seems to greatly extend the applicability of chaotic control.

Commentary by A.I. Mees

This is an excellent example of what we are trying to achieve at this workshop. It takes a controls approach but is inspired by some ideas from dynamical systems and manages to achieve something that neither group seems to have done on its own.

Specifically, it uses state-of-the-art, but nevertheless fairly standard, control engineering methods to design a feedback controller to stabilize a fixed point in the chaotic Hénon system, but modifies the controller using the insight from dynamical systems theory that one should do nothing at all when the system is away from the controllable region, since ergodicity will eventually bring the state back into that region. The targeting techniques described elsewhere in this volume could be used to improve the time to reach the region, but since the methods used in this paper can achieve a good-sized controllable region with small control effort, there is less need to worry about targeting methods; nevertheless, this is, as the author points out, an area for further research.

The control of the bouncing ball both as a theoretical study and a practical example is very satisfying. The control is not quite the one analyzed theoretically since a switch to a high driving frequency is needed to give the ball enough acceleration to get into the desired state, but the lesson still seems to be the same as in the Hénon example, namely that applying zero control effort when the system is outside a certain region makes it possible to design a successful controller. The paper doesn't say, but I understand that leaving the controller switched on (presumably saturated) when the state is outside the controllable region actually interferes with the desired behavior and makes control worse or impossible.

Perhaps this sort of combination of the ideas of both control engineers and dynamical systems theorists will lead to new understandings of how to control at least some kinds of nonlinear systems.

The Dynamics of Evolutionary Stable Strategies

Yosef Cohen
Department of Fisheries and Wildlife
University of Minnesota, St. Paul, MN 55108, USA

Thomas L. Vincent
Department of Aerospace and Mechanical Engineering
University of Arizona, Tucson, AZ 85721, USA

Abstract

We use the so-called G-function to determine the number of coexisting species with evolutionary stable strategies (ESS). We show that a simple model, with a single resource (carrying capacity) can result in an ESS with more than one (and potentially many) coexisting species. We also show how the adaptive landscape (the shape of the G-function) changes in the pursuit of the evolutionary stable strategy. Next, we use a populations model with a random mating system, and trace the strategy dynamics to ESS from arbitrary initial conditions. We show that with appropriate parameter values, the dynamics lead to chaotic trajectories of the species populations *and* their evolutionary stable strategies. This indicates that under appropriate conditions, ESS are not necessarily fixed; to be ESS, they must vary (chaotically) with time. The approach we take relies on the fundamental definitions of ESS, and simulation models. This allows applications of ESS analysis to moderately complex models.

1 Introduction

Maynard-Smith's (Maynard-Smith 1982) application of mathematical game theory stimulated novel approaches to the study of evolutionary theory. A mathematical game is defined by rules, players, strategies, and payoffs. In the evolutionary game the rules are "defined" by Nature, the "players" are the individual organisms, the "strategies" are the values of (heritable) traits, and the "payoffs" are the fitnesses. An evolutionary game can give rise to evolutionary stable strategies (ESS). An ESS is a strategy such that if all members of a biotic community attain it (through the evolutionary process), then no mutant strategy could persist with the community under the influence of natural selection (Maynard-Smith 1982).

This game theoretic approach underwent substantial development within the last decade. See for example review by Hines (Hines 1987). One specific approach, which was originally developed by Vincent and Brown (Vincent and Brown 1984) and Vincent et al. (Vincent 1993) gives the necessary conditions for ESS in multi-player continuous games. In here, we limit our attention to this approach. By using the so-called ESS maximum principle (as stated below), this theory is constructive in the sense that it provides a recipe for calculating candidates solutions for ESS in terms of population dynamics, given that the evolutionary strategies are fixed at their evolutionary stable values. The theory also provides the tools for calculating the number of existing species at the ESS. Because necessary conditions require that ESS exists, the theory provides no information about how ESS is actually (if at all) reached from arbitrary initial conditions. Recently Vincent et al. (Vincent 1993) extended the theory to include strategy dynamics.

Here we further develop the ideas of different strategy dynamics that were proposed in Vincent et al. (Vincent 1993). More specifically, we extend the theory to allow simulations of strategy dynamics, and thus provide the tools for exploring moderately complex (and analytically intractable) models of systems undergoing evolution toward ESS. We also explore the interaction between phenotypic variance and ESS, and show that consideration of strategy dynamics can lead to strategies that are not fixed, and in fact are chaotic. These results require rethinking of the definition of ESS.

2 The Theory in a Nutshell

The theory was developed for both discrete and continuous models, with parallel results (Vincent and Fisher 1988). We discuss the discrete version only. The theory requires subtle notations, and we begin with these. We shall denote vectors with boldface characters and often neglect to explicitly indicate dependence on time; e.g., we may write x instead of $x(t)$. It should be clear from the context when time is involved.

We consider a community of N individuals. The community is composed of r species, and each species has x_i individuals. Thus, $N = \sum_{i=1}^{r} x_i$. Each species is identified by a unique set of values of some adaptive (and heritable) traits. This set of traits, called strategies, is denoted by

$$\mathbf{u}^i = [u_1^i, \ldots, u_{s_i}^i] \,, \; s_i \geq 1 \,, \; i = 1, \ldots, r$$

where s_i is the set of strategies for species i. We denote the collection of all strategies for all species by

$$\tilde{\mathbf{u}} = [\mathbf{u}^1, \ldots, \mathbf{u}^r] \,.$$

We seek those strategies (and their values), $\mathbf{u}^i$, which are ESS. One may think of the strategies as parameters of a model. We consider a model with discrete generations. The dynamics of each of the species population is

$$x_i(t+1) = x_i(t)[1 + H_i(\tilde{\mathbf{u}}(t), \mathbf{p}(t), N(t))] \quad [1]$$

where $\mathbf{p} = [p_1, \ldots, p_r]$ is the vector of frequencies $p_i = \dfrac{x_i}{N}$. The term in the square brackets in [1] determines the changes in the number of individuals of species i, through generations. From [1], the fitness function is

$$H_i(\tilde{\mathbf{u}}, \mathbf{p}, N) \,.$$

To be able to determine the number of species in the community, r, it is expedient to rearrange $\tilde{\mathbf{u}}$ into 2 distinct groups: those strategies that form (together) an ESS and those that do not. The former is denoted by $\mathbf{u}^c = [\mathbf{u}^1, \ldots, \mathbf{u}^\mu]$ and the latter by

$\mathbf{u}^m = [\mathbf{u}^{\mu+1}, \ldots, \mathbf{u}^r]$. The vector $\mathbf{u}^c$ is called the coalition vector, and $\mathbf{u}^m$ is called the mutant vector (Brown and Vincent 1987). The term coalition is borrowed from game theory (Owen 1982). Let

$$x_c = \sum_{i=1}^{\mu} x_i; x_m = \sum_{i=\mu+1}^{r} x_i; p_c = \frac{x_c}{N}; p_m = \frac{x_m}{N},$$

then

$$\sum_{i=1}^{r} p_i = 1 \text{ and } N = x_c + x_m.$$

For $\mathbf{p} = [p_1, \ldots p_r]$ we re-write [1] as

$$\left. \begin{array}{rcl} p_i(t+1) & = & \dfrac{p_i[1 + H_i(\tilde{\mathbf{u}}, \mathbf{p}, N)]}{\bar{H}} \\ N(t+1) & = & N\bar{H} \end{array} \right\}$$

where $\bar{H}$, the average fitness of all the individuals in the community, is

$$\bar{H} = 1 + \sum_{i=1}^{r} p_i H_i(\tilde{\mathbf{u}}, \mathbf{p}, N).$$

With the assumption that an equilibrium point for system [1] exists, we must have

$$H_i(\bar{\mathbf{u}}, \mathbf{p}^*, N^*) \equiv; i = 1, \ldots, \mu; \sum_{i=1}^{r} p_i^* = 1; p_i^* = 0; i = \mu+1, \ldots, r$$

where $*$ denotes equilibrium conditions. Note that the equilibrium assumption is not required (Vincent and Brown 1987); we use it to simplify the presentation.

An evolutionary stable strategy is defined as a strategy such that, if all members of a population adopt it, then no mutant strategy could invade the population under the influence of natural selection (Maynard-Smith 1982). To develop a method for finding the ESS, the concept of a fitness generating function is used (Vincent and Brown 1984). This function is defined as follows:

A function $G(\mathbf{u}, \tilde{\mathbf{u}}, \mathbf{p}, N)$ is a fitness generating function for system [1] if $G(\mathbf{u}^i, \tilde{\mathbf{u}}, \mathbf{p}, N) = H_i(\tilde{\mathbf{u}}, \mathbf{p}, N)$.

The definition says that if G is a fitness generating function, then substitution of a particular strategy $\mathbf{u}^i$ for $\mathbf{u}$ in G gives H_i. This definition simplifies the task of finding ESS: we no longer need to consider each species separately. We can simply calculate the fitness for a particular species when we need it. Because the ESS concept applies to the whole community, the introduction of the dummy variable $\mathbf{u}$ in G allows examining the fitness of all species in the community simultaneously. With the introduction of the G-function we can give a recipe for calculating ESS. The recipe is given by the following theorem.

ESS Maximum Principle (Vincent and Brown 1984; Vincent and Fisher 1988): Let $G(\mathbf{u}, \tilde{\mathbf{u}}, \mathbf{p}, N)$ be a generating function for system [1]. If $\mathbf{u}^c$ is an ESS such that $\{\mathbf{p}(t)\}$ is a monotone increasing sequence for all $t > 0$ and if $\mathbf{p}^*$ and N^* are asymptotically stable, then as $t \to \infty$, $G(\mathbf{u}, \tilde{\mathbf{u}}, \mathbf{p}, N)$ must take on a global maximum with respect to $\mathbf{u}$ at $\mathbf{u}^1, \ldots, \mathbf{u}^\mu$.

From the definition of the G-function, and the system [1], it follows from the equilibrium requirement that all maxima must equal zero.

With this theorem, the task of finding candidate solutions for an ESS is reduced to maximizing the G-function. Since the theorem states necessary conditions for an ESS, it is assumed that ESS exists. This may not be true. Thus, the solutions generated by the maximum principle are candidate solutions and their validity should be confirmed. This can be done, for example, by running a computer-coded model of the system [1] with strategy values that were generated by the maximum principle. Every few generations, a 'mutant' strategy (a strategy with a value different from the ESS value) is introduced. The frequency of these mutant strategies should go to zero with time. A more detailed discussion (than the one given here) of the G-function and the maximum principle can be found elsewhere (Vincent and Brown 1984; Brown and Vincent 1987; Vincent and Fisher 1988).

Recall that species are defined as those having a unique set of values for their strategy vector. Thus, the ESS maximum principle provides the number of species (r) that coexist under the ESS solution: r is simply the number of maxima found for the G-function. The maximum principle also requires that the G-function be maximized. If we were to maximize the fitness functions, as opposed to the G-function, we ignore the requirement that at ESS, all fitnesses must be maximized simultaneously. Thus, when density dependent mechanisms do not operate within the community, maximizing the G-function and the fitness functions is equivalent. If density dependent mechanisms do operate within the community, then maximizing the fitness functions amounts to invoking the group selection argument.

3 ESS of the Lotka-Volterra Competition Model

To demonstrate and further discuss the results above, we modify the L-V competition model (Brown and Vincent 1987; Vincent et al 1993) and examine the consequences of deriving ESS conditions according to the maximum principle in a community where competition only determines the ESS. The model, although naive, captures some of the important features of the approach discussed above. The results from this model will also be used to verify some of the results on strategy dynamics (below). For simplicity, we assume that the strategy is a scalar: there is a single trait that is maximized by all species of the community, and that all species possess the same G-function. A species is identified by having a unique value of that trait. The system model is

$$x_i(t+1) = x_i(t) \frac{R}{k(u^i)} \left[k(u^i) - \sum_{j=1}^{r} \alpha(u^i, u^j) x_j \right] \quad . \qquad [2]$$

Note that (i) the competition function $\alpha(\cdot)$ is symmetric; (ii) its value is determined by the strategies of all of the species in the community; (iii) the carrying capacity $k(\cdot)$ is a function of the strategy value for species i; it does not depend on strategy values of other species; (iv) by modeling k as a function of the strategy u^i, we admit that the carrying capacity is a function of both the environment and the strategy (for example, u^i may reflect the efficiency of utilizing food items by various phenotypes); and (v) the number of species in the community, r, is a parameter, and its value is determined by the ESS solution.

With the definitions in Section 2 the fitness function for species i is

$$H_i(\tilde{\mathbf{u}}, \mathbf{p}, N) = \frac{R}{k(u^i)} \left[k(u^i) - N \sum_{j=1}^{r} \alpha(u^i, u^j) p_j \right],$$

and the G-function is

$$G(u, \tilde{\mathbf{u}}, \mathbf{p}, N) = \frac{R}{k(u)} \left[k(u) - N \sum_{j=1}^{r} \alpha(u, u^j) p_j \right]. \qquad [3]$$

To see this, replace u in [3] with u^i to get [2]. Next, let:

$$k(u) = K \exp \left\{ -\frac{u^2}{2\sigma_k^2} \right\} \qquad [4]$$

and

$$\alpha(u, u^j) = 1 + \exp \left\{ -\frac{(u - u^j + \beta)^2}{2\sigma_a^2} \right\} - \exp\{\gamma\}. \qquad [5]$$

There are specific values of u^i and u^j which maximize the function $\alpha(u, u^j)$. The β and γ parameters in [5] assure that this maximum does not occur when $u^i = u^j$. Other parameters of interest are: R - the growth rate; σ_k - the "penalty" for deviation from the value of u which maximizes the carrying capacity; σ_α - the penalty for deviation from the value of u which maximizes the function α; and K - a scalar for the carrying capacity.

4 How Many Species Can Coexist at ESS

We use the following values for the parameters in [2]-[5]:

$$\beta = 2; \gamma = -\frac{1}{2}; R = 0.25; 2\sigma_k^2 = 8; 2\sigma_\alpha^2 = 8; K = 100; r = 1.$$

To find a candidate ESS solution, we first assume that the ESS is a coalition of 1 (i.e., $r = 1$ in [3]). The ESS maximum principle requires that:

$$\left. \begin{array}{rcl} G(u^1, u^1) & = & 0 \\ \dfrac{\partial}{\partial u} G(u^1, u^1) & = & 0 \end{array} \right\} \qquad [6]$$

The partial derivative in [6] is first taken with respect to the dummy variable u, and then u^1 is substituted for u. Equation [6] gives $u^{1*} = 1.213$ and $x^* = 83.199$. Examination of the G-function (Fig. 1) verifies that this solution satisfies the ESS maximum principle. With some algebra we can show that second order conditions are also satisfied. This solution may also be obtained via simulation (Brown and Vincent 1987). If the model [2]-[5] is run with the ESS parameter values and every few generations a random mutant (i.e., with a strategy value different from u^{1*}) is introduced, then all mutants will eventually become extinct.

By increasing the value of the parameter $2\sigma_k^2$ in [4], the penalty on deviating from the ESS value of u diminishes; i.e., $k(u)$ in [4] becomes flatter. This may allow

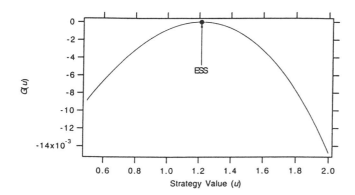

Figure 1: ESS solution for the G-function of the L-V competition model given in [2]–[5]. Parameter values are: $\beta = 2$; $\gamma = -\frac{1}{2}$; $R = 0.25$; $2\sigma_k^2 = 8$; $2\sigma_\alpha^2 = 8$; $K = 100$; $r = 1$. The dot locates the ESS solution: the global maximum ($G = 0$) occur at $u^{1*} = 1.213$.

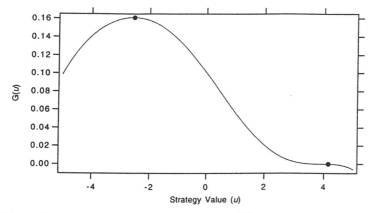

Figure 2: An attempt to solve for an ESS with [6] for the L-V competition model given in [2]–[5]. Parameter values are: $\beta = 2$; $\gamma = -\frac{1}{2}$; $R = 0.25$; $2\sigma_k^2 = 25$; $2\sigma_\alpha^2 = 8$; $K = 100$; $r = 1$. Note the 2 local maxima, shown as dots. Since both maxima do not occur at $G = 0$, the solution is not an ESS, but indicates that for $r = 2$ it may.

different species to invade an existing biotic community, and persist with it as an ESS coalition of more than one species. For $2\sigma_k^2 = 25$ (with the other parameter values unchanged) we find that, under the assumption of a coalition of $1 (r = 1)$, the ESS candidate solution obtained using [6] does not satisfy the ESS maximum principle (Fig. 2). However, examining a coalition of 2 we find that a candidate solution (Fig. 3) which satisfies the ESS maximum principle is

$$u^{1*} = 3.129; u^{2*} = -0.240; p_1^* = 0.565; p_2^* = 0.435; N^* = 90.346 \ .$$

We used these values to run the model [2]–[5] in simulations with mutants (different values for the strategy) introduced at random times. In all cases, the mutants eventually went extinct. This suggests that the candidate ESS solution is indeed ESS.

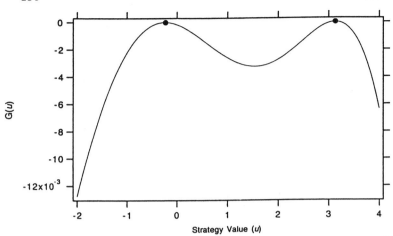

Figure 3: ESS solution for the G-function of the L-V competition model given in [2]–[5]. Parameter values are: $\beta = 2$; $\gamma = -\frac{1}{2}$; $R = 0.25$; $2\sigma_k^2 = 25$; $2\sigma_\alpha^2 = 8$; $K = 100$; $r = 2$. The dots indicate the ESS solution; i.e., the global maxima ($G = 0$) occur at $u^{1*} = 3.129$ and $u^{2*} = -0.240$.

Next, we set $2\sigma_k^2 = 40$. This should allow more species to co-exist at ESS. Figure 4 indicates that a coalition of 1 ($r = 1$ in [3]) is not possible (the maximum does not occur at $G = 0$). The existence of 2 local maxima (Fig. 4), indicates that a coalition of 2 species may give an ESS candidate solution. Thus, we try to solve [6] with $r = 2$ in [3]. This gives 3 local maxima (Fig. 5), none of which occur at $G = 0$. We therefore try a solution with $r = 3$ which gives 3 global maxima (Fig. 6) which occur at $G = 0$. Thus, the candidate ESS solution is

$$u^{1*} = 4.214; u^{2*} = 0.769; u^{3*} = -3.102$$
$$p_1^* = 0.396; p_2^* = 0.480; p_3^* = 0.125; N^* = 99.445$$

This candidate solution was verified by simulations, and we conclude that it is an ESS solution.

The statement "a flatter function of carrying capacity, with respect to some adaptive trait, allows more species to coexist" is intuitively clear. How to quantify it is not. The ESS maximum principle provides the appropriate tools. We have shown how a simple model of a single species population that relies on a single resource can give rise to multiple species. Other models behave similarly. For example, Brown (Brown 1990) used the L-V predator-prey model to show that the introduction of a predator into a single prey community can cause speciation of both the prey and predators at ESS.

5 Strategy Dynamics

The ESS maximum principle states necessary conditions which must be satisfied at an ESS. Sufficiency and existence remain open. The theory does not provide a mechanism to determine how ESS is reached from arbitrary initial conditions. Until recently, applications of the theory were limited to simple, and analytically tractable models. The theory does not address genetic and behavioral mechanisms which may

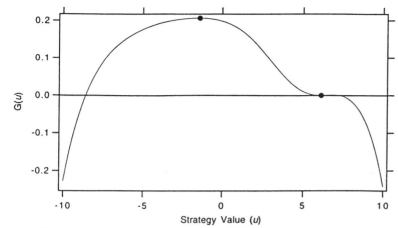

Figure 4: An attempt to solve for an ESS with [6] for the L-V competition model given in [2]–[5]. Parameter values are: $\beta = 2$; $\gamma = -\frac{1}{2}$; $R = 0.25$; $2\sigma_k^2 = 40$; $2\sigma_\alpha^2 = 8$; $K = 100$; $r = 1$. Note the 2 local maxima, shown as dots. Since both maxima do not occur at $G = 0$, the solution is not an ESS, but indicates that for $r = 2$ it may.

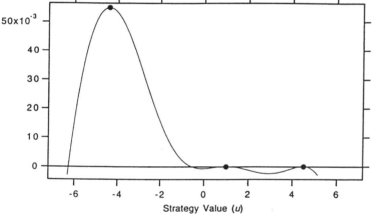

Figure 5: An attempt to solve for an ESS with [6] for the L-V competition model given in [2]–[5]. Parameter values are: $\beta = 2$; $\gamma = -\frac{1}{2}$; $R = 0.25$; $2\sigma_k^2 = 40$; $2\sigma_\alpha^2 = 8$; $K = 100$; $r = 2$. Note the 3 local maxima, shown as dots. Since none of the maxima occur at $G = 0$, the solution is not an ESS, but indicates that for $r = 3$ it may.

affect the ESS solution. The definition of ESS is in terms of a fixed point in the strategy space. Simply put, the potentially rich behavior of the evolution of strategies toward (and at) the ESS is not examined. The theory does not distinguish between those extinctions that occur because the strategies converge and those that occur because the strategies were eliminated. For example, when the dynamics begin with 2 different initial values of a strategy (i.e., 2 species), both strategies may converge to the ESS values, or one may go extinct. In both cases the final outcome is a single ESS strategy (i.e., a single species). Some of these shortcomings may be addressed by introducing strategy dynamics within the context of population dynamics and the

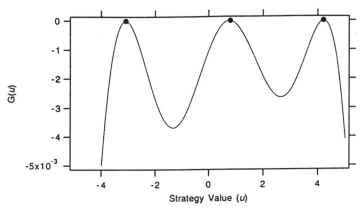

Figure 6: ESS solution for the G-function of the L-V competition model given in [2]-[5]. Parameter values are: $\beta = 2$; $\gamma = -\frac{1}{2}$; $R = 0.25$; $2\sigma_k^2 = 40$; $2\sigma_\alpha^2 = 8$; $K = 100$; $r = 3$. The dots indicate the ESS solution; i.e., the global maxima ($G = 0$) occur at $u^{1*} = 4.214$, $u^{2*} = 0.769$ and $u^{3*} = -3.102$.

pursuit of ESS. Strategy dynamics has been previously discussed by Roughgarden (Roughgarden 1983) in connection with coevolution and was introduced in an ad hoc way for use with the ESS maximum principle by Vincent (Vincent 1990).

Elsewhere (Vincent, 1993) we show that the strategy dynamics may be given by

$$u^i(t+1) = u^i(t) + \frac{(\sigma^i h^i)^2}{1 + G(u^i, \tilde{\mathbf{u}}, \mathbf{p}, N)} \frac{\partial}{\partial u}[G(u^i, \tilde{\mathbf{u}}, \mathbf{p}, N)] \quad [7]$$

where σ^i and h^i are the phenotypic variance and the heritability coefficients–of the strategy–respectively. The derivation of [7] relies on 2 potentially restricting assumptions: (i) the frequency distribution of phenotypes in the population is symmetric, and (ii) the variance of this distribution is small.

To trace the strategy dynamics, and verify that they lead to the expected ESS, we use [7] with the following initial conditions and parameter values:

$$\beta = 2; \gamma = -\frac{1}{2}; R = 0.25; 2\sigma_k^2 = 25; 2\sigma_\alpha^2 = 8; K = 100; r = 4$$
$$(\sigma^i h^i)^2 = 2, 3, 4, 5; u_0^i = 5, 4, 0, -3; x_i(0) = 10, 20, 30, 40 \ .$$

With these parameter values we should get 2 species coexisting at ESS with the values shown in Fig. 3. The ESS dynamics (Figs. 7 and 8) eventually lead to the expected values of ESS (Fig. 3). The ESS values reached via the dynamics agreed with the analytical results to the 10-th decimal digit after 300 iterations. From Figs. 7 and 8 we see that there are 2 ESS strategies:

$$p_1^* + p_2^* + p_3^* = 0.565 \quad \text{and} \quad p_4^* = 0.435 \ .$$

Thus, the initial 4 phenotypes coevolve into 2 phenotypes, and depending on initial conditions, the identity of some phenotypes is lost. Figure 7 also verifies that the speed of evolution depends on the phenotypic variance and the coefficient of heritability; e.g., examination of the numerical values confirmed that u_t^4 converged to the ESS solution faster than u_t^1. We return to this point below.

The Dynamics of ESS   287

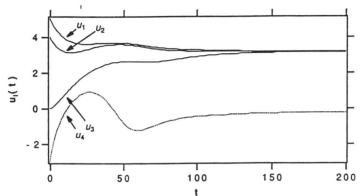

Figure 7: Strategy dynamics [2]–[5] according to [7] with the following initial conditions for 2 species and parameter values: $\beta = 2$; $\gamma = -\frac{1}{2}$; $R = 0.25$; $2\sigma_k^2 = 25$; $2\sigma_\alpha^2 = 8$; $K = 100$; $r = 4$ $(\sigma^i h^i)^2 = 2, 3, 4, 5$; $u_0^i = 5, 4, 0, -3$; $x_i(0) = 10, 20, 30, 40$. Note how 4 species (4 different initial conditions for strategy values) lead to 2 strategies, which correspond to the ESS values shown in Fig. 3.

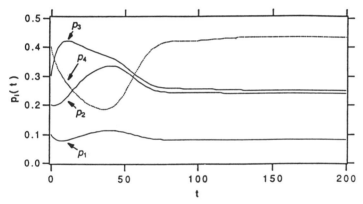

Figure 8: The dynamics [2]–[5] according to [7] of relative population size with the initial conditions for 4 species and parameter values: $\beta = 2$; $\gamma = -\frac{1}{2}$; $R = 0.25$; $2\sigma_k^2 = 25$; $2\sigma_\alpha^2 = 8$; $K = 100$; $r = 4$ $(\sigma^i h^i)^2 = 2, 3, 4, 5$; $u_0^i = 5, 4, 0, -3$; $x_i(0) = 10, 20, 30, 40$. Although 4 populations are traced, $p_1(t)$, $p_2(t)$ and $p_3(t)$ eventually become a single species; i.e., u_t^1, u_t^2 and u_t^3 converge to a single ESS value (Fig. 7), which is given in Fig. 3. $p_4(t)$ remains as a separate species: its u_t^4 remains unique.

Figures 7 and 8 show the dynamics of the populations and their strategies. Because the strategies are in fact parameters of the G-function, the G-function itself evolves with time: the adaptive landscape, $G_t(\cdot)$, is changing with time, and eventually stabilizes at ESS. To see this, we use only 2 phenotypes with the following initial conditions as in Figs. 7 and 8:

$$\beta = 2; \gamma = -\frac{1}{2}; R = 0.25; 2\sigma_k^2 = 25; 2\sigma_\alpha^2 = 8; K = 100; r = 2$$
$$(\sigma^i h^i)^2 = 2, 3; u_0^i = 5, 4; x_i(0) = 10, 20 \ .$$

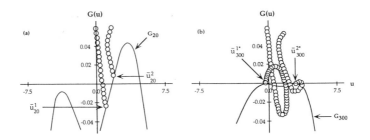

Figure 9: Strategy dynamics [2]–[5] according to [7] with the following parameter: $\beta = 2$; $\gamma = -\frac{1}{2}$; $R = 0.25$; $2\sigma_k^2 = 25$; $2\sigma_\alpha^2 = 8$; $K = 100$; $r = 2$ $(\sigma^i h^i)^2 = 2, 3$; $u_0^i = 5, 4$; $x_i(0) = 10, 20$. The dots in trace the trajectory of $G_t(\cdot)$ at u_t^1 and u_t^2. G itself (the adaptive landscape) is shown for $\{u_{20}^1, u_{20}^2\}$ and $\{u_{300}^1, u_{300}^2\}$ in the left and right panels, respectively. The ESS values corresponds to those shown in Figs. 3 and 7.

The dots in Fig. 9 trace the trajectory of $G_t(\cdot)$ at u_t^1 and u_t^2. G_t itself (the adaptive landscape) is shown for $\{u_{20}^1, u_{20}^2\}$ and $\{u_{300}^1, u_{300}^2\}$ in Figs. 9a and 9b, respectively. The ESS values correspond to those shown in Figs. 3 and 7.

The derivation of [7] uses a Taylor Series approximation of G with the assumptions that (i) the distribution of u around its mean is symmetric and (ii) the values of $(\sigma^i h^i)^2$ are small. In simulations with the model in Section 3 and with the parameter values in Figs. 7 and 8 and with $(\sigma^i h^i)^2 > 5$ (in which case the Taylor Series approximation deteriorates), none of the phenotypes reached the ESS solution obtained in Figs. 7 and 8. To relax assumptions (i) and (ii), and yet achieve ESS values, we use a different approach to trace strategy dynamics.

To simplify the development, we use scalar strategies. A species population, $x_i(t)$, is defined as a collection of its phenotype populations, $y_{ij}(t)$; thus, $x_i(t) = \sum_j y_{ij}(t)$.
A phenotype is identified by its unique strategy value. The strategy value of a species is the mean of the strategy values of its phenotypes. Thus, the phenotypic variance and mean can be used to parameterize the species strategy. Let $\bar{u}_t^i$ be the mean strategy value for species i at time t. The strategy value itself has some frequency distribution among the phenotypes, and we denote this frequency by $q_t^i(u)$. With small Δ and large M, the potential range of values that u may take is

$$-M\Delta u, -(M-1)\Delta u, \ldots, -\Delta u, 0, \Delta u, \ldots, (M-1)\Delta u, M\Delta u .$$

M must be large enough so that the entire potential range of strategy values is included. Because species are identified by the mean of the values of strategies of their phenotypes, and because the number of individuals having a specific strategy value may change with time (due to differential mortality, for example), the mean strategy value (which identifies a species) evolves with time. Let $y_{ij}(t)$ denote the number of phenotypes of species i whose strategy value is $j\Delta u$. Then, by definition:

$$q_t^i(j\Delta u) = \frac{y_{ij}(t)}{x_i(t)} \qquad [8]$$

$$\bar{u}_t^i = \sum_{j=-M}^{M} q_t^i(j\Delta u) j\Delta u \qquad [9]$$

and

$$(\sigma_t^i)^2 = \sum_{j=-M}^{M} [j\Delta u - \bar{u}_t^i]^2 q_t^i(j\Delta u) . \qquad [10]$$

These definitions, with the dynamics equation:

$$y_{ij}(t+1) = y_{ij}(t)[1 + G_t(j\Delta u, \tilde{\mathbf{u}}, \mathbf{p}, N)] \qquad [11]$$

completely determine the dynamics of: phenotype and species populations, the frequency distribution of phenotypes within a species, and the moments of this distribution. In fact, we used the assumption that $q_t^i(u)$ is symmetric and σ_t^i is small to derive [7] [Vincent, 1993]. Note that we insist on denoting the dummy variable u by $j\Delta u$. We do this to emphasize the relations between phenotype populations, and their strategy values.

Equation [8] indicates that the frequency distribution of phenotypes within a species evolves with time: the relative number of individuals (of a species) with a particular strategy value may change. These changes depend on the number of individuals in the community, N, the frequency of individuals of a species, $\mathbf{p}$, the strategy value of all other individuals, $\tilde{\mathbf{u}}$, and the particular value 'chosen' by the individual. Equation [11] uses the assumption of a single G-function for all individuals, and also emphasizes the fact that it depends on time; i.e., the adaptive landscape evolves. The assumption of a single G can be relaxed (Brown, 1987). To simplify notation, we shall write $G_t(j\Delta u)$ instead of $G_t(j\Delta u, \tilde{\mathbf{u}}, \mathbf{p}, N)$.

6 ESS Dynamics with a Specific Mating System

We now develop a specific model. With it, we wish to: (i) clarify the ideas above; (ii) show how behavior–in particular mating behavior–can be incorporated in the dynamics of ESS; (iii) show how one might go about incorporating biological reality in modeling ESS; and (iv) examine the consequences of the definitions [8]-[10] to the definition of ESS, the rate at which populations arrive at ESS, and how this rate depends on phenotypic variance. We use the following assumptions:

(i) The system dynamics are determined by [11]. There are r species populations. Each species has $2M + 1$ potential phenotypes. Because we choose large M and small Δu, all potential strategy values for all species are included: for each species, the dummy variable u can take on $2M + 1$ discrete values, indexed by $j\Delta u$, $j = -M, -(M-1), \ldots, -1, 0, 1, \ldots, M-1, M$.

(ii) Mortality during the mating season is negligible. We therefore assume a short mating season (relative to the generation time).

(iii) The G-function (e.g., natural selection through, say, differential mortality) operates during the time interval $t-1$ to t^-. We use t^- to indicate the instant before the mating season of the $t-1$ generation, and t to denote the beginning of the t-th generation.

(iv) The genetic makeup of an individual defines its phenotype, and during the mating season there is a complete mixing of genes to produce an average phenotype. Thus, without an external variation on the genetic structure of the population, we get a single phenotype, whose strategy is $\bar{u}_t^i$ (see [8]) and whose population size is given in [11].

(v) Due to random mutations, there is a redistribution of phenotypes around the mean in (iv). This redistribution occurs with a fixed variance, and is effected by $Q_t^i(u)$. Thus, after the mating season we have a new generation whose population size, $x_i(t) = \sum_j y_{ij}(t)$, is determined by [11]. The distribution of phenotypes in the new generation is determined by $y_{ij}(t) = Q_t^i(j\Delta u)x_i(t)$.

(vi) $Q_t^i(j\Delta u)$ is parameterized by its mean, $\bar{u}_t^i$, and its constant variance, σ^i.

These assumptions can be relaxed at the expense of proliferation in notations and complex simulations. The process is summarized in Fig. 10. Note that $Q_t^i(u)$ is governed by the species mating behavior, and can be chosen arbitrarily. One can, for example, choose to exclude from reproduction those phenotypes whose strategy values are 'too far' from the mean strategy value, or have some other peculiar value. Below, we choose a normal distribution for $Q_t^i(u)$.

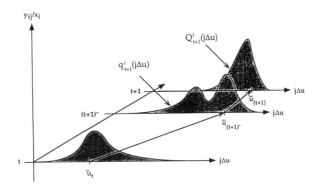

Figure 10: Strategy dynamics of [8]-[11] with the model explained in §6. Note that we are tracing the dynamics of $\bar{u}_t^i$, with its value changing between t and $(t+1)^-$, because of changes in $y_{ij}(t)$ and thus $x_i(t)$. The value of $\bar{u}_t^i$ is not changing between $(t+1)^-$ and $(t+1)$, but the frequencies of phenotypes do because of the redistribution of $y_{ij}(t)$ due to mating (i.e., due to $Q_t^i(j\Delta u)$). Note that the distribution of $y_{ij}(t)$ is spread over all of the potential values of the strategy because $j = -M\Delta u, -(M-1)\Delta u, \ldots, (M-1)\Delta u$ for large M and small Δu.

7 Setting the Initial Conditions

We assume that the species mating behavior causes the values of the adaptive parameter $j\Delta u$ to be distributed normally among j phenotypes of the i-th species. Thus,

$Q_t^i(j\Delta u) \sim N(\bar{u}_t^i, \sigma^i)$. The notation emphasizes the fact that the species mean strategy value evolves with time, but its variance does not. To specify initial conditions we must choose values for the population of species i, $x_i(0)$, its mean strategy value, $\bar{u}_0^i$, and the number of species, r. Next, we set:

$$y_{ij}(0) = Q_0^i(j\Delta u)x_i(0); i = 1, \ldots, r \qquad [12]$$

where

$$Q_0^i(j\Delta u) = \frac{1}{\sqrt{2\pi\sigma^i}} \exp\left\{-\frac{1}{2}\left(\frac{j\Delta u - \bar{u}_0^i}{\sigma^i}\right)^2\right\}; i = 1, \ldots, r; j = -M, \ldots, M \ . \qquad [13]$$

(recall that $y_{ij}(t)$ is the number of individuals of species i whose strategy value is $j\Delta u$). Note that r may change with time by either convergence of means of the strategy values of different species, or by $y_{ij}(t) = 0$ for all j and for $t > t_n > 0$. t_n is the first time that all phenotypes of a species disappear.

8 The Dynamics

Let the dynamics of $y_{ij}(t)$ be governed by the discrete L-V model [2]-[5], with appropriate substitution of $y_{ij}(t)$ for $x_i(t)$ according to [11]. Thus, the strategy value 'selected' by a particular phenotype is influenced by all phenotypes in the community (intra- and inter-specific).

At time t, we have $y_{ij}(t)$, and $\bar{u}^i(t)$. Based on assumptions (i) - (vi), and [8]-[11] we have:

$$x_i(t) = \sum_{j=-M}^{M} y_{ij}(t^-) \qquad [14]$$

$$y_{ij}(t) = Q_t^i(j\Delta u)x_i(t); Q_t^i(j\Delta u) \sim N(\bar{u}_t^i, \sigma^i); i = 1, \ldots, r; j = -M, \ldots, M \qquad [15]$$

$$q_t^i(j\Delta u) = \frac{y_{ij}(t^-)}{x_i(t)} \qquad [16]$$

$$\bar{u}_t^i = \sum_{j=-M}^{M} q_t^i(j\Delta u)j\Delta u \qquad [17]$$

$$(\sigma_t^i)^2 = \sum_{j=-M}^{M} [j\Delta u - \bar{u}_t^i]^2 q_t^i(j\Delta u) \ . \qquad [18]$$

$$y_{ij}(t+1)^- = y_{ij}(t)[1 + G_t(j\Delta u)] \ . \qquad [19]$$

Equations [14]-[19] provide a recipe for calculating all of the quantities of interest. It is important to realize that the absence of a particular strategy value (phenotype) is achieved by $y_{ij}(t) = 0$; thus, we are free to choose a large M. This guarantees that all potential values of the strategy are considered. The process we model here implies that as the strategy values tend to ESS, we get $\lim_{t\to\infty} q_t^i(u) = Q^i(u)$ where $Q^i(u)$ is the stationary ESS distribution of strategy values among phenotypes; it is parameterized by $\bar{u}^{i*}$ and σ^i. We return to this point below.

9 The G-function

From [19], and [3]-[5] we have:

$$G_t(j\Delta u) = \frac{R}{k(j\Delta u)} \left[k(j\Delta u) - \sum_{i=1}^{r} \sum_{j=-M}^{M} \alpha(j\Delta u, u^{ij}) y_{ij}(t) \right] . \quad [20]$$

Similar to [4] and [5], we write

$$k(j\Delta u) = K \exp\left\{ -\frac{1}{2} \left(\frac{j\Delta u}{\sigma_k} \right)^2 \right\} \quad [21]$$

and

$$\alpha_t(j\Delta u, u^{ij}) = 1 + \exp\left\{ -\frac{1}{2} \left(\frac{j\Delta u - u^{ij} + \beta}{\sigma_\alpha} \right)^2 \right\} - \exp\{\gamma\} . \quad [22]$$

where u^{ij} is the strategy value of y_{ij}. It is important to note that since the sum on j is taken over all possible values of the strategy, individuals of a particular phenotype compete against all phenotypes of all species by virtue of the value of y_{ij}.

10 Coalitions with a Fixed-point ESS

We now verify (numerically) that the model above leads to ESS. We use the ESS maximum principle (as given in [6]) and the results in Section 4 (Fig. 1) as benchmarks. We set the parameter to:

$$\beta = 2; \gamma = -\frac{1}{2}; R = 0.25; 2\sigma_k^2 = 8; 2\sigma_\alpha^2 = 8; K = 100; r = 1; \sigma^i = 0.5$$
$$x_1(0) = 1; \bar{u}_0^1 = -1$$

(note that since we consider densities, populations are positive real numbers). The trajectories of both $x_1(t)$ and $\bar{u}_t^1$ shown in Fig. 11a converge to the ESS values shown in Fig. 1. Figure 11b shows the dynamics of the distribution of the strategy among the phenotypes from the initial to the final values. These determine $q_t^i(j\Delta u)$ as shown in [16].

For $2\sigma_k^2 = 25$ we showed that a coalition of 2 exists (Figs. 3 and 7). We obtain similar results based on the model above (Figs. 12 and 13). We also obtained the same ESS values from the dynamical process for a coalition of 3. With larger coalitions we cannot verify results analytically, and the search for ESS solutions requires a large number of runs from various initial conditions. In general, as the number of coexisting species increases, specifying the initial conditions, that end with ESS, becomes more difficult.

11 Phenotypic Variance, ESS, and the Time It Takes to Reach It

One of the necessary conditions for evolution is the presence of genetic variance. This variance must be reflected in phenotypic variance. The strategy dynamics, as given

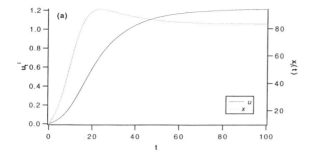

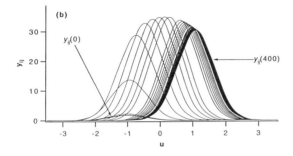

Figure 11: Strategy dynamics with phenotypic variance. The model is explained in §6 and [8]–[22]. Parameter values are: $\beta = 2$; $\gamma = -\frac{1}{2}$; $R = 0.25$; $2\sigma_k^2 = 8$; $2\sigma_\alpha^2 = 8$; $K = 100$; $r = 1$; $\sigma^i = 0.5$; $x_1(0) = 1$; $\bar{u}_0^1 = -1$ (a) Strategy dynamics (compare to Fig. 1); (b) the distribution of phenotypes with respect to strategy values at time 0, and after 400 iterations; these determine $q_0^i(j\Delta u)$ and $q_{400}^i(j\Delta u)$ according to [16].

in [7] and [8]–[12], indicate that without variance there is no evolution. To explore the effect of the phenotypic variance–that may be caused by factors independent of the process dynamics, such as random mutations (the σ^i of $Q_t^i(u)$)–on ESS, we increased σ^i from the value used above (in Figs. 11-13). We assumed that σ^i were equal for all species. There are 2 competing consequences of increasing σ^i: the time to reach a stable strategy (T^*) becomes shorter; however, the values of the ESS $(\bar{u}^{i*})$ will, in general, be different from those predicted by the ESS maximum principle (Fig. 14). We consider that $\bar{u}_t^i = \bar{u}_{T^*}^i = \bar{u}^{i*}$ when $|\bar{u}^i(t+1) - \bar{u}^i(t)| \leq 10^{-9}$ for the first time. [In our models the convergence to ESS or stable strategies is uniformly asymptotic.] Species with a large variance evolve quickly. They "pay" a penalty however: their steady state strategies would not be ESS in the presence of others with small variance (Fig. 14). They thus become vulnerable to extinction (in constant environments). Species with small variance evolve slowly and their steady state strategies will be close to those predicted by the ESS maximum principle. However, in changing environments they could become vulnerable to extinction in the presence of species with large variance. These (trivial) conclusions highlight the implications–of the interactions between environmental changes and phenotypic variance–to the evolutionary process. The quantification of such conclusions in the context of ESS

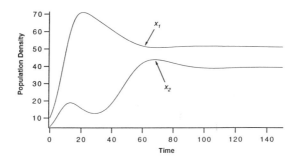

Figure 12: Population dynamics with phenotypic variance. The model is explained in §6 and [8]–[22]. Parameter values are: $\beta = 2$; $\gamma = -\frac{1}{2}$; $R = 0.25$; $2\sigma_k^2 = 25$; $2\sigma_\alpha^2 = 8$; $K = 100$; $r = 2$; $\sigma^i = 0.5$; $x_i(0) = 10$, 1; $\bar{u}_0^i = -1$, -1; $i = 1$, 2. Compare to Figs. 3 and 7.

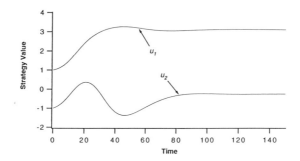

Figure 13: Strategy dynamics with phenotypic variance. The model is explained in §6, [12] through [20]. Parameter values are: $\beta = 2$; $\gamma = -\frac{1}{2}$; $R = 0.25$; $2\sigma_k^2 = 25$; $2\sigma_\alpha^2 = 8$; $K = 100$; $r = 2$; $\sigma^i = 0.5$; $x_i(0) = 10$, 1; $\bar{u}_0^i = -1$, -1; $i = 1$, 2. Compare to Figs. 3 and 7.

dynamics, however, is not trivial.

The conclusions drawn from Fig. 14 also emphasize the fact that it is advantageous for species in changing (constant) environments to maintain a large (small) phenotypic variance. Yet, in drawing conclusions from observations, we must distinguish between the process independent phenotypic variance, σ^i, and the variance effected by the dynamics. σ^i can be detected when species are close to their ESS strategies. In other words, unless a population is close to its ESS strategy values, it is difficult to separate $q_t^i(u)$ from $Q_t^i(u)$. As populations get closer to their ESS values, the difference between the post-mating distribution of phenotypes, $Q_t^i(u)$, and their pre-mating distribution $q_t^i(u)$ should diminish; and so is the difference between $\bar{u}_t^i$ and $\bar{u}_{t+1}^i$. Note that $Q_t^i(u)$ depends on t through $\bar{u}_t^i$: the mating system redistributes the frequency of phenotypes around $\bar{u}_t^i$ with the time independent variance σ^i. One

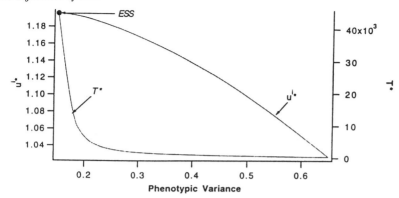

Figure 14: The interaction between the time it takes to reach a stable 'ESS' solution, T^*, and the 'ESS' value $\bar{u}^{i*}$, as a function of phenotypic variance. We put ESS in quotes because these values differ from that predicted by the ESS maximum principle; yet, in simulations, $\bar{u}^{i*}$ persisted over time, with all mutations ($\bar{u}^j \neq \bar{u}^{i*}$) eventually disappearing (i.e., $x_j(t) = 0$).

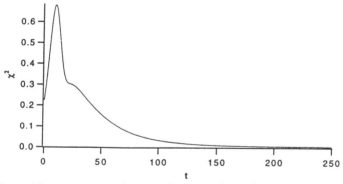

Figure 15: The difference between the pre- and post-breeding distributions of phenotypes ($Q_t^i(u)$ and $q_{t-}^i(u)$) for the model simulated in Fig. 11. This difference is calculated according to [23].

way to measure this diminishing difference is with the χ^2 score

$$\sum_{i=1}^{r} \sum_{j=-M}^{M} \frac{(q_t^i(j\Delta u) - Q_t^i(u))^2}{Q_t^i(u)}.$$ [23]

This difference, according to our model, is shown in Fig. 15. These differences among the distributions can then be used as indicators of how close a population might be to its 'ESS'. We put ESS in single quotes to emphasize the fact that although the stable strategies that are achieved with phenotypic variance (e.g., Fig. 14) persist and do not allow mutants to persist, they do not correspond to the ESS values generated by the maximum principle through [6].

12 Chaos

Because of the complexity of the evolutionary process, one can hardly expect to see a fixed-point ESS in nature. Yet, as a "utopian concept", a fixed-point ESS may be used as a reference: a point where an evolutionary dynamical system moves toward, but never reaches. To examine the potentially complex behavior of the dynamics of ESS, we changed some of the parameters in the L-V model to those given in Table 1. With $R = 3.082$, for example, the distributions of $y_{ij}(0)$ and $y_{ij}(200,000)$ are shown in Fig. 16. It is interesting to note that with these parameter values after about 100 iterations both $\bar{u}_t^1$ and $x_1(t)$ fluctuate. Their trajectories for $43,000 \leq t \leq 43,400$ are shown in Fig. 17.

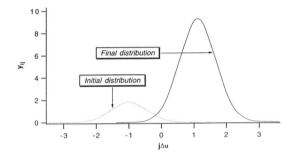

Figure 16: The distributions of $y_{ij}(0)$ and $y_{ij}(200,000)$ with $R = 3.082$ and other parameter values shown in Table 1 according to the model in §6 and [8]–[22].

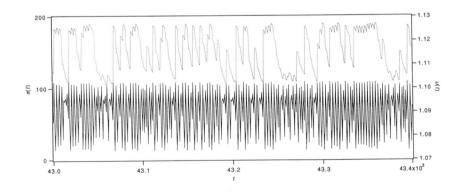

Figure 17: The trajectories of $x_1(t)$ (solid) and $\bar{u}_t^1$ (dotted) for $43,000 \leq t \leq 43,400$, with parameter values shown in Table 1, and $R = 3.082$ according to the model in §6 and [8]–[22]. Note the high-intermixed with low-frequencies in the fluctuations of $\bar{u}_t^{1*}$.

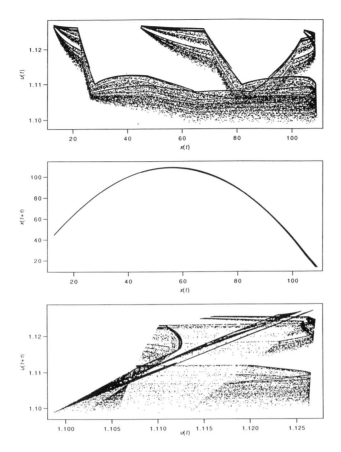

Figure 18: The attractor of the trajectories shown in Fig. 17. Top panel: $\bar{u}_t^1$ vs $x_1(t)$; middle panel: $x_1(t+1)$ vs $x_1(t)$; and bottom panel: $\bar{u}_{t+1}^1$ vs $\bar{u}_t^1$. Attractor ranges are as in Fig. 17. The number of iterations shown is 200,000. Parameter values are given in Table 1 with $R = 3.082$. The system spends more time in the darker (denser) regions of the attractor than in the lighter regions. These are created by the high-intermixed with low-frequencies in the fluctuations of $\bar{u}_t^{1*}$.

The ESS definition does not require a fixed point in the strategy space (a set of unchanging strategy values), as does the ESS maximum principle given here. Thus, an ESS may exist under conditions not satisfied by the ESS maximum principle. For example, simulations indicate an attractor and potentially cycles in both $\bar{u}_t^1$ and $x_1(t)$. To check if the results we achieve through simulations (Figs. 16 and 17) are an ESS, we injected at random times during the simulations new mutants. These mutants followed the same dynamics of the existing species $x_1(t)$, except that their mean strategy values ($\bar{u}_t^i; i > 1$) were fixed to a value chosen randomly from the set of values within the range shown in Fig. 17. None of the 100 mutants thus introduced

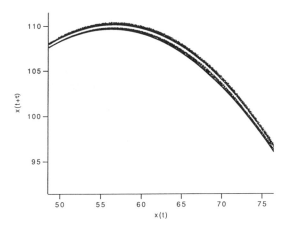

Figure 19: Magnification of a specific region of the attractor shown in the middle panel of Fig. 18.

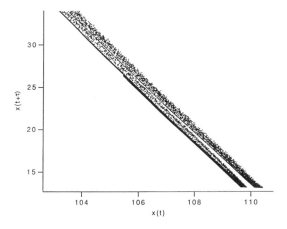

Figure 20: Magnification of a specific region of the attractor shown in Fig. 19.

persisted with $x_1(t)$ for more than 200 generations. Thus, the simulations indicate that fluctuations in $\{\bar{u}_t^1, x_1(t)\}$ are necessary for the ESS, which are no longer a fixed point, but rather an attractor. It is interesting to look at the fluctuations of $q_t^i(u)$, and $[q_t^i(t) - Q_t^i(t)]$. Such analysis is (perhaps) forthcoming.

The implications of these results indicate that we need to extend the ESS maximum principle. The simulations indicate that in some models, fluctuations are necessary, and that during the evolutionary process, we can expect to observe strategies fluctuating with time. Without such fluctuations, no persistence is achieved. These fluctuations are produced from a deterministic model, and no stochastic effects are

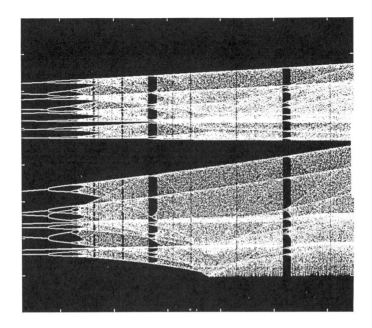

Figure 21: Bifurcation diagram of $0 \leq x_1(t) \leq 150$ (left to right) for $0.25 \leq R \leq 3.5$ (top to bottom). Parameter values are given in Table 1.

necessary to produce them. The trajectories of $\{\bar{u}_t^1, x_1(t)\}$ appear to be chaotic (Fig. 18). A sequence of magnifications of the attractor (Figs. 19 and 20) confirms that the attractor is fractal. The chaotic behavior of $\{\bar{u}_t^1, x_1(t)\}$ is further confirmed by the bifurcation diagram (Fig. 21). This diagram–explained, for example in (May 1976)–shows how the number of cycles of the $x_1(t)$ trajectory change with the value of the parameter R. When the value of the parameter R was changed to 2.999, a different attractor emerged (Fig. 22). Magnification of the attractor (i.e., u vs x in Fig. 23) revealed again a fractal structure. Within this new attractor, one can clearly see the attractor that was produced with $R = 3.082$ (top panel, Fig. 18). This phenomenon is called crises, and has been previously reported (Grebogi et al. 1982). In crisis, the shape of the attractor changes abruptly at some specific values of the parameters.

The implications of these results to the evolutionary process as modeled here are that in addition to the seemingly random–but in fact chaotic–fluctuations of the ESS strategies and the accompanying populations, we can expect sudden changes in the shape of the attractor. Such sudden changes can be interpreted as 'jumps', or macro-evolution. Also, note the darker regions in the attractors shown in Figs. 18, 20, and 22: The system spends more time in these than in other regions. This means that as we trace the dynamics, we may observe small changes in the values of the strategies and in fluctuations of populations for a while. Then, an abrupt change occurs (the system moves to a different region of the attractor). For example, observe the trajectory of $\bar{u}_t^{1*}$ (the broken line in Fig. 17): small amplitude with high

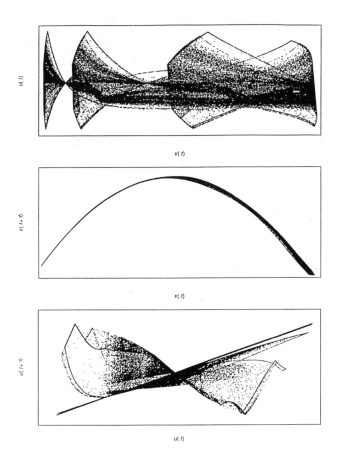

Figure 22: The attractor of the model in §6 and [8]–[22]. Top panel: $\bar{u}_t^1$ vs $x_1(t)$; middle panel: $x_1(t+1)$ vs $x_1(t)$; and bottom panel: $\bar{u}_{t+1}^1$ vs $\bar{u}_t^1$. Attractor ranges are as in Fig. 17. The number of iterations shown is 200,000. Parameter values are given in Table 1 with $R = 2.999$.

frequencies for a while at high values of the strategy are broken with sudden large amplitudes with low frequencies. Note that we denote the species strategy by $\bar{u}_t^{1*}$. This notation indicates that the strategy value is not constant (and hence the time subscript); yet it is 'ESS' in the sense that mutant strategies do not coexist with $\bar{u}_t^{1*}$ (hence the *). For some parameters values, this mix of high and low frequency fluctuations in $\bar{u}_t^{1*}$ (and possibly $x_1(t)$) may last many generations. In such cases we may mistakenly interpret the strategy dynamics to be moving towards an ESS; i.e., we may think that we are observing $\bar{u}_t^1$, while in fact we are observing $\bar{u}_t^{1*}$.

Our analysis is numerical and probably has nothing to do with 'reality' (whatever that means). Yet, our results indicate that we can expect the concept of ESS to produce a rich behavior of population trajectories and the values of their adap-

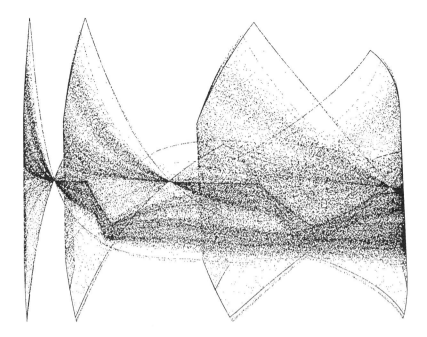

Figure 23: Magnification of the top panel in Fig. 22. Note that the attractor shown in the top panel of Fig. 18 is 'buried' within the attractor here.

tive parameters. It seems that the maximum principle should be extended to time-dependent ESS sets (attractors), with each attractor (a subset of the strategy set) defining a species. Within the region of the strategy set that is occupied by the attractor, the 'ESS' strategies may be found anywhere and the ability to forecast where deteriorates with time. This ability depends on the impossible determination of the precise location of the system. Thus, the 'ESS' we describe here captures some of the rich behavior of the evolutionary process we may observe in Nature. One may use a more statistical approach to account for variance and fluctuations in ESS values. This may be useful, but is sure to lose the interesting consequences of unpredictable deterministic processes.

The results here (Figs. 16-23) mean that: (i) the specific values of the ESS, at a particular time, may be sensitive to initial conditions; (ii) long-term changes in ESS traits may be unpredictable; (iii) large inter-generation differences (with regard to the ESS strategy) can be expected; (iv) complex strategy dynamics can arise from simple models; (v) sudden changes in the 'ESS' strategy values are possible; (vi) fluctuations in these values may be an integral part of (and even necessary for) the ESS. These conclusions also indicate that seemingly evolving populations (with regard to some traits) may actually be in the 'ESS' set; that is, rather than evolving, they are on the attractor. Furthermore, apparent random changes (with generations), in the values of some heritable traits, may be driven by deterministic evolutionary mechanisms.

13 Conclusions

In closing, we wish to point out some of the difficulties and limitations in our approach, and speculate on potentially fruitful extensions to the theory presented here. The theory assumes that all individuals (phenotypes) have a perfect 'knowledge' of their environment and the strategies undertaken by other organisms. The updating of strategies is simultaneous, and occurs for all members of the community instantaneously. We do not consider the effects of learning and uncertainty in the pursuit of the ESS. The derivation of the G-function can be tricky and in some cases impossible. Some of these assumptions are ridiculous: there is no biological system without some amount of delay built in, and some amount of uncertainty. It is also difficult to see how all strategies can be updated simultaneously. The ESS maximum principle should be generalized to include sets, with fixed points as a special case. It seems that sequential updating, and adaptive learning may play an important role in further developments in the theory. If adaptive learning is considered, state-dependent strategies would have to be considered.

The framework for modeling the dynamics of ESS, as prescribed in Section 5 and [8]-[22] can be modified to include the effects of mating behavior on the dynamics and the ESS strategy values. In particular, it is important to understand how a particular mating behavior leads to the redistribution of phenotypes (and perhaps genes) in the population. It is also interesting to develop a continuous parallel of the discrete models discussed here. Finally, one needs to keep in mind that all of our results presented in Section 6 depend on a specific model. As with all models, the conclusions we draw from ours are specific; their generality, and credibility are limited, and can be established with future work on a variety of other models of the dynamics of ESS.

References

[1] Brown, J. and T.L. Vincent. 1987. A theory for the evolutionary game. Theoret. Popul. Biol. 31: 140-166.

[2] Brown, J. and T.L. Vincent. 1987. Coevolution as an evolutionary game. Evolution 41: 66-79

[3] Brown, J.S. 1990. Community organization under predator-prey coevolution. Pp. 263-288 in T. L. Vincent, A. I. Mees and L. S. Jennings ed. Dynamics of Complex Interconnected Biological Systems. Birkhaüser, Boston.

[4] Grebogi, C., E. Ott and J.A. Yorke. 1982. Chaotic attractors in crisis. Phy. Rev. Lett. 48: 1507-1510.

[5] Hines, W.G.S. 1987. Evolutionary stable strategies: a review of basic theory. Theoret. Pop. Biol. 31: 195-272.

[6] May, R.M. 1976. Theoretical Ecology: Principles and Applications. W.B. Saunders Company, Philadelphia.

[7] Maynard-Smith, J. 1982. Evolution and the Theory of Games. Cambridge University Press, Cambridge, UK.

[8] Owen, G. 1982. Game Theory. Academic Press, Inc., Orlando.

[9] Roughgarden, J. 1983. The theory of coevolution. Pp. 33-64 in D. J. Futuyma and M. Slatkin ed. Coevolution. Sinaüer, Sunderland, Massachussetts.

[10] Vincent, T.L. 1990. Strategy dynamics and the ESS. Pp. 236-262 in T. L. Vincent, A. I. Mees and L. S. Jennings ed. Dynamics of Complex Interconnected Biological Systems. Birkhaüser, Boston.

[11] Vincent, T.L. and J.S. Brown. 1984. Stability in an Evolutionary Game. Theoret. Popul. Biol. 26: 408-427.

[12] Vincent, T.L. and J.S. Brown. 1987. Evolution under nonequilibrium dynamics. Math. Model. 8: 766-771.

[13] Vincent, T.L., Y. Cohen, and J.S. Brown. 1993. Evolution via strategy dynamics. Theoret. Pop. Biol.

[14] Vincent, T.L. and M.E. Fisher. 1988. Evolutionary stable strategies in differential and difference equation models. Evol. Ecol. 2: 321-377.

Commentary by W.J. Grantham

Some researchers in population biology believe that chaos in a population system is ridiculous. That school believes that any observed "random" fluctuations must be caused by random effects in the environment. The paper by Cohen and Vincent helps to dispel this bias by showing that, at least for a certain simulated population system, chaos may not only occur in a deterministic population model, but also that it can occur under a game theoretic optimal strategy for maximum fitness and an increasing level of complexity in the population. Thus the search for "equilibrium" behavior should include not only fixed points and periodic orbits, but also the possibility of chaotic strange attractors.

Nitrogen Cycling and the Control of Chaos in a Boreal Forest Model

John Pastor
Natural Resources Research Institute
University of Minnesota
Duluth, MN 55812, USA

Yosef Cohen
Dept. of Fisheries and Wildlife
University of Minnesota
St. Paul, MN 55108, USA

Abstract

The cycling of nitrogen — a growth limiting nutrient — between plants and soils is a strong feedback in forest ecosystems. The rate by which nitrogen is released from decaying plant litter controls its supply rate to plants with which they can fix carbon through photosynthesis; the decay rate of litter in turn is controlled by its carbon chemistry, which varies among species. Here, we present a well-validated forest ecosystem model that simulates this feedback and its interaction with competing plant species and climate. We show, using spectral analysis and the Grassberger-Procaccia algorithm, that under climates typical of northern hardwoods the model produces a periodic attractor but under cooler climates typical of boreal forests the output is a strange attractor. We further show that these attractors are governed by this feedback — the chaotic behavior of the boreal forest disappears when the assumption that plants are limited by nitrogen is relaxed. We then augment (by fertilization) or decrease (by leaching or harvesting) the amount of nitrogen that is cycling through the forest to determine if we can control this behavior. Augmenting the supply of nitrogen to the plants increased productivity, but also increased variance of model output, essentially increasing the scaling region of the attractor. Decreasing nitrogen availability decreased variance, but also decreased productivity. This suggests that increasing productivity by fertilization may come at the expense of long-range predictability. If sustainability of production implies long range predictability, then it appears that increases in production are not sustainable.
1.

1 Introduction

Feedbacks between organisms and resources in ecosystems are highly non-linear, thus precluding direct experimental tests via general linear models of all but the simplest of systems. For lack of better alternatives, insights into ecosystem functioning must rely on simulation models. This has practical as well as theoretical consequences.

Global climate change (Global Climate Change Committee 1988) and the accelerating rate of harvesting of forest ecosystems (USDA 1982) accentuate the need for understanding long-term ecosystem processes, but global climate change is an uncontrolled, unreplicated experiment and large scale manipulations of forest ecosystems are not often possible. Recent developments in non-linear systems theory (Packard, et al. 1980; Takens 1981; Farmer 1982; Swinney 1983; Schuster 1984; Grassberger 1986; Procaccia 1988; Abraham, et al. 1989; Sugihara and May 1990) point out that mathematical models in general, and ecosystem models in particular, can produce results whose long term predictions are governed by multiple periodicities or are even chaotic. Whether these predictions reflect real-world processes or are simply an artifact of model construction is a fundamental problem in the applications of mathematical theory to complex biological systems and natural resource policy aimed at controlling the output of biological systems.

There are a growing number of examples of periodic and chaotic behavior of simple biological models, and even some real world systems (May 1974, 1976, May and Oster 1976, Sugihara and May 1990, Hanski et al. 1993, Turchin 1993, Hastings et al. 1993, Ellner and Turchin 1995, Constantino et al. 1995, Kareiva 1995). For the most part, these examples show that shifts in demographic parameters such as fecundity or mortality can alter behavior of real or simulated populations from periodic oscillations to aperiodic and deterministic chaos.

These examples derive mainly from population studies, where the supply of limiting resources is greatly simplified either mathematically (e.g., K in Lotka-Volterra or logistic equations — May 1974) or experimentally (e.g., flour beetle microcosms — Constantino et al. 1995). In the context of the ecosystem, fecundity and mortality depend strongly on rates of supply of limiting resources. Feedbacks between the populations and the resources arise where organisms control the supply rate of those resources and in particular where the ability to control supply rates differ among species. We will examine here how feedbacks between plant species and soils inherent in the cycling of nitrogen, a limiting nutrient to plant growth, cause periodic and chaotic behavior of a forest ecosystem model. First, we present a brief explanation of the nature of such feedbacks.

2 Nitrogen Cycling and Feedbacks in Forest Ecosystems

The rate by which plants can grow by fixing carbon through photosynthesis is controlled partly by the rate that nitrogen — a limiting nutrient — is released from decaying plant litter; the amount of nitrogen released by decomposition is in turn controlled by the amount of litter and its carbon chemistry, which varies among species. This reciprocal feedback loop between the carbon and nitrogen cycles results in some interesting dynamics, as we shall show. The cycle of nitrogen in intact forests is relatively closed — whatever nitrogen is released from decaying litter and humus is rapidly taken up by plants, and consequently very little leaches from most forests (Bormann and Likens 1979). Inputs of nitrogen are also very low, except for inputs by symbioses of certain plants such as alder with bacteria that can fix atmospheric nitrogen, inputs in acid precipitation (esssentially a dilute solution of nitric and sulfuric acids), and inputs in fertilizer.

The feedback between plants and soils as nitrogen and carbon cycle in the forest is therefore an important regulator of production. This feedback in turn is constrained by climate, through the length of the growing and decomposing season and through the temperature and precipitation during this season. With colder climates typical of northern boreal forests, the availability of nitrogen decreases and there is a premium for plants to use nitrogen efficiently during growth. Such efficient use can come about in two ways: 1) rapid growth during the short periods of nitrogen availability, as exhibited by many deciduous species such as aspen (Populus tremuloides) and paper birch (Betula papyrifera); 2) slow growth and conservation of nitrogen by evergreen conifers such as spruce (Picea glauca) and balsam fir (Abies balsamea). These two strategies are correlated with other traits, important ones being the decay rate of litter and the ability to survive in the shade cast by taller trees. Fast growing deciduous species in northern forests have nitrogen rich, easily decomposible litter but are unable to tolerate dense shade. While they occupy a site, nitrogen availability is enhanced, but they cannot perpetuate occupancy unless some disturbance causes mortality of taller trees and opens the canopy, allowing their seedlings to receive full sunlight. The slower growing conifers, on the other hand, depress nitrogen availability because their litter is nitrogen poor and slow to decompose, but they can occupy sites for several generations because their seedlings tolerate dense shade.

3 The Model

Such interactions between plant traits and the cycle of nitrogen between plants and soils lead to complex long-term behaviors. We explore these behaviors using a forest ecosystem model, called LINKAGES.

Like many of its kind (Botkin, et al. 1972; Shugart and West 1977; Shugart 1984) the LINKAGES model (Pastor and Post 1985; Pastor and Post 1986; Pastor and Post 1988) begins with a priori assumptions about how climate and soils affect tree growth and recruitment. A detailed description and validations are given in Pastor and Post (1986).

The model simulates the birth, growth, and death of individual trees in a 0.01 ha forest plot. Predictions are usually reported as means of many independent runs with stochastic variation of input parameters (Fig. 1).

The model includes a soil moisture simulator (Thornthwaite and Mather 1957) and also simulates the effects of soil nitrogen limitation on tree growth (Weinstein, et al. 1982; Pastor and Post 1985; Pastor 1986; Pastor and Post 1986). Decomposition of annual cohorts of litter and humus and nitrogen release from them are functions of actual evapotranspiration as well as litter lignin:nitrogen ratio, which varies among species (McClaugherty et al. 1985, Moore 1984).

The responses of trees to resource availabilities are determined by species-specific response curves. The response curves scale the effects of availabilities of each resource from 0 (resource prohibitively low) to 1.0 (resource at optimal levels). These response curves are non-linear (i.e., either asymptotic or quadratic), and appear to be responsible for complex behaviors to be discussed below. Temperature and soil moisture response functions are quadratic functions, while the soil nitrogen and light response functions are saturating. These response functions limit both fecundity and growth of individual trees. Recruitment of each species is reduced from maximum to the extent that light, moisture, and degree-days are less than optimal. Diameter growth

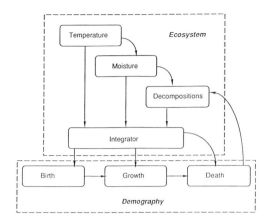

Figure 1: Block diagram of the LINKAGES model.

of each tree is reduced from maximum by these factors and soil nitrogen, whichever is most limiting. The probability of mortality increases with tree age and slow growth due to stresses imposed by low resource availabilities.

LINKAGES belongs to a class of models which have been extensively used to explore the interaction between demographic and ecosystem processes as driven by resource availability and climate. A surprising conclusion is that changes in competitive advantages of different tree species under different climates can either amplify or obscure changes in forest biomass, productivity, and species composition (Solomon and Shugart 1984, Pastor and Post 1988). Because trees affect the availability of their resources through shading or decomposition of their litter (McClaugherty et al. 1985, Moore 1984), changes in species composition resulting from climate change produce secondary responses in forest resources that in turn cause lags, hysteresis, and diverging characteristics (such as species composition) as the forest changes (Pastor and Post 1988, Cohen and Pastor 1991).

These forest ecosystem models had been validated by simulating changes in climate along altitudinal gradients. The models' predictions were then tested against measured changes in ecosystem properties. The models correctly predicted the altitude at which hardwoods are replaced by conifers in New England when driven by climate data at different elevations (Botkin, et al. 1972). Species distribution in the Australian Alps were correctly predicted from climate data (Shugart and Noble 1981). The models have also been validated against soil moisture gradients (Solomon and Shugart 1984; Pastor and Post 1986). Other applications of these models to the interpretation of vegetation response to late Quaternary climate changes were reviewed (Solomon and Webb 1985). These predictions, which had been used to validate the models, are not claimed to be accurate. They essentially produce output trajectories within bounds which had been observed (i.e., within an attractor).

The above considerations lend credibility to the LINKAGES (and similar) models. Yet, some important questions remain: What mechanisms are responsible for the apparent ecosystem dynamics? How sensitive are the model predictions to initial conditions? Are there any fundamental ecosystem traits that can be inferred and verified from the model as a dynamical system? The theory of non-linear dynamical systems can provide answers to some of these questions.

4 Complex Behaviors of the LINKAGES Model

Early in the application of this model to problems of projecting forest growth, it became apparent that the model exhibited complex behaviors, such as bifurcations and oscillations. The plant-soil feedback inherent in the nitrogen cycle was implicated as a cause of these behaviors. For example, Pastor et al. (1987) showed that simulated boreal forests show oscillations of occupancy by different species when the feedback between plants and soil nitrogen was allowed to operate, but asymptotically stable dominance by one species when nitrogen availability was kept constant. Pastor and Post (1988) showed that, with climate warming projected with the increased loading of radiatively active trace gasses such as CO_2, the forests of northern North America will show a bifurcated response sensitive to small differences in soil water holding capacity. On clay soils that can retain water, productivity increases with warming because the species that can occupy such sites have litter that decays rapidly and increases nitrogen availability. On sandy soils, productivity decreases because the species that can occupy such droughty soils have litter that is difficult to decay and depress nitrogen availability. Yet, the difference in water holding capacity between a sand and a clay is relatively small — only $8cm$ of water per meter of soil depth. Cohen and Pastor (1991) repeated these climate warming simulations including observed serial correlations in temperature of 46 and 89 months. When superimposed on gradually changing mean temperatures, these serial correlations increased the divergent responses of forests observed in the Pastor and Post (1988) model runs.

These results are significant for several reasons. First, they indicate that the model is qualitatively sensitive to small changes in driving variables such as temperature and soil moisture, particularly when these driving variables operate by setting thresholds on fecundity and mortality of different species. Second, they indicate that the model is quantatively and qualitatively sensitive to secondary effects inherent in the nitrogen feedback, which tend to amplify the small changes in driving variables.

Such secondary responses give a strong indication that the model dynamics may be chaotic under some domains. This raises the possibility that a large and complicated ecosystem model (and perhaps the ecosystems themselves) are low dimensional: the systems are controlled by few important variables which produce complicated, and seemingly stochastic outputs.

Accordingly, we began examining the output of LINKAGES for complex behaviors, particularly with regard to the feedback between carbon and nitrogen as they cycle between plants and soils. The reciprocal feedback between carbon and nitrogen cycles can be positive when, because of easily decomposible carbon compounds, litter decays rapidly and enhances nitrogen availaability and hence plant fixation of more carbon through photosynthesis; the feedback is negative when litter decays slowly because of high concentrations of recalcitrant carbon compounds, thereby depressing nitrogen availability and subsequently the rate of carbon fixation by plants. The sign of the feedback is likely to be reversed at the borders of ecosystems comprised of species that strongly differ in their effect on carbon and nitrogen cycles. Such is the case at the border between northern hardwood forests — with deciduous species that grow rapidly and have litter that decays rapidly — and the boreal forest - with coniferous species that grow slowly and have litter that decays slowly. This boundary in the Lake Superior region is defined by several climatic conditions, such as the winter position of Arctic fronts and mean annual temperature (Bryson 1966).

Cohen and Pastor (submitted) ran the model for three climates that span this boundary: 1) northeastern Minnesota, now at the boundary between these two forests; 2) northeastern Minnesota $+2°$ C, which enhances growth of northern hardwoods and inhibits growth of boreal conifers; 3) northeastern Minnesota $-2°$ C, which inhibits growth of northern hardwoods and, by freeing them from competition for light and nitrogen, enhances the growth of boreal conifers. Model runs were made with the nitrogen cycle intact and with the nitrogen feedback deactivated by assuming that nitrogen does not limit growth (i.e., by setting the nitrogen response function always equal to 1.0). Runs were made for 20,000 years to generate a sufficiently long time series to detect presence or absence of periodicities. In these runs, monthly temperature and precipitation vary stochastically and were generated from normal distributions whose mean and variance are defined by the historical record over the past 100 years. In addition, probability of birth and mortality also vary stochastically (Pastor and Post 1986).

We analyzed the power spectral densities (Priestly 1981) of 20,000 years of model output of total production for each of the climates, with and without the nitrogen feedback intact. Total production integrates life history characteristics with ecosystem properties such as the nitrogen cycle. Others (Emanual et al. 1978) have used a similar approach in analyzing previous versions of the model that did not include nitrogen cycling.

Under the coldest climate typical of boreal forests, there were no strong dominating periods, but removing the nitrogen feedback produced three dominating periods in total production (Cohen and Pastor, submitted; Fig. 2). Under current climate of northeastern Minnesota at the borel-northern hardwood border, there was one dominating period with the nitrogen feedback intact and two with the nitrogen feedback removed. In contrast, under the warmest climate characteristic of northern hardwoods, there were two dominant periods with the feedback intact and only one with the feedback removed. Thus the effect of the nitrogen cycle is to dampen or filter oscillations in the boreal forest and to increase the number of oscillations in northern hardwoods.

The lack of strong dominating periodicities in the boreal forest simulations could reflect the stochastic simulation of some of the variables, or it could reflect an infinite number of unstable periodic orbits that constitute a strange attractor (Ott et al. 1990). To examine this, Cohen and Pastor (submitted) ran the simulations again but with all stochastic effects removed except for probability of mortality with increasing age. After verifying that productivity is stationary about a mean, differential time interval plots (Babloyantz 1989) were constructed from a series of the time intervals, R_i, between successive minima of $x(t)$. Next, $\Delta R_{i+1} = R_{i+2} - R_{i+1}$ was plotted against $\Delta R_i = R_{i+1} - R_i$. For a limit cycle, the DTI plot will appear as a single point at the origin since all R_i will be equal. A periodic trajectory with additive noise will appear as a cloud of points scattered homogeneously around the origin. This approach is similar to the interspike interval approach presented by Sauer in this volume. Our time series is clearly non-homogeneously structured (Figure 3).

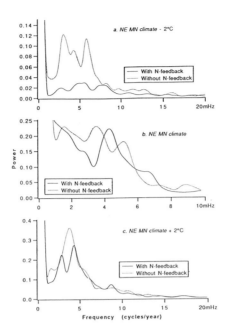

Figure 2: Power spectral density of a 20,000 year run of the model ouput for production rate (t/ha per year) with (a) Northeastern Minnesota climate cooled by 2° C; (b) Northeastern Minnesota climate at present; and (c) Northeastern Minnesota climate warmed by 2° C. Note the effect that turning on and off the nitrogen feedback has on the number of dominating frequencies within each climate regime. Also note the lack of strong dominating frequencies under the coolest (boreal) climate when the nitrogen feedback is intact.

This indicates that the dynamics underlying the trajectory of production rate is not periodic, but much more complex. It suggests the possible existence of an attractor on non-integer dimension.

Next we calculated the slopes of the correlation integral (Grassberger and Procaccia 1983) to determine the dimension of the attractor. To do this, we first calculated the autocorrelation function of the time series of productivity and found zero autocorrelation at a time interval between 49 and 50 years delays. We therefore used 49 year delays in constructing the embedding matrix. The slopes of the correlation integral were calculated for embedding dimensions 2 through 10, where each successive dimension represents an increment of 49 year delay from the previous dimension. A scaling region emerges with dimension between 5 and 7. We also scrambled the time series output and recalculated the correlation integral. This removes any hidden periodicities from the data. If the process is chaotic, there should be no change in the correlation integral after scrambling, and there was not.

Finally, we examined the attractor itself (Fig. 4). The attractor demonstrates a

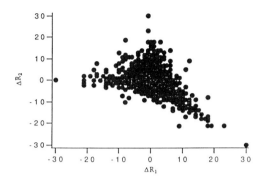

Figure 3: Differential time interval plot of model output for Northeastern Minnesota climate cooled by 2° C with the nitrogen feedback intact and with all stochastic effects removed except for probability of mortality. Note the distinct structure indicating non-random behavior.

structure in that: 1) it does not fill its space uniformly, as would a random walk time series; 2) magnification of the attrctor reveals "microstructures" of small arcs rather than a uniform distribution of points. The attractor is bounded between 2 and 8 tons of production per ha per year, which are reasonable ranges of production reported in the literature for these forests (Cannell 1982). Within these bounds, the trajectory of the dynamics of production on 49 year time steps is not predictable.

We therefore conclude that there is a strong possibility of chaotic behavior of a medium dimensional attractor when the model is run (nearly) deterministically under climatic conditions typical of the boreal forest. We do not suggest that we have conclusively shown that the model represents a low dimensional chaotic system that is an analogue for the real forest. The points we wish to make are: 1) with the nitrogen cycle turned off, the model exhibits periodicities, but with the nitrogen feedback intact and with deterministic input the DTI plots and the attractor reveal structures and a scaling region emerges; 2) if the simulation of the nitrogen cycle allows one to validate the model against measurements of a wide variety of ecosystem properties in these forests (Pastor and Post 1986), then chaotic behavior in long term because of the nitrogen feedback is also a possibility.

5 The Nitrogen Cycle and the Control of Chaos in Boreal Forests

The previous results suggest that the internal cycle of nitrogen in a boreal forest produces a medium dimensional, chaotic attractor. This suggests that chaos can be controlled or even eliminated by augmenting or decrementing nitrogen in the forest.

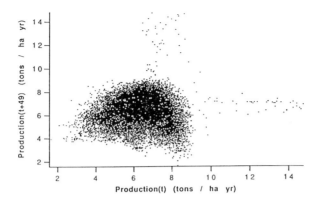

Figure 4: The attractor of the model, with lag of 49 years (i.e., at an interval where autocorrelation = 0.0). The "rays" of points emanating from the body of the attractor are from the first 200 years when the mean is not stationary. Note that the body of the attractor after the first 200 years is bounded between 3 and 8 t/ha of production per year, which is the range generally reported for boreal forests (see text).

The amount of nitrogen that cycles and its effect on productivity would then be determined more by inputs to and outputs from the system rather than the internal cycle. Thus, we implement the following control on the nitrogen cycle:

$$\dot{x} = f(x) + g(k)$$

where $\dot{x}$ is the rate of change in productivity, $f(x)$ represents the non-linear feedback between production and soils inherent in the nitrogen cycle (essentially the model) and $g(k)$ is a function representing an augmentation of production for a fixed amount of nitrogen, k, added or subtracted each year for plant uptake. Essentially $g(k)$ is the nitrogen response function solved for a given input of nitrogen k and added to production each year.

Accordingly, we repeated the above runs for the boreal forest with stochastic effects removed, but this time incrementing available nitrogen by 5 and 50 kg per ha per year, and decrementing it by these amounts. An incrementation of nitrogen would be analogous to applying fertilizer to augment the supply from decay of litter. A decrement of nitrogen would be analogous to removals by leaching before the plant could obtain it.

The effect that augmenting or decrementing nitrogen supply has on the time series of productivity is shown in Fig. 5 and on the attractor in Fig 6. Augmenting nitrogen supply increased mean productivity and decrementing it decreased it, although productivity did not go to zero (a point to which we shall return shortly). This is not surprising, as the supply of nitrogen, through the species response curves, limit plant growth.

What is more interesting is the effect it has on the shape of the attractor. Aug-

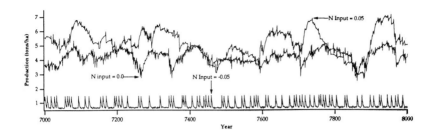

Figure 5: Time series of model output for Northeastern Minnesota cooled by 2° C with various levels of augmenting or decrementing nitrogen in the system (in t/ha per year).

menting nitrogen supply increased the size of the attractor, essentially increasing variance around the mean. Decreasing nitrogen supply decreased the spread of points and under the highest rates of removal produced an attractor which is a small square with empty spaces. Furthermore, with increased nitrogen supply the attractor becomes more and more asymmetric. This is shown by examining the position of the attractors relative to the line defining equal production in year t and year $t + 49$ (the diagonal in Fig. 6). With increasing nitrogen supply, there is a higher density of points below this line than above it. This means that decreases in productivity over a 49 year time interval become more common than increases in productivity with augmentations of nitrogen, despite the fact that increased nitrogen should increase growth rates. And yet, the time series remains stationary (Fig. 5). These apparent paradoxes can be resolved by noting, from the time series in Fig. 5, that declines in productivity over 49 years are usually small, but increases are large jumps. Essentially, a few relatively rare increases in productivity cancel many small decreases, and the series remains stationary.

Further simulations by increasing and decreasing nitrogen supply by 500 kg/ha per year had no further effect on the attractor. This dimishing effect with further enhancements is attributed to the saturating function of plant growth in response to nitrogen — beyond 50 kg/ha of augmentation + supply from the soil, the plants are no longer limited by nitrogen and there is no further effect.

Why is there no further effect with more removals and why does productivity not decline to zero even at higher rates of removal? The answer lies in the fact that new individuals (seedlings) are established with an internal supply of nitrogen -seedling recruitment is an external supply of nitrogen to the system. This supply, although very small, is sufficient to support plant growth for a few years, whereupon requirements exceed internal supply and the plants die. This results in the "square" attractor in which the truncated sides represent inital productivity when seedlings are established and final productivity when they die because of lack of nitrogen supply from external sources. The empty spaces in the square are the result of incrementing growth by one year time steps — analogous to adding new rings of growth in a real seedling.

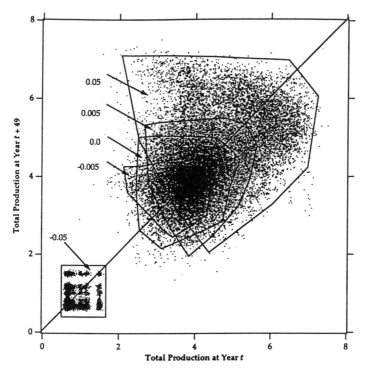

Figure 6: The effect of augmenting or decrementing nitrogen in the system on the shape of the attractor. Note the increased variance of the system (i.e., size of attractor) as nitrogen is added, and the increasing asymmetry of points around the 1:1 diagonal.

We are currently examining the effect of augmenting or decrementing nitrogen supply on the size of the scaling region as above. However, it appears possible that augmenting nitrogen supply by a fixed amount did not eliminate chaotic behavior, as we originally expected. Instead, the nitrogen simply was taken up by the plants and incorporated into the nitrogen feedback responsible for the chaotic behavior. Therefore, while augmenting nitrogen increased mean productivity in the simulations, it also increased variance about the mean. Adding nitrogen — even in large amounts — does not preclude declines in productivity or even very low productivities equal to that under more nitrogen stressed conditions because of the attractor that the nitrogen cycle itself generates. Thus increased mean productivity by continuously adding a constant amount of fertilizer is gained at the expense of predictability.

An alternative method of controlling chaotic behavior is not to augment nitrogen by a constant amount, but to add just enough fertilizer each year so that the total amount of available nitrogen for the plants from both fertilizer and decomposition is equivalent to the amount required to obtain, say, 95% of maximum growth. The amount of nitrogen required for this amount of growth can be calculated from the nitrogen response function. Thus, the control in this case would be implemented by adjusting productivity, x, through the N response function, $f(\cdot)$, according to the difference between nitrogen required for 95% growth and the actual nitrogen available from the soil and fertilizer at time t:

$$x_{0.95} - x(t) = f(N_{x=0.95}) - f(a(t) + N_{soil}(t))$$

In this case, the amount of N augmented in fertilizer (a) varies with time such that $(a(t) + N_{soil}(t))$ always equals the amount required to achieve 95% of growth and hence minimize $x_{0.95} - x(t)$.

How to achieve this control in a practical sense is not yet clear. The most straightforward method is to continuously monitor the amount of nitrogen released from the soil using standard soil analysis methods (i.e., Pastor et al. 1984, Binkley et al. 1985) and adjust fertilizer applications accordingly. However, this could be expensive to implement. Alternatively, one could employ a targeting algorithm a described by Ott et al. (1990; "OGY" control) and tested experimentally by Shinbrot et al. (1992) to steer the system toward a prescribed targeted nitrogen cycling and hence productivity that is on a stable periodic attractor. Unfortunately, given current theories of forest ecosystems as well as stochastic noise in the real system due to weather, etc., one cannot predict whether the system is within a small neighborhood of a stable target, a condition that is required by the OGY algorithm. In these cases, it may be possible to use targeting algorithms presented by Kostelich (this volume), Teo and Rehbock (this volume), and Glass (this volume) to design a control strategy based on the time series of model output of productivity and nitrogen availability, with or without stochastic noise superimposed.

It is clear that managers of complex ecological systems (foresters, fisheries managers, wildlife managers) need to be more cognizant of non-linear feedbacks in order to consistently achieve a targeted output of services from those systems (Vincent et al. 1990). As demand for outputs from ecological systems increases (e.g., timber supply; USDA, Forest Service 1982), so does the need for precise predictions of those outputs. However, given our current understanding of these systems and methods to control chaotic behavior, it is equally unclear what should be the best method to achieve a desired and predictable output. The fertilization example shown here clearly indicates that simply augmenting the supply of a limiting nutrient indiscriminantly of the system dynamics might increase mean productivity (output) but only at the expense of predictability. In fact, the underlying chaotic behavior itself is enhanced. More precise and delicate methods of control of the nitrogen cycle are clearly required if we are to sustain target levels of productivity in complex ecological systems.

Acknowledgements

This research was supported by a grant from the National Science Foundation's Ecosystem Studies and Computational Biology Programs, which also provided travel support to attend the workshop "Control and Chaos" where these results were presented. The continued support of this agency is greatly appreciated.

References

[1] Abraham, N.B., A.M. Albano, A. Passamante and P.E. Rapp. 1989. *Measures of Complexity and Chaos*. New York, Plenum Press.

[2] Babloyantz, A. 1989. *Some remarks on nonlinear data analysis.* Pages 51-62 in N.B. Abraham, A.M. Albano, A. Passamante and P.E. Rapp (editors). Measures of Complexity and Chaos. New York, Plenum Press.

[3] Binkley, D., J. Aber, J. Pastor, and K. Nadelhoffer. 1986. *Nitrogen availability in some Wisconsin forests: comparisons of resin bags and on-site incubations.* Biology and Fertility of Soils 2: 77-82.

[4] Bormann, F.H. and G.E. Likens. 1979. *Pattern and Process in a Forested Ecosystem.* Springer-Verlag, New York.

[5] Botkin, D.B., J.F. Janak and J.R. Wallis. 1972. *Some ecological consequences of a computer model of forest-growth.* Journal of Ecology 60: 849-872.

[6] Bryson, R. 1966. *Air masses, streamlines, and the boreal forest.* Geographical Bulletin 8: 228-269.

[7] Cannell, M.G.R. 1982. *World Forest Biomass and Primary Production Data.* London, Academic Press.

[8] Cohen, Y. and J. Pastor. 1991. *The responses of a forest model to serial correlations of global warming.* Ecology 72: 1161-1165.

[9] Cohen, Y. and J. Pastor. submitted. *Cycles, randomness, chaos, and sustainability in a forest ecosystem model.* American Naturalist.

[10] Constantino, R.F., J.M. Cushing. B. Dennis, andd R.A. Desharnals. 1995. *Experimentally induced transitions in the dynamic behavior of insect populations.* Nature 375: 227-230.

[11] Ellner, S. and P. Turchin. 1995. *Chaos in a noisy world: new methods and evidence from time-series analysis.* American Naturalist 145: 343-375.

[12] Emanuel, W.R., H.H. Shugart, and D.C. West. 1978. *Spectral analysis and forest dynamics: The effects of perturbations on long-term dynamics.* Pages 193-206 in H.H. Shugart (ed.) Time Series and Ecological Processes. SIAM, Philadelphia, PA.

[13] Farmer, J.D. 1982. *Dimension, fractal measures and chaotic dynamics.* Pages 228-246 in H. Haken (editor) Evolution of Order and Chaos. Berlin, Springer-Verlag.

[14] Fraser, A.M. and H.L. Swinney. 1986. *Independent coordinates for strange attractors from mutual information.* Physical Review A 33: 1134-1140.

[15] Global Climate Change Committee. 1988. *Toward an Understanding of Global Change.* Washington, D.C., National Academy Press.

[16] Grassberger, P. 1986. *Estimating the fractal dimensions and entropies of strange attractors.* Pages 291-311 in A.V. Holden (editor) Chaos. Princeton, N.J., Princeton University Press.

[17] Grassberger, P. and I. Procaccia. 1983. *Measuring the strangeness of strange attractors.* Physica D 9: 189-208.

[18] Hanski, I., P. Turchin, E. Korpimki, and H. Hentonnen. 1993. *Population oscillations of boreal rodents: regulation by mustelid predators leads to chaos.* Nature 364: 232-235.

[19] Hastings, A., C. L. Hom, S. Ellner, P. Turchin, and H.C.J. Godfrey, Jr. 1993. *Chaos in ecology: Is Mother Nature a strange attractor?* Annual Reviews of Ecology and Systematics 24: 1-34.

[20] Kareiva, P. 1995. *Predicting and producing chaos.* Nature 375: 189-190.

[21] May, R.M. 1974. *Biological populations with non-overlapping generations: stable points, stable cycles, and chaos.* Science 186: 645-647.

[22] May, R.M. 1976. *Simple mathematical models with very complicated dynamics.* Nature 261: 459-467.

[23] McClaugherty, C.A., J. Pastor, J.D. Aber and J.M. Melillo. 1985. *Forest litter decomposition in relation to soil nitrogen dynamics and litter quality.* Ecology 66: 266-275.

[24] Moore, T.R. 1984. *Litter decomposition in a subarctic spruce-lichen woodland, eastern Canada.* Ecology 65: 299-308.

[25] Ott, E., C. Grebogi, and J.A. Yorke. 1990. Physical Review Letters 64: 1196.

[26] Packard, N.H., J.P. Crutchfield, J.D. Farmer and R.S. Shaw. 1980. *Geometry from a time series.* Physical Review Letters 45: 712-716.

[27] Pastor, J., J.D. Aber, C.A. McClaugherty, and J.M. Melillo. 1984. *Aboveground production and N and P cycling along a nitrogen mineralization gradient on Blackhawk Island, Wisconsin.* Ecology 65: 256-268.

[28] Pastor, J. and W.M. Post. 1985. *Development of a Linked Forest Productivity-Soil Process Model.* Oak Ridge National Laboratory.

[29] Pastor, J. and W.M. Post. 1986. *Influence of climate, soil moisture, and succession on forest carbon and nitrogen cycles.* Biochemistry 2: 3-27.

[30] Pastor, J., R.H. Gardner, V.H. Dale, and W.M. Post. 1987. *Successional changes in soil nitrogen availability as a potential factor contributing to spruce dieback in boreal North America.* Canadian Journal of Forest Research 17: 1394-1400.

[31] Pastor, J. and W.M. Post. 1988. *Response of northern forests to CO_2-induced climate change.* Nature 334: 55-58.

[32] Priestly, M.B. 1981. *Spectral Analysis and Time Series.* New York: Academic Press.

[33] Procaccia, I. 1988. *Universal properties of dynamical complex systems: the organization of chaos.* Nature 333: 618-623.

[34] Schuster, H.G. 1984. *Deterministic Chaos: An Introduction.* Weinheim, Federal Republic of Germany, Physik Verlag.

[35] Shinbrot, T., C. Grebogi, E. Ott, and J.A. Yorke. 1993. *Using small perturbations to control chaos.* Nature 363: 411-417.

[36] Shugart, H.H. 1984. *A Theory of Forest Dynamics.* New York, Springer-Verlag.

[37] Shugart, H.H. and I.R. Noble. 1981. *A computer model of succession and fire response of the high altitude Eucalyptus forest of the Brindabella Range, Australian Capital Territory.* Australian Journal of Ecology 6: 149-164.

[38] Shugart, H.H. and D.C. West. 1977. *Development of an Appalachian deciduous forest succession model and its application to assessment of the impact of the chestnut blight.* Journal of Environmental Management 5: 161-179.

[39] Solomon, A.M. and H.H. Shugart. 1984. *Integrating forest-stand simulations with paleoecological records to examine long-term forest dynamics.* Pages 333- 356 in G.I. Agren (editors) State and Change of forest ecosystems — Indicators in current research. Uppsala, Sweden, Swedish University of Agricultural Science Report Number 13.

[40] Solomon, A.M. and T. Webb. 1985. *Computer-aided reconstruction of Late-Quarternary landscape dynamics.* Annual Reviews of Ecology and Systematics 16: 63-84.

[41] Sugihara, G. and R.M. May. 1990. *Nonlinear forecasting as a way of distinguishing chaos from measurement error in time series.* Nature 344: 734-741.

[42] Swinney, H.L. 1983. *Observations of order and chaos in nonlinear systems.* Physica D 7: 3-15.

[43] Takens, F. 1981. *Detecting strange attractors in turbulence.* Pages 315-381 in D.A. Rand and L.S. Young (editors) Dynamical Systems and Turbulence. Berlin, Springer-Verlag.

[44] Thornthwaite, C. and J.R. Mather. 1957. *Instructions and tables for computing potential evapotranspiration and the water balance.* Publications in Climatology 10: 183-311.

[45] Turchin, P. 1993. *Chaos and stability in rodent population dynamics: evidence from time-series analyses.* Oikos 68: 167-172.

[46] USDA, F.S. 1982. *An Analysis of the timber situation in the United States, 1952-2030.* USDA, Forest Service Resource, Washington, D.C.

[47] Vincent, T.L., A.I. Mees, and L.S. Jennings (editors). 1990. *Dynamics of Complex Interconnected Biological Systems.* Mathematical Modelling No. 6, Birkhüser, Boston.

[48] Weinstein, D.A., H.H. Shugart and D.C. West. 1982. *The long-term nutrient retention properties of forest ecosystems: A simulation investigation.* Oak Ridge National Laboratory, Oak Ridge, TN, USA.

Commentary by T. Sauer

Pastor and Cohen present a compelling case for the applicability of nonlinear dynamics methodology to forest ecosystem modeling. The Grassberger-Procaccia correlation integral and the return plots individually are weak evidence. This is not surprising, given that the possession of interesting, weak evidence is the canonical position ecological modelers have found themselves in with regards to nonlinearity and chaos; see (Ellner and Turchin, 1995) for a compendium of current work. Added to these measures is the fact that varying temperature inputs to the LINKAGES model results in different periodicities as measured in the power spectrum, which points toward bifurcation behavior characteristic of a nonlinear system. The intervals between minima of nitrogen production levels shows interesting return plot structure as well. The total picture is one that is intriguing, and worthy of further study.

The type of dynamics present in the ecosystem model is pivotal, for the kind of external loading that could be applied as a control protocol is entirely dependent on this. Cohen and Pastor find that artificially increasing nitrogen levels in the model can increase productivity at the expense of increased variance, which is a qualitative fact of importance to policy management.

Self-organization Dynamics in Chaotic Neural Networks

Masataka Watanabe
Department of Quantum Engineering and Systems Science
The University of Tokyo
7-3-1 Hongo Bunkyo-ku Tokyo 113, Japan

Kazuyuki Aihara
Department of Mathematical Engineering and Information Physics
The University of Tokyo
7-3-1 Hongo Bunkyo-ku Tokyo 113, Japan

Shunsuke Kondo
Department of Quantum Engineering and Systems Science
The University of Tokyo
7-3-1 Hongo Bunkyo-ku Tokyo 113, Japan

Abstract

We examine roles of deterministic chaos in artificial neural networks with self-organization dynamics which has the ability to switch automatically between learning and retrieving modes in accordance with the given input pattern. We demonstrate that the chaotic dynamics works as means to learn new patterns and increases the memory capacity of the neural network.

1 Introduction

Freeman and his colleagues have been investigating roles of chaos in the biological neural systems with interesting physiological experiments [1], in which rabbits with an array of 64 electrodes implanted permanently on the lateral surface of the left olfactory bulb were used. From the experiments, odorant-specific information was found to exist in the spatial patterns of the amplitude of the waveform of EEG potentials [1]. It was reported [1] that when a rabbit was given a known odorant, the EEG signals became a limit cycle and when given an unknown odorant, it became chaotic. Moreover, as the rabbit learned the unknown odorant, the response changed from periodic to chaotic. They proposed the following hypothesis on the basis of the experiments that deterministic neural chaos is utilized as means for ensuring continual access to previously learned sensory patterns and also for learning new sensory patterns.

We investigate the latter part of the hypothesis on learning new patterns in relation with the stability-plasticity problem [25] in modeling of neural networks; a network capable of autonomously adapting in real time must make a decision, namely either recognize the input as a previously learned pattern (stable mode) or newly learn it as a new pattern (plastic mode) according to the given input. The conventional neural autoassociators can only retrieve stored patterns according to the pre-installed memory.

Our network model is a single layer mutually connected neural network with three major extensions, which realizes automatic switching of learning and retrieving modes [21, 22]. The first extension is to use the input pattern as a continuous bias so that the network is capable of recognizing whether the input pattern is a known one or not. The second is that the synaptic connections among neurons have incessant plasticity based upon a spatio-temporal learning rule which we propose in section 3. This property works to store the input pattern automatically when the network recognizes it as unknown. The final extension is that Chaotic Neurons [17] are used for our network. This neuron model was proposed to qualitatively describe chaotic response observed in squid giant axons under periodic current stimulation [13, 11, 2]. We use the chaotic neurons to investigate the role of deterministic chaos in learning and retrieving patterns.

2   Chaotic Neuron Model

Artificial neural networks are usually composed of simple elements of artificial neurons modeling biological neurons. A typical neuron model is a simple threshold element transforming a weighted spatial summation of the inputs into the output through nonlinear output function with threshold. However, it is criticized from the viewpoint of neurophysiology that real neurons are far more complicated than such simple threshold elements. One of the typical characteristics which real neurons have but the conventional artificial neurons lack is chaotic behavior experimentally observable in a single neuron [13, 11, 2]. ODE models of excitable nerve membranes as the Hodgkin-Huxley equations [19] and the FitzHugh-Nagumo equations [24, 3] are too complicated as elements of large-scale artificial neural networks even though these equations can reproduce the experimentally observed chaos very well [13, 11, 15]. Under these circumstances the following simple discrete-time neuron model with chaotic dynamics was proposed by Aihara et al. [17, 11]:

$$x_i(t+1) = f\left(\sum_{r=0}^{t} k^r A_i(t-r) + \sum_{j=1}^{N} W_{ij} \sum_{r=0}^{t} k^r x_j(t-r) - \alpha \sum_{r=0}^{t} k^r x_i(t-r) - \theta\right) \quad (1)$$

where $x(t+1)$ is the output of the ith neuron which takes an analog value between 0 and 1 at discrete time $t+1$; f is an analog function, e.g. the logistic function $f(y) = 1/(1 + e^{-y/\varepsilon})$ with the steepness parameter ε; $A_i(t-r)$ is the strength of the external input to the neuron at discrete time $t-r$; W_{ij} is the synaptic weight from the jth neuron to the ith neuron; k is the temporal decay factor; α is a positive scaling parameter which adjusts the relative strength of refractoriness and θ is the threshold. This model shows chaotic dynamics similar to the response observed in real neurons.

We define $I^{neurons}$, $I^{external}$ and I^{total} on eq.(1) as follows:

$$I^{neurons} = \sum_{j=1}^{N} W_{ij} \sum_{r=0}^{t} k^r x_j(t-r) \qquad (2)$$

$$I^{external} = \sum_{r=0}^{t} k^r A_i(t-r) \qquad (3)$$

$$I^{total} = \sum_{j=1}^{N} W_{ij} \sum_{r=0}^{t} k^r x_j(t-r) + \sum_{r=0}^{t} k^r A_i(t-r) = I^{neurons} + I^{external} \qquad (4)$$

3 Spatio-temporal Learning Rule

The Hebbian rule, which increases the synaptic weight W_{ij} when both of the pre-synaptic neuron j and the post-synaptic neuron i are simultaneously excited, is given in the following equation [23]:

$$\Delta W_{ij} = \eta a_i x_j \qquad (5)$$

where $x_j(t)$ is the output of the jth neuron at discrete time t, which takes a value between 0 (non-firing) and 1(full-firing) ; a_i is the activation potential of the ith neuron; η is the constant of proportionality representing the learning rate.

Sejnowski suggested that a decrease in synaptic strength should take place in order for plastic neuronal circuits to operate effectively [27]. A modified rule with such an extension is given as follows:

$$\Delta W_{ij} = \eta a_i (x_j(t) - \Theta) \qquad (6)$$

where Θ is a threshold value to judge whether the pre-synaptic neuron is in a high firing state or in a low firing state (includes non-firing). It should be noted that both of these two rules use only instantaneous information at the synapse. Assuming that effects of previous inputs to the synapse remain with exponential time decay and modeling the property with a leaky integrator at the synapse, we obtain the following equation:

$$\Delta W_{ij} = \eta h(a_i) g\left(\sum_{r=0}^{t} k^r x_j(t-r) - \Theta\right) \qquad (7)$$

where the functions g and h are assumed as follows:

$$g(x) = \begin{cases} 1, & \text{if } x \geq 0, \\ 0, & \text{if } x < 0, \end{cases} \qquad (8)$$

$$h(x) = \begin{cases} 1, & \text{if } \theta^+ < x < b^+, \\ -1, & \text{if } b^- < x < \theta^-, \\ 0, & otherwise. \end{cases} \qquad (9)$$

Learning thresholds θ^+, θ^- provide a non-plastic region in the middle for a read-only synaptic state. This property is quite consequential to our model. We also set learning bounds b^+, b^- to terminate the learning regions(Figure 1).

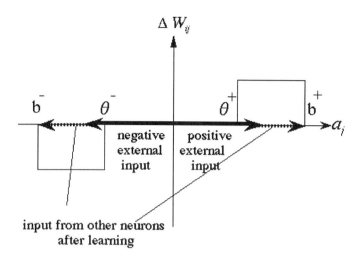

Figure 1: A spatio-temporal learning rule which uses temporal summation together with spatial summation. In the case that $\sum_{r=0}^{t} k^r x_j(t-r) - \Theta > 0$

4 The Network Model

We describe the network model in this section. The model is a single-layer mutually connected neural network with switching ability between learning and retrieving modes. The network structure is shown in Figure 2. Each neuron is connected to all other neurons in the network and also receives a external input corresponding to the input pattern.

The first extension to the conventional single-layer mutually connected network is that the input pattern is given as a continuous bias rather than as an initial state of the network, although most of the usual neural autoassociators use the input only as the initial state of the network. In the conventional case the network usually falls into a fixed-point attractor in the state space. The stable fixed-point may be either one of the stored patterns or a spurious memory state. On the contrary, when we use input as a continuous bias in our network model, the behavior of the network differentiates depending on the distance between the input pattern and the stored patterns. If the distance is small(known input), the network state converges to the corresponding stored pattern. If the distance is large, on the other hand, and if the input pattern is almost orthogonal to the previously learned patterns, the network holds the input pattern.

The second extension to the conventional network is that the learning rule is used continuously. As stated above, the network can distinguish between known and unknown patterns. We can adjust parameter values so that the activation potential of a neuron which receives the correct input is outside of the learning boundaries and that of a neuron which receives the incorrect input is in the non-plastic region when the input is contaminated by a small amount of noise but still close enough to a known pattern. Therefore, besides the transient time, there is no change to the synaptic weights and the memorized pattern is retrieved. When an unknown input is given, neurons' action potentials lie in plastic regions between the learning thresholds and boundaries and the synaptic weights change to store this new pattern.

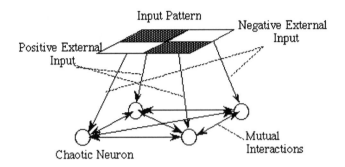

Figure 2: network model

The final extension is that the chaotic neurons are used for our network model [17, 11]. As the model of chaotic neural networks include conventional neural network models with changing the parameter values, we can compare the results and investigate roles of deterministic chaos in associative neurodynamics and learning.

5 Behavior of the Network According to the Given Input

5.1 When given unknown input

When an unknown input is given to the network, inputs from other neurons sum up to 0 if it is orthogonal to the previously stored patterns. As a result the total input is nearly equal to the external input strength. While the temporally summed output of a neuron which receives positive external input becomes larger than Θ, that of a neuron which receives negative external input becomes smaller than Θ. Input from other neurons at time $t + 1$ is given as follows:

$$
\begin{aligned}
I_i^{neurons}(t+1) &= \sum_{j \in P^+}^{N} (W_{ij}(t) + \Delta W_{ij}) \sum_{r=0}^{t} k^r x_j(t-r) + \\
&\quad \sum_{j \in P^-}^{N} W_{ij}(t) \sum_{r=0}^{t} k^r x_j(t-r) \text{ for } i \in P^+
\end{aligned}
\tag{10}
$$

$$
I_i^{neurons}(t+1) = \sum_{j \in P^+}^{N} (W_{ij}(t) - \Delta W_{ij}) \sum_{r=0}^{t} k^r x_j(t-r) +
$$

$$\sum_{j \in P^-}^{N} W_{ij}(t) \sum_{r=0}^{t} k^r x_j(t-r) \text{ for } i \in P^- \qquad (11)$$

where P^+ and P^- are sets of indexes of neurons receiving positive and negative inputs, respectively. Since the output of neuron $i(i \in P^-)$ is almost 0, inputs from other neurons becomes

$$I_i^{neurons}(t+1) > I_i^{neurons}(t) \text{ for } i \in P^+ \qquad (12)$$

$$I_i^{neurons}(t+1) < I_i^{neurons}(t) \text{ for } i \in P^- \qquad (13)$$

If we assume that the external input is kept constant, the total input to a neuron satisfies the following inequalities:

$$I_i^{total}(t+1) > I_i^{total}(t) \text{ for } i \in P^+ \qquad (14)$$

$$I_i^{total}(t+1) < I_i^{total}(t) \text{ for } i \in P^- \qquad (15)$$

Consequently, the new pattern is stored in the network if the input pattern is kept for a while.

5.2 When given known input

When a known input is given to the network, inputs from other neurons and the external input have same signs and the total input becomes,

$$I_i^{neurons} \approx positive\ learning\ bound(b^+) \text{ for } i \in P^+ \qquad (16)$$

$$I_i^{neurons} \approx negative\ learning\ bound(b^-) \text{ for } i \in P^- \qquad (17)$$

Consequently, the corresponding pattern is quickly retrieved without significant changes of the synaptic connections.

6 Simulation Results on Automatic Switching of Learning and Retrieving Modes

In this section, we show simulation results with the network composed of 81 neurons. The parameters of the chaotic neural networks are set as follows:

$$k = 0.2,\ \varepsilon = 3.0,\ \alpha = 103.68,\ \theta^+ = 60,\ \theta^- = -60,\ b^+ = 180,\ b^- = -180. \qquad (18)$$

Examples of the input patterns are given in Figure 3.

6.1 Learning and Retrieving

6.1.1 Learning an unknown pattern

A simulation result of learning an unknown input is given in Figure 4. In the following Figures 4-7, each square represents an activation potential of a neuron receiving positive external input and each cross represents an activation potential of a neuron receiving negative external input. Until time step 5, the network is retrieving a known

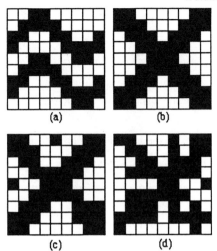

Figure 3: Examples of input patterns used in simulation

pattern (a) in Figure 3 and the new pattern (b) in Figure 3 is given at time step 5. Figure 4 shows that the activation potentials of the neurons converge to the strength of the external input at time step 8, which strength is in plastic regions between the learning thresholds and the learning bounds. When the activation potentials lie in the regions, the change of the synaptic weights due to the learning rule works to store the new pattern as demonstrated in Figure 4.

6.1.2 Retrieving a known pattern

Figure 5 shows the simulation result of retrieving a known pattern. Until time step 10 the network is retrieving a known pattern (a) in Figure 3 and another known pattern (b) in Figure 3 is given at time step 10. Figure 5 shows that the activation potentials of the neurons which receive the reversed external input go beyond the learning bounds on the opposite side quickly. There is a little change to the synaptic weights while the values of the activation potentials pass through the plastic regions between the learning thresholds and the learning bounds.

6.1.3 Correct retrieving a known pattern with a small amount of noise

Figure 6 shows the behavior of the activation potentials when given a known pattern with noise(c) in Figure 3. Since the amount of noise is still small enough, the network recognizes it as a known pattern with noise rather than as a new pattern to learn. The activation potentials of neurons which receives the incorrect external input (noise) converges to the read-only synaptic state(between the negative learning threshold and the positive learning threshold). Figure 6 shows that the signs of the activation potentials are same as those of the memorized pattern, even though they receive the external input with the opposite sign. This is due to larger input from the consistuent neurons than the external input. As a result, the memorized pattern is correctly retrieved from the noisy input.

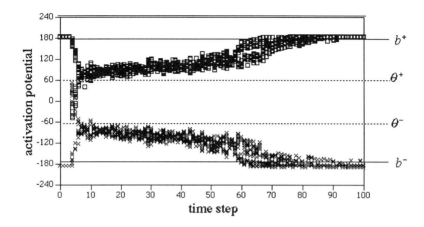

Figure 4: Learning an unknown pattern; time series of activation potentials of all neurons in the network

6.1.4 Incorrect retrieving a known pattern with more noise incorrectly retrieved

When there is larger noise to the given input pattern as shown in Figure 3 (d), the network recognizes it as a known pattern but cannot retrieve it correctly. Figure 7 shows that the activation potential of a neuron receiving the incorrect input is in the wrong side. This is because the input from the other neurons is smaller than the external input. The network fails to retrieve but still there is no change to the synaptic weight since the activation potentials of neurons which receive the incorrect external input is in the read-only synaptic region.

6.2 Threshold distance on learning as a new pattern

Simulations were carried out to investigate the threshold value of distance between a input pattern and the closest stored pattern where the network's behavior dramatically changes from retrieving to learning. Figure 8 is the result of Monte-Carlo simulations. The horizontal axis is the distance of the input pattern from one of the stored pattern and the vertical axis is the rate of learning as a new pattern. Figure 8 shows that there is a region where all the patterns are recognized as a known pattern and a region where almost all patterns are newly learned. In between, there is a kind of fuzzy region where retrieving and learning co-exist.

6.3 Comparing the capacity with conventional networks

In this section we compared the capacity of conventional networks (composed of neurons with step or sigmoidal output functions) and the chaotic neural network

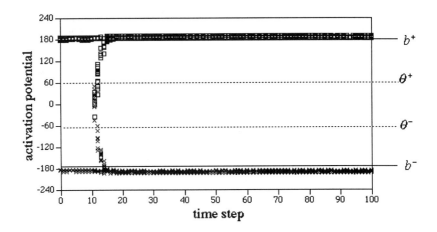

Figure 5: Retrieving a known pattern; time series of activation potentials of all neurons in the network

when different patterns are successively given to the networks. The horizontal axis of Figure 9 is the robustness factor R of the network which is defined as follows:

$$R = I^{neurons}/I^{external} \tag{19}$$

The robustness factor R is a parameter which determines the network's ability to correctly retrieve known patterns from ones with noise. The greater the value the more robust to noise. This value must be greater than 1 to guarantee least robustness. Figure 9 indicates that the network composed of neurons with step output function cannot learn the second pattern when R is more than 1.1. Figure 9 also shows that the capacity of the chaotic neural networks to store patterns successively is more than those of the conventional neural networks with step or sigmoidal output functions.

7 Discussions

We have proposed a new learning rule to cope with the stability-plasticity problem in neural autoassociators. According to the simulations, networks composed of chaotic neurons show more capacity than networks with conventional neuron models with step or sigmoidal output functions. This result is ascribed to the ability of the chaotic neural networks to jump out from attractions in the state space. The upper limit on the number of learnable patterns is determined by the failure of jumping out of previously built attractors(memory) when given an unknown pattern. It is interesting that this result seems to support the Freeman's hypothesis that chaos is utilized as means to learn new sensory patterns although further analysis on the dynamical structure still remain to be a future problem.

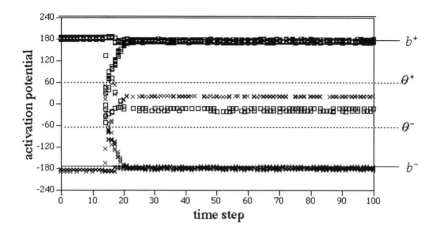

Figure 6: Correct retrieving of a noisy known pattern; time series of activation potentials of all neurons in the network

The learning rule proposed in this article can be extended to store mutual-correlation of neighboring patterns in the sequence as well as auto-correlation due to effects of leaky integrators in the synapses [20].

We have shown a neural mechanism to utilize chaotic dynamics for learning in neural networks. Generally speaking, on the contrary to the conventional view of control in biological systems with the "homeostasis" concept, recent researches on biological systems show that nonlinear dynamics with possibly chaotic fluctuations may play important functional roles in biological systems [14, 18, 5, 8]. In other words, not only homeostasis but also "homeodynamics" should be considered to understand the biological control principle [6, 7, 9, 12]. On the other hand, the concept of homeostasis has been influencing studies of control in engineering [26]. It may be also worthy of intensive studies to seek for possible "homeodynamical control" mechanisms in biological systems and their applications [7, 9, 12]. Studies on the inverse of chaos control [4] and engineering applications of chaotic fluctuations [16] may be stepping into this direction to some extent. It seems important to the present authors not to ignore possibilities that real biological systems adopt various homeodynamical control mechanisms which "harness" vitality of deterministic chaos without necessarily killing it [12, 10, 11].

References

[1] Skarda C.A. and Freeman W.J. How brains make chaos in order to make sense of the world. *Behavioral and Brain Science*, 10:161–195, 1987.

[2] Matsumoto G., Aihara K., Takahashi N., Yoshizawa S., and Nagumo J. Chaos

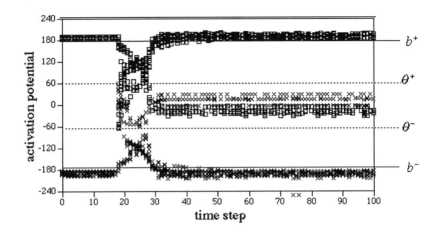

Figure 7: Incorrect retrieving of a noisy known pattern; time series of activation potentials of all neurons in the network

and phase locking in normal squid axons. *Phys. Lett. A*, 123(4):162–166, 1987.

[3] Nagumo J., Arimoto S., and Yoshizawa S. An active pulse transmission line simulating nerve axon. *Proc. IRE*, 50:2061–2070, 1962.

[4] Schiff S. J., Jerger K. Duong., Chang T., Spano M. L., and Ditto W. L. Controlling chaos in the brain. *Nature*, 370:615–620, 1994.

[5] West B. J. *Fractal Physiology and Chaos in Medecine*. World Scientific, Singapore, 1990.

[6] Aihara K. Biochaotic information and its applications to engineering. *Denshi Kohgyo Geppoh*, 34(1):30–39, 1992.

[7] Aihara K. Chaos: Towards applications. *Mathematical Sciences*, 30(6):5–10, 1992.

[8] Aihara K. *Chaos in Neural Systems*. Tokyo Denki University Press, Tokyo, 1993.

[9] Aihara K. Nonlinear engineering: Towards science and technology in the 21st century. *Mathematical Sciences*, 31(9):5–7, 1993.

[10] Aihara K. *Applied Chaos and Applicable Chaos*. Science-sha, Tokyo, 1994.

[11] Aihara K. Chaos in neural response and dynamical neural network models: Toward a new generation of analog computing. In Yamaguchi M., editor, *Towards the Harnessing of Chaos*, pages 83–98. Elisevier, Amsterdam, 1994.

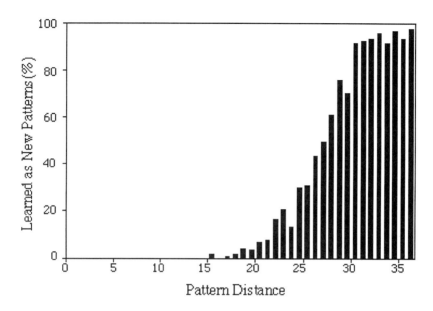

Figure 8: A relationship between pattern distances and rates of learning as a new pattern

[12] Aihara K. Life, chaos and engineering. *Mathematical Sciences*, 33(3):5–10, 1995.

[13] Aihara K. and Matsumoto G. Chaotic oscillations and bifurcations in squid giant axons. In A. V. Holden, editor, *Chaos*, pages 257–269. Manchester University Press, Princeton University Press, Manchester, Princeton. NJ, 1986.

[14] Aihara K. and Matsumoto G. Forced oscillations and routes to chaos in the hodgkin-huxley axons and squid giant axons. In Degn H., Holden A. V., and Olsen L. F., editors, *Chaos in Bilogical Systems*, pages 121–131. Plenum Press, New York, 1987.

[15] Aihara K., Matsumoto G., and Ikegaya. Periodic and non-periodic responses of a periodically forced hodgkin-huxley oscillator. *J. Theor. Biol.*, 109:249–269, 1984.

[16] Aihara K. and Katayama R. Chaos engineering in japan. Communications of the ACM(in press).

[17] Aihara K., Takabe T., and Toyoda M. Chaotic neural networks. *Phys. Lett. A*, 144(6,7):333–340, 1990.

[18] Goldberger L., Rigney D. R., and West B. J. Chaos and fractals in human physiology. *Scientific American*, 262(2):34–41, 1990.

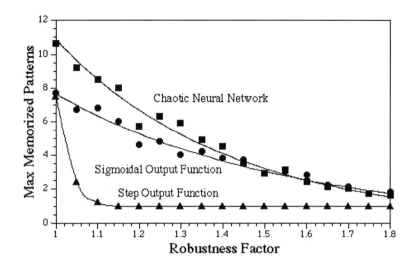

Figure 9: Comparing the capacities of the networks to store patterns successively

[19] Hodgkin A. L. and Huxley A. F. A quantitative description of membrane current and its application to conduction and excitation in nerve. *J. Physiol. (Lond.)*, 117:500–544, 1952.

[20] Watanabe M., Aihara K., and Kondo S. Learning and retrieving of spatio-temporal patterns in chaotic neural networks. in preparation.

[21] Watanabe M., Aihara K., and Kondo S. Automatic learning in chaotic neural networks. *Trans. IEICE*, J78-A(6):686–691, 1995.

[22] Watanabe M., Aihara K., and Kondo S. Automatic learning in chaotic neural networks. In *1994 IEEE Symposium on Emergin Technologies and Factory Automation*, pages 245–248, 1995.

[23] Hebb D. O. *The Organaization of Behavior*. Wiley, New York, 1949.

[24] FitzHugh R. Mathematical models of excitation and propagation in nerve. In H. P. Schwan, editor, *Biological Engineering*, pages 1–85. McGraw-Hill, New York, 1969.

[25] Grossberg S. Competitive learning: From interative activation to adaptive resonance. *Cognitive Science*, 11:23–63, 1987.

[26] Utusnomiya T. *Biological Control and Information Systems*. Asakura Shoten, Tokyo, 1978.

[27] Sejnowski T.J. Storing covariance with nonlinearly interactin neurons. *J. Math. Biol.*, 4:303–321, 1977.

Commentary by J. Pastor

Most of this conference was concerned with ways that we as humans can control systems that exhibit chaotic behavior. This paper addresses perhaps the opposite problem: how can chaos be used to learn about external systems and better control our responses to them? Watanabe and coauthors present a mathematical model of a neural network that shows how chaotic behavior of the sequential firing of neurons can be harnessed to learn new patterns. Presumably, this allows the organism to better control its environment, or at least its response to its environment. The chaotic behavior arises in the model from setting the output of each neuron during some time interval to be a logistic function of the sum of all inputs from the environment and other neurons minus the sum of the decays of the signal through the network and a learning threshold. These summations can be viewed as a sort of "mass balance" accounting of information transfer throughout the system. Control is achieved by adjusting the learning threshold to maximize the probability that a pre-synaptic neuron j and a post-synaptic neuron i are both firing in response to changes in external inputs. The organism thus "learns" a new pattern by synchronizing the firing of all neurons in the neural network in respone to the new pattern. This learning happens faster if the neural network is in a chaotic state before being presented with a new pattern. This is a very nice model because all the parameters have real physical or biological meaning. It is thus very amenable to experimental testing. One is tempted to speculate on whether the minds of creative people such as artists and musicians (perhaps even scientists?) are in a chaotic state when they learn or discover new patterns.

Commentary by T. Sauer

This article describes a successful approach to the stability-plasticity problem of neural network design. An intelligent system must be capable of deciding between two cases when given an externally-presented pattern: (1) (classification) the pattern is known, or close to a previously-stored pattern, or (2) (learning) the pattern is not already known, and should be stored. The protocol described in the article, where learning is used only for intermediate values of neural activation, is a means to this end.

Watanabe and co-authors construct their network from neurons that are chaotic in some parameter regimes. The idea is that this design leads to avoidance of periodic or quasiperiodic entrainments that would restrict access to parts of state space in which memories are stored. This possibility is very appealing to researchers in dynamical systems, and may become an organizing principle for investigations into information storage in distributed nonlinear systems. Watanabe's article is a step toward exploiting this promising direction.